TOPOLOGY — HAWAII

TOPOLOGY
HAWAII

HONOLULU
12 – 18 AUGUST 1990

Editor: **Karl Heinz Dovermann**
Department of Mathematics
University of Hawaii

World Scientific
Singapore • New Jersey • London • Hong Kong

Published by

World Scientific Publishing Co. Pte. Ltd.

5 Toh Tuck Link, Singapore 596224

USA office: 27 Warren Street, Suite 401-402, Hackensack, NJ 07601

UK office: 57 Shelton Street, Covent Garden, London WC2H 9HE

British Library Cataloguing-in-Publication Data
A catalogue record for this book is available from the British Library.

TOPOLOGY — HAWAII
Proceedings of the Topology Conference

ISBN-13 978-981-02-0683-3
ISBN-10 981-02-0683-6

PREFACE

The conference *Topology–Hawaii* was held on the campus of the University of Hawaii at Manoa, Honolulu, Hawaii, August 12–18, 1990. Researchers from all areas in topology, as well as related fields, were invited, and this was reflected in the participants of the symposium. It was an international meeting with participants from every country with an active research program in mathematics. We planned the conference at this time and in this location because of the up-coming International Congress of Mathematicians, which was held a week later in Kyoto, Japan.

The conference was supported by a grant from the National Science Foundation, and the Vice President for Research and Graduate Education and the Dean of Natural Sciences of the University of Hawaii. Additional support was provided by the Soros Foundation – Soviet Union, Oceanic Cablevision, and Hawaiian Electric Industries INC. In the name of all of those who received support for their travel, we would like to thank all of the above.

The proceedings contain papers related to lectures given at the meeting. All papers were refereed and are in final form. We would like to express our gratitude to the authors and the many referees.

The participants and, in particular the lecturers, contributed to the success of the conference and we are most grateful to them. Special thanks are due to Professors Wu Chung Hsiang and Ted Petrie who shared the responsibility as members of the organizing committee and Professor Michael Freedman who helped during the preparation. Thank yous also go to the organisors of the special sessions, Professors C. Allday, R. Brown, R. Cohen, W. Kucharz, I. Madsen, R. Oliver, R. Piccinini, and R. Skora. The conference would have been impossible without the support of several people on the staff at the university, Professor R. Little and D. Y. Suh, P. Goldstein, A. Young, K. Miike and N. Pagador. They took care of an endless number of tasks until they were completely exhausted.

TABLE OF CONTENTS

TOPOLOGY — HAWAII

ELEMENTARY ABELIAN p-GROUP ACTIONS ON LENS SPACES

CHRISTOPHER ALLDAY

University of Hawaii

ABSTRACT. Actions of elementary abelian p-groups are studied on spaces having mod. p cohomology isomorphic to that of the standard mod. p lens spaces (including the action of the Steenrod algebra). A best possible upper bound for the size of the smallest orbit is obtained.

1. INTRODUCTION

An elementary abelian p-group, or p-torus, is a group isomorphic to $(\mathbb{Z}/(p))^r$, i.e. the r-fold product of $\mathbb{Z}/(p)$ with itself, where p is prime and $r \geq 0$. When studying p-torus actions the following two invariants are of interest.

(1.1) Defintions. (1) For a topological space X, the free p-rank of X, denoted $\mathrm{frk}_p(X)$, is defined to be the largest r such that $(\mathbb{Z}/(p))^r$ can act freely on X.

(2) Let X be a topological space, let $G \cong (\mathbb{Z}/(p))^r$ and let $\Phi : G \times X \longrightarrow X$ be an action of G on X. Let $\mathrm{rk}(\Phi) = s$, where $\max\{|G_x| : x \in X\} = p^{r-s}$. Then the p-rank (or Hsiang p-rank) of X, denoted $\mathrm{rk}_p(X)$, is defined to be $\max \mathrm{rk}(\Phi)$, where Φ ranges over all p-torus actions on X.

Clearly $\mathrm{rk}_p(X) \geq \mathrm{frk}_p(X)$.

We shall usually assume that X satisfies conditions which make the Localization Theorem ([Hs] or [Q2]) valid.

(1.2) Definition. A topological space X is said to satisfy condition (LT) if X is paracompact and either X is finitistic or $\mathrm{cd}_p(X) < \infty$, where cd_p means sheaf cohomology dimension over the field $\mathbb{F}_p$ of integers modulo p. (See [Bre] and [Q2].) (Clearly a finite-dimensional CW-complex satisfies condition (LT) on both accounts.)

We also make the following defintions for brevity.

(1.3) Defintions. (1) If X and Y are topological spaces, then $X \underset{p}{\sim} Y$ means that $H^*(X; \mathbb{F}_p) \cong H^*(Y; \mathbb{F}_p)$ as graded $\mathbb{F}_p$-algebras via an isomorphism preserving the action of the Steenrod algebra.

(2) For a positive integer n, $\mathrm{ord}_p(n) = \max\{j : p^j \text{ divides } n\}$.

(3) L_p^{2n+1}, for an odd prime p, denotes the standard lens space $S^{2n+1}/\mathbb{Z}/(p)$.

The first theorem is due essentially to Quillen ([Q1]).

(1.4) Theorem. *Let X be a space satifying condition (LT), and suppose that $X \underset{2}{\sim} \mathbb{R}P^n$. Then $\mathrm{rk}_2(X) \le \mathrm{ord}_2(n+1)$. Furthermore $\mathrm{rk}_2(\mathbb{R}P^n) = \mathrm{ord}_2(n+1)$.*

The second theorem is the analogue of the first for odd primes. We shall prove it by a method used by T. Skjelbred ([Sk]) to treat complex projective spaces.

(1.5) Theorem. *Let X be a space satisfying condition (LT), let p be an odd prime, and suppose that $X \underset{p}{\sim} L_p^{2n+1}$. Then $\mathrm{rk}_p(X) \le 1 + \mathrm{ord}_p(n+1)$. Furthermore $\mathrm{rk}_p(L_p^{2n+1}) = 1 + \mathrm{ord}_p(n+1)$.*

(1.6) Remark. In the situation of Theorem (1.4), $\mathrm{frk}_2(X) \le \mathrm{frk}_2(\mathbb{R}P^n) = 0, 1, 2$, respectively, when n is even, 1 modulo 4, -1 modulo 4, respectively. (The free action of $(\mathbb{Z}/(2))^2$ on $\mathbb{R}P^{4m-1}$ comes from the free action of the quaternion 8-group on S^{4m-1}.) Equally it is well-known that, in the situation of Theorem (1.5), $\mathrm{frk}_p(X) \le \mathrm{frk}_p(L_p^{2n+1}) = 1$. (See Section 5.)

In Section 2 we review some of the results of [Q1] and apply them to Theorem (1.4). In Section 3 we discuss Theorem (1.5), proving the inequality part. In Section 4 we give a simple example to realize the equalities in Theorems (1.4) and (1.5). And we make some related comments in Section 5.

2. QUILLEN'S RESULTS

Let $G = (\mathbb{Z}/(2))^r$. Then $H^*(BG; \mathbb{F}_2) = \mathbb{F}_2[t_1, \dots, t_r]$, where the polynomial generators $t_1, \dots, t_r$ all have degree 1. For a homogeneous element $f \in H^*(BG; \mathbb{F}_2)$, let $S_{-1}(f) = f$, $S_0(f) = Sq^1 f$, $S_1(f) = Sq^2 Sq^1 f$, and in general, $S_m(f) = Sq^{2^m} Sq^{2^{m-1}} \cdots Sq^2 Sq^1 f$.

Now a principal result of [Q1] is the following.

(2.1) Theorem [Q1]. *Let $G = (\mathbb{Z}/(2))^r$, and let $f \in H^2(BG; \mathbb{F}_2)$ with $f \ne 0$. Then there is an integer $m \ge -1$ such that*

(1) *$f = S_{-1}(f), S_0(f), \dots, S_m(f)$ is a regular sequence; and*
(2) *for all $n \ge -1$, $S_n(f) \in (f, S_0(f), \dots, S_m(f))$, the ideal generated by f, $S_0(f), \dots, S_m(f)$.*

In particular, letting $J = (f, S_0(f), \dots, S_m(f))$, then J is unmixed of height $m + 2$. (See [M] or [Q1] for defintions and relevant results from commutative algebra.)

(2.2) Corollary. *Let $G = (\mathbb{Z}/(2))^r$, let X be a space such that $X \underset{2}{\sim} \mathbb{R}P^n$, and let $\Phi : G \times X \longrightarrow X$ be an action of G on X. Let $X_G = (EG \times X)/G$ be the Borel construction, and $X \overset{i}{\longrightarrow} X_G \overset{\pi}{\longrightarrow} BG$ the associated fibre bundle. Let $u \in H^1(X; \mathbb{F}_2)$ be the generator (and so $H^*(X; \mathbb{F}_2) = \mathbb{F}_2[u]/(u^{n+1})$). And let $f \in H^2(BG; \mathbb{F}_2)$ be the transgression of u in the Leray-Serre spectral sequence of π. Finally let $J = (f, S_0(f), \dots, S_m(f))$, where $m \ge -1$ is the integer given by Theorem (2.1). Then $2^{m+1} \le n$, and $H_G^*(X; \mathbb{F}_2) := H^*(X_G; \mathbb{F}_2) \cong (R/J \otimes \mathbb{F}_2[w])/(h(w))$, where $R = H^*(BG; \mathbb{F}_2) = \mathbb{F}_2[t_1, \dots, t_r]$, w has degree 2^{m+2}, and*

$h(w)$ is a monic polynomial in w with coefficients in R/J and having degree k in w where k is the least positive integer such that $k2^{m+2} > n$. (In particular, if $2^{m+2} > n$, then $H^*_G(X;\mathbb{F}_2) \cong R/J$.) And $J = \ker[\pi^* : R \longrightarrow H^*_G(X;\mathbb{F}_2)]$. Furthermore $m + 1 < \mathrm{ord}_2(n+1)$.

Proof. Except for the last statement everything follows immediately from Theorem (2.1), the defining property of a regular sequence, and the properties of the spectral sequence, including the way in which transgression commutes with the action of the Steenrod algebra.

To see the last statement let $j = \mathrm{ord}_2(n+1)$. If $n = 2^j - 1$, then the result is clear. So suppose $n = 2^j s - 1$, where $s > 1$ and 2 does not divide s. Let $v = u^{2^j}$, and let $i = 2^j + 1$. Then $0 = d_i(v^s) = d_i(v)v^{s-1} = S_{j-1}(f)v^{s-1}$. Hence $S_{j-1}(f) \in (f, S_0(f), \ldots, S_{j-2}(f))$. Thus $m \le j - 2$. (If $X = \mathbb{R}P^n$, then this follows also because u^{2^j} is a Stiefel-Whitney class; and Stiefel-Whitney classes survive to ∞ in the spectral sequence: see [A, P], Section 5.4.) $\square$

(2.3) Corollary. *With the notation and conditions of Corollary (2.2) suppose in addition that X satisfies condition (LT). Then $\mathrm{rk}(\Phi) \le \mathrm{ord}_2(n+1)$. Furthermore all the maximal isotropy subgroups have the same rank.*

Proof. If X satisfies condition (LT), then it follows from the Localization Theorem (see [Hs]), that the maximal isotropy subgroups are in one-one correspondence with the minimal prime ideals of $\ker \pi^* = J$, an isotropy subgroup $K \subseteq G$ corresponding to the prime ideal $PK := \ker[H^*(BG;\mathbb{F}_2) \longrightarrow H^*(BK;\mathbb{F}_2)]$. Clearly the height of PK, $\mathrm{ht}\, PK$, is the corank of K in G. Hence $\mathrm{rk}\, \Phi = \mathrm{ht}\, J = m + 2 \le \mathrm{ord}_2(n+1)$. And all the maximal isotropy subgroups have the same rank since J is unmixed. $\square$

(2.4) Remarks. (1) Corollary (2.3) proves Theorem (1.4) except for the assertion that $\mathrm{rk}_2(\mathbb{R}P^n) = \mathrm{ord}_2(n+1)$, which will be shown by example in Section 4. (See also [Sk].)

(2) In [Q1], Quillen shows how to compute the number of maximal isotropy subgroups using a version of Bézout's Theorem. The number depends upon the type of the quadratic form f. He shows also that $m + 2 = \mathrm{rk}\, \Phi \le r/2$, $\frac{1}{2}(r+1)$, $\frac{1}{2}(r+2)$, respectively, according as f is real, complex, quaternionic, respectively.

3. LENS SPACES

In this section we shall suppose that p is an odd prime; and we shall develop the analogues of the results of Section 2 following Tezuka and Yagita ([T, Y]).

Let $G = (\mathbb{Z}/(p))^r$. Then $H^*(BG;\mathbb{F}_p) = \bigwedge(s_1, \ldots, s_r) \otimes \mathbb{F}_p[t_1, \ldots, t_r]$, where each s_i has degree 1, each t_i has degree 2, and $t_i = \beta s_i$ for $1 \le i \le r$, where β is the Bockstein operation. If X is a space such that $X \underset{p}{\sim} L_p^{2n+1}$, then $H^*(X;\mathbb{F}_p) = \bigwedge(u) \otimes \mathbb{F}_p[v]/(v^{n+1})$, where $\deg(u) = 1$, $\deg(v) = 2$ and $v = \beta u$. We shall assume for the rest of this section that G and X are of these forms.

Now suppose that G is acting on X, and consider the Leray-Serre spectral sequence (with coefficients in $\mathbb{F}_p$) of the Borel construction bundle $X \overset{i}{\longrightarrow} X_G \overset{\pi}{\longrightarrow} BG$.

Then u transgresses to a class in $H^2(BG; \mathbb{F}_p)$, which we may write as $\lambda + \phi$, where λ is linear in $t_1, \ldots, t_r$ and $\phi \in \bigwedge^2(s_1, \ldots, s_r)$. ϕ can be thought of as an alternating bilinear form on G; and we can choose a generating set for G (and hence the dual basis $s_1, \ldots, s_r$) to make the expression for $\lambda + \phi$ as simple as possible. Since a vector space with a non-singular alternating bilinear form has a symplectic basis starting (or ending) with any given non-zero vector, one easily has the following.

(3.1) Lemma. *Given $\lambda + \phi$ as above, there is a generating set $s_1, \ldots, s_r$ for $H^*(BG; \mathbb{F}_p)$ so that $\lambda + \phi$ has one of the following five forms:*

(1) $\lambda + \phi = t_{2m} + \sum_{i=1}^{m} s_i s_{m+i}$ (where $r \geq 2m$);

(2) $\lambda + \phi = t_{2m+1} + \sum_{i=1}^{m} s_i s_{m+i}$ (where $r > 2m$);

(3) $\lambda = 0$ and $\phi = \sum_{i=1}^{m} s_i s_{m+i}$ (where $r \geq 2m$);

(4) $\phi = 0$ and $\lambda = t_1$;

(5) $\lambda = 0$ and $\phi = 0$.

(3.2) Remark. We shall refer to these cases frequently below; but we are only interested in cases (1)–(3). When X satisfies condition (LT), then case (5) arises if and only if the fixed point set $X^G \neq \emptyset$; and so the rank of the action is 0. In case (4) let $K \cong (\mathbb{Z}/(p))^{r-1}$ be the subgroup of G on which $s_1 = 0$. Then for the action of K on X we are in case (5); and so $X^K \neq \emptyset$ if X satisfies condition (LT). Thus, if X satisfies condition (LT), then case (4) arises only if the rank of the action is 1. Since we are concerned with actions of maximal rank, we consider only cases (1)–(3), therefore.

Let $R = \mathbb{F}_p[t_1, \ldots, t_r]$, the polynomial part of $H^*(BG; \mathbb{F}_p)$. And for a subgroup $K \subseteq G$, let R_K be the polynomial part of $H^*(BK; \mathbb{F}_p)$. Inclusion of K into G induces a restriction homomorphism $R \longrightarrow R_K$, and one sets $PK = \ker[R \longrightarrow R_K]$. PK is then a prime ideal in R generated by homogeneous linear polynomials. (For the particular K in Remark (3.2), $PK = (t_1)$.) The height of PK, ht PK, which coincides with the minimal number of generators, is equal to the corank of K: i.e. ht $PK = j$ if and only if $K \cong (\mathbb{Z}/(p))^{r-j}$.

(3.3) Definition. A homogeneous ideal $J \subseteq R$ is said to be a Hsiang-Serre ideal if every minimal prime ideal of J is of the form PK for some subgroup $K \subseteq G$: i.e. if every minimal prime is linear.

The name Hsiang-Serre was chosen in part because of the following theorem of Serre.

(3.4) Theorem ([S]). *Let $J \subseteq R = \mathbb{F}_p[t_1, \ldots, t_r]$ be a homogeneous ideal. Then J is Hsiang-Serre if J is invariant under the Steenrod algebra (i.e. for any $a \in J$ and $j \geq 0$, $\mathcal{P}^j a \in J$, where $\mathcal{P}^j$ is the Steenrod operation). (This holds also for $p = 2$.)*

Returning to the action of G on X let $\pi_0^* = \pi^*|R : R \longrightarrow H_G^*(X; \mathbb{F}_p)$. Then $\ker \pi_0^*$ is invariant under the Steenrod algebra; and hence it is Hsiang-Serre. When

X satisfies condition (LT), then it follows from the Localization Theorem (as shown in [Hs]) that the subgroups of G corresponding to the minimal primes of $\ker \pi_0^*$ are precisely the maximal isotropy subgroups. Hence $\mathrm{ht}\,\ker \pi_0^*$ is equal to the rank of the action (when X satisfies (LT)).

In Section 2 Quillen's theorem gave a complete description of $\ker \pi_0^*$ (which is the same as $\ker \pi^*$ there); and it is easy to obtain analogues of Quillen's results for odd primes (as was done in [T, Y]), and they give some interesting information about $\ker \pi_0^*$ as we shall now describe.

Before stating some of the results of Tezuka and Yagita we need to describe some of the elements in $\ker \pi^*$. Recall that in the spectral sequence $u \in H^1(X; \mathbb{F}_p)$ transgresses to $\lambda + \phi \in H^2(BG; \mathbb{F}_p)$, and by Lemma (3.1) we may assume that

$$\phi = \sum_{i=1}^{m} s_i s_{m+i},$$ where $2m$ is the rank of ϕ as an alternating bilinear form. Now

$$v = \beta u \text{ transgresses to } -\beta\phi = -\sum_{i=1}^{m}(t_i s_{m+i} - t_{m+i} s_i) = \phi_0, \text{ say. And, applying}$$

further Steenrod operations, for all $j \geq 0$, v^{p^j} transgresses to $\mathcal{P}^{p^{j-1}} \ldots \mathcal{P}^p \mathcal{P}^1 \phi_0 =$

$$-\sum_{i=1}^{m}(t_i^{p^j} s_{m+i} - t_{m+i}^{p^j} s_i) = \phi_j, \text{ say. } \phi_j \in \ker \pi^*; \text{ and so } \beta\phi_j = -\sum_{i=1}^{m}(t_i^{p^j} t_{m+i} - t_{m+i}^{p^j} t_i) \in \ker \pi_0^*.$$ (Indeed, by Kudo's Transgression Theorem (see, e.g., [McC]), $d_i(\phi_{j-1} \otimes v^{p^j - p^{j-1}}) = -\beta\phi_j$, where $i = 2p^{j-1}(p-1) - 1$.) Let $B_j = \beta\phi_j$ for $j \geq 1$.

Now the result of Tezuka and Yagita which concerns us is the following.

(3.5) Theorem ([T, Y]). **Let** $\phi = \sum_{i+1}^{m} s_i s_{m+i}$ **as above. Then**

(1) $B_1, \ldots, B_m$ **is a regular sequence (where B_i is as defined above);**
(2) **letting** $J = (B_1, \ldots, B_m)$, **the ideal in** $R = \mathbb{F}_p[t_1, \ldots, t_r]$ **generated by** $B_1, \ldots, B_m$, $B_i \in J$ **for all** $i \geq 1$;
(3) J **is Hsiang-Serre,** $J = \sqrt{J}$, **and the number of minimal prime ideals of J is** $N(p, m) := (p+1)(p^2+1)\cdots(p^m+1)$; **and**
(4) **letting** PK_i, $1 \leq i \leq N(p, m)$, **be the minimal primes of J, the subgroups K_i are the maximal subgroups on which the alternating bilinear form ϕ vanishes. (So the K_i are the maximal subgroups such that ϕ restricts to 0 in $H^*(BK_i; \mathbb{F}_p)$.)**

(3.6) Lemma. *With the notation of Theorem (3.5) and above,* $\sqrt{J + (\lambda)} = J + (\lambda)$; *and $J + (\lambda)$ is generated by a regular sequence.*

Proof. In case (3) of Lemma (3.1) the result is part of Theorem (3.5).

In case (2), $J + (\lambda) = (B_1, \ldots, B_m, t_{2m+1})$. The sequence $B_1, \ldots, B_m, t_{2m+1}$ is regular, and the minimal primes of $J + (\lambda)$ are $PK_i + (t_{2m+1})$ for $1 \leq i \leq N(p, m)$. And clearly $\sqrt{J + (\lambda)} = J + (\lambda)$.

In case (1), $\lambda + \phi = t_{2m} + \sum_{i=1}^{m} s_i s_{m+i}$.

$$B_j = -\sum_{i=1}^{m}(t_i^{p^j} t_{m+i} - t_{m+i}^{p^j} t_i)$$

$$= -\sum_{i=1}^{m-1}(t_i^{p^j} t_{m+i} - t_{m+i}^{p^j} t_i) - t_{2m}(t_m^{p^j} - t_{2m}^{p^j-1} t_m)$$

$$= B_j' - t_{2m}(t_m^{p^j} - t_{2m}^{p^j-1} t_m),$$

say. Hence $J + (\lambda) = (B_1', \ldots, B_m', t_{2m})$. By Theorem (3.5), $B_1', \ldots, B_{m-1}'$ is a regular sequence and $B_m' \in (B_1', \ldots, B_{m-1}')$. Thus $J + (\lambda) = (B_1', \ldots, B_{m-1}', t_{2m})$; and the result follows as for case (2). $\square$

(3.7) Remark. In cases (1) and (3), $\text{ht}(J + (\lambda)) = m$; and in case (2), $\text{ht}(J + (\lambda)) = m + 1$. Also, in cases (2) and (3), the number of minimal primes is $N(p, m)$; and in case (1), the number is $N(p, m - 1)$.

(3.8) Corollary. *With the notation used above,*

$$J + (\lambda^p) \subseteq \ker \pi_0^* \subseteq J + (\lambda); \text{ and } \sqrt{\ker \pi_0^*} = J + (\lambda).$$

Proof. We know that $J \subseteq \ker \pi_0^*$; and $\lambda^p \in \ker \pi_0^*$, since, in the spectral sequence, $d_2((\lambda + \phi)^{p-1} \otimes u) = (\lambda + \phi)^p = \lambda^p$.

Now suppose $PK \supseteq J + (\lambda)$. So B_1 and λ restrict to 0 in $H^*(BK; \mathbb{F}_p)$. But $B_1 = 0$ only if $\phi = 0$. So $\lambda + \phi$ restricts to 0 in $H^*(BK; \mathbb{F}_p)$. Thus the spectral sequence of $X \longrightarrow X_K \longrightarrow BK$ collapses; and so $\ker \pi_0^* \subseteq PK$. $\square$

Now we have a good grip on the spectral sequence for $X \longrightarrow X_G \longrightarrow BG$ when $G = (\mathbb{Z}/(p))^r$ and $X \underset{p}{\sim} L_p^{2n+1}$. Nevertheless it does not seem clear that we can easily give a proof of the inequality part of Theorem (1.5) analogous to the proof of Theorem (1.4). The problem is this: suppose v^{p^j} survives to E_∞; then $B_{j+1} = 0$ in $E_i^{*,0}$, where $i = 2p^j(p-1) + 1$; but how do we know that B_{j+1} is not killed earlier in the spectral sequence by some class lower than and further to the right of $\phi_j \otimes v^{p^j(p-1)}$? Anyhow there is a simple way to prove the inequality part of Theorem (1.5), which was used by T. Skjelbred ([Sk]) to show that $\text{rk}_p(\mathbb{C}P^n) = \text{ord}_p(n+1)$.

(3.9) Proof of inequality part of Theorem (1.5). Let $K \subseteq G$ be a maximal isotropy subgroup; and let $F_1, \ldots, F_s$ be the components of X^K. Now it is an easy exercise (see [Bre], Chap. VII, Exercise 8) to show that for each i, $1 \leq i \leq s$, $F_i \underset{p}{\sim} L_p^{2n_i+1}$ for some $n_i \geq 0$ such that $\sum_{i=1}^{s}(n_i+1) = n+1$. G/K acts on X^K; and the action must be free since K is a maximal isotropy subgroup. Let $H_i \subseteq G/K$ be the subgroup which leaves F_i invariant. So H_i acts freely on F_i; and hence $\text{rk}\,H_i \leq 1$. Let $F_{i_1}, \ldots, F_{i_t}$

be a complete set of representatives of the orbits of the action of G on the set $\{F_1, \ldots, F_s\}$. Let $|K| = p^j$. Then the order of the orbit of F_i is p^{r-j} or p^{r-j-1} according as rk $H_i = 0$ or 1. And all components in one orbit are homeomorphic. Hence, adding up $\dim_{\mathbb{F}_p} H^*(X^K; \mathbb{F}_p)$ in two ways, we have

$$2(n+1) = \sum_{\alpha=1}^{t} 2(n_{i_\alpha} + 1) p^{r-j-\epsilon_\alpha} ,$$

where $\epsilon_\alpha = 0$ or 1 according as rk $H_{i_\alpha} = 0$ or 1. So p^{r-j-1} divides $n+1$: i.e. $r - j - 1 \leq \mathrm{ord}_p(n+1)$. Finally, of course, $r - j$ is the rank of the action. $\square$

(3.10) Summary. Let $G = (\mathbb{Z}/(p))^r$, p an odd prime, and let X be a space satisfying condition (LT) such that $X \underset{p}{\sim} L_p^{2n+1}$. Let $\Phi : G \times X \longrightarrow X$ be an action of G on X. Then in this section we have shown the following:

(1) $\mathrm{rk}(\Phi) \leq 1 + \mathrm{ord}_p(n+1)$; and so $\mathrm{rk}_p(X) \leq 1 + \mathrm{ord}_p(n+1)$.
(2) All the maximal isotropy subgroups have the same order.
(3) If $\mathrm{rk}(\Phi) = \rho$, then the number of maximal isotropy subgroups is $N(p, \rho - 1)$ in cases (1) and (2) of Lemma (3.1), and $N(p, \rho)$ in case (3).
(4) If $\mathrm{rk}(\Phi) = \rho$, then $\rho \leq r/2$ in cases (1) and (3), and $\rho \leq \frac{1}{2}(r+1)$ in case (2).

4. THE STANDARD ACTION

In this section we shall describe an action of a p-torus on L_p^{2n+1} having rank $1 + \mathrm{ord}_p(n+1)$. The action is an obvious one, and presumably it has been known in various contexts for a long time. (In the case $p = 2$ it is presumably equivalent to the action given in [Sk].)

We begin by describing a (not quite extra-special) p-group $\widetilde{G}$ of order p^{2m+2} where $m \geq 1$. $\widetilde{G}$ has generators $x_1, \ldots, x_m$, $y_1, \ldots, y_m$ and g. In $\widetilde{G}$ there is a central element b of order p. There are the following relations: $x_i^p = 1$ and $y_i^p = 1$ for $1 \leq i \leq m$; $g^p = b$; $[x_i, x_j] = 1$ and $[y_i, y_j] = 1$ for $1 < i < m$ and $1 \leq j \leq m$, where $[x, y] = xyx^{-1}y^{-1}$; $[x_i, y_j] = 1$ for $1 \leq i \leq m$, $1 \leq j \leq m$ and $i \neq j$; $[x_i, y_i] = b$ for $1 \leq i \leq m$; $[x_i, g] = 1$ and $[y_i, g] = 1$ for $1 \leq i \leq m$. In particular, $\widetilde{G}$ is an extension $1 \longrightarrow K \longrightarrow \widetilde{G} \longrightarrow G \longrightarrow 1$, where K is the central cyclic subgroup generated by b, and $G \cong (\mathbb{Z}/(p))^{2m+1}$.

Now we shall describe a faithful unitary representation of $\widetilde{G}$ on $\mathbb{C}^{p^m}$. Put the coordinates $z_1, \ldots, z_{p^m}$ into blocks (bks) as follows.

0-bks: $B_{01} = (z_1), \ldots, B_{0p^m} = (z_{p^m})$.

1-bks: $B_{11} = (z_1, \ldots, z_p), B_{12} = (z_{p+1}, \ldots, z_{2p}), \ldots, B_{1p^{m-1}} = (z_{p^m-p+1}, \ldots, z_{p^m})$.

$\vdots$

j-bks: $B_{j1} = (B_{j-11}, \ldots, B_{j-1p}), \ldots, B_{jp^{m-j}} = (B_{j-1p^{m-j+1}-p+1}, \ldots, B_{j-1p^{m-j+1}})$.

$\vdots$

m-bk: $B_{m1} = (B_{m-11}, \ldots, B_{m-1p})$.

Now let x_j act by cyclically permuting the $(j-1)$-blocks: i.e.

$$x_j B_{jk} = x_j(B_{j-1pk-p+1}, B_{j-1pk-p+2}, \ldots, B_{j-1pk})$$
$$= (B_{j-1pk-p+2}, \ldots, B_{j-1pk}, B_{j-1pk-p+1}).$$

Let y_j act as follows:

$$y_j B_{jk} = (B_{j-1pk-p+1}, \omega B_{j-1pk-p+2}, \ldots, \omega^{p-1} B_{j-1pk}),$$

where $\omega = e^{2\pi i/p}$.

Finally let $gz = g(z_1, \ldots, z_{p^m}) = \theta z = (\theta z_1, \ldots, \theta z_{p^m})$, where $\theta = e^{2\pi i/p^2}$. So $bz = \omega z$.

It is easy to check that this is indeed a well-defined faithful unitary representation of $\widetilde{G}$. Furthermore, for $k \geq 1$, we can extend the representation to $\mathbb{C}^{kp^m}$ as $\mathbb{C}^{p^m} \otimes \mathbb{C}^k$, with the trivial representation on $\mathbb{C}^k$.

For brevity let $N = kp^m$, and let $V = \mathbb{C}^N$ with the above action of $\widetilde{G}$. Let $S(V)$ be the unit sphere. So $S(V) \approx S^{2N-1}$. Now $K = \langle b \rangle \cong \mathbb{Z}/(p)$ acts freely on $S(V)$. Let $L_K(V) = S(V)/K$. So $L_K(V) \approx L_p^{2N-1}$; and $G = \widetilde{G}/K \cong (\mathbb{Z}/(p))^{2m+1}$ acts on $L_K(V)$.

Viewing BK as $E\widetilde{G}/K$, G acts freely on BK and $BK/G = B\widetilde{G}$. Hence $(BK)_G \simeq B\widetilde{G}$: i.e. the Borel construction for the free G-action on BK is homotopy equivalent to $B\widetilde{G}$.

Now we have the following diagram of Borel construction bundles.

$$
\begin{array}{ccccc}
L_K(V) & \xleftarrow{\ q\ } & S(V)_K & \xrightarrow{\ \pi_1\ } & BK \\
\downarrow & & \downarrow & & \downarrow \\
L_K(V)_G & \longleftarrow & (S(V)_K)_G & \longrightarrow & (BK)_G \\
\downarrow & & \downarrow & & \downarrow \pi \\
BG & \xleftarrow{\ id\ } & BG & \xrightarrow{\ id\ } & BG
\end{array}
$$

π_1 is the bundle map with fibre $S(V)$, and q is the homotopy equivalence $S(V)_K \longrightarrow S(V)/K$.

Let $s \in H^1(BK; \mathbb{F}_p)$ correspond to b, and let $s_1, \ldots, s_{2m}, s_{2m+1} \in H^1(BG; \mathbb{F}_p)$ correspond to the images of $x_1, \ldots, x_m, y_1, \ldots, y_m, g$, respectively. Now by [McL], Chapter XI, sections 9 and 10, in the Lyndon-Hochschild-Serre spectral sequence for the extension $1 \longrightarrow K \longrightarrow \widetilde{G} \longrightarrow G \longrightarrow 1$, s transgresses to $-(t_{2m+1} + \sum_{i=1}^{m} s_i s_{m+i})$. Hence in the Leray-Serre spectral sequence for π, s also transgresses to

$$-(t_{2m+1} + \sum_{i=1}^{m} s_i s_{m+i}),$$

or some non-zero multiple of it. Since the fibre of π_1 is $S(V)$, $\pi_1^*(s) \neq 0$. So $(q^*)^{-1}\pi_1^*(s) \neq 0$. Hence there is a generator $u \in H^1(L_K(V); \mathbb{F}_p)$

which transgresses to $t_{2m+1} + \sum\limits_{i-1}^{m} s_i s_{m+i}$ in the Leray-Serre spectral sequence of $L_K(V)_G \longrightarrow BG$. Hence, by the results of Section 3 (see especially Remark (3.7)), the rank of the action of G on $L_K(V)$ is $m+1$. And $m+1 = 1 + \mathrm{ord}_p(N)$, if p does not divide k.

(4.1) Summary. $\mathrm{rk}_p(L_p^{2n+1}) = 1 + \mathrm{ord}_p(n+1)$. A similar, but easier, argument with $p = 2$ shows that $\mathrm{rk}_2(\mathbb{R}P^n) = \mathrm{ord}_2(n+1)$. When $p = 2$, $\widetilde{G}$ is constructed as above but without the element g: so $\widetilde{G}$ is extra-special in this case. (See [Sk].)

(4.2) Remark. A. Adem in [A] uses constructions similar to those above together with a theorem of his on exponents in equivariant integral Tate cohomology to detect integral cohomology classes of extra-special p-groups having large exponent (i.e. additive order). We shall now indicate briefly how to do the same for the group $\widetilde{G}$, avoiding Tate cohomology, using a more elementary result of Browder ([Bro]), or, independently, Gottlieb ([G]), instead of Adem's theorem in [A].

Let $x \in H^{2N-1}(L_K(V); \mathbb{Z})$ be a top class, and let $i : L_K(V) \longrightarrow L_K(V)_G$ be the inclusion of the fibre. Since the rank of the action of G on $L_K(V)$ is $m+1$, by Browder's result on exponents ([Bro]) or Gottlieb's on fibre number ([G]), $p^{m+1}x \in \mathrm{im}\, i^*$, but $p^m x \notin \mathrm{im}\, i^*$. With the notation of the diagram above, let $y = q^*(x)$. Since q is a homotopy equivalence, letting $i_1 : S(V)_K \longrightarrow (S(V)_K)_G$ be the inclusion of the fibre, we have $p^{m+1}y \in \mathrm{im}\, i_1^*$ and $p^m y \notin \mathrm{im}\, i_1^*$. Since K is acting freely on $S(V)$, letting $i_2 : S(V) \longrightarrow S(V)_K$ be the inclusion of the fibre and $z \in H^{2N-1}(S(V); \mathbb{Z})$ an appropriate choice of the top class, by Browder and Gottlieb (or, in fact, by more elementary considerations), $pz \in \mathrm{im}\, i_2^*$ and $z \notin \mathrm{im}\, i_2^*$. Hence, letting i_3 be the inclusion of the fibre in $S(V) \longrightarrow (S(V)_K)_G \longrightarrow (BK)_G$, it follows that $p^{m+2}z \in \mathrm{im}\, i_3^*$ and $p^{m+1}z \notin \mathrm{im}\, i_3^*$. But it is easy to see that $S(V) \longrightarrow (S(V)_K)_G \longrightarrow (BK)_G$ is homotopy equivalent to $S(V) \overset{j}{\longrightarrow} S(V)_{\widetilde{G}} \longrightarrow B\widetilde{G}$. So $p^{m+2}z \in \mathrm{im}\, j^*$ and $p^{m+1}z \notin \mathrm{im}\, j^*$. Thus, letting $c = \delta z \in H^{2N}(S(V)_{\widetilde{G}}, S(V); \mathbb{Z})$, c has exponent p^{m+2}. Let $\phi : (S(V)_{\widetilde{G}}, S(V)) \longrightarrow (B\widetilde{G}, *)$ be the map of pairs where $*$ is a base point in $B\widetilde{G}$. Then it follows from the Leray-Serre spectral sequence of ϕ, that $\phi^* : H^i(B\widetilde{G}, *; \mathbb{Z}) \longrightarrow H^i(S(V)_{\widetilde{G}}, S(V); \mathbb{Z})$ is an isomorphism for $i \leq 2N$. Hence we obtain a class $a_k \in H^{2N}(B\widetilde{G}; \mathbb{Z})$ of exponent p^{m+2} (where the subscript k comes from $N = kp^m$). By construction, a_k is the transgression of z in the spectral sequence of $S(V)_{\widetilde{G}} \longrightarrow B\widetilde{G}$; and so a_k is the Euler class of the representation V. Since $V = \mathbb{C}^{p^m} \otimes \mathbb{C}^k$, $a_k = a_1^k$. So we have $a_1 \in H^{2p^m}(B\widetilde{G}; \mathbb{Z})$ such that a_1^k has exponent p^{m+2} for all $k \geq 1$.

5. COMMENTS

We begin with the following easy lemma.

(5.1) Lemma. Let X be a space satisfying condition (LT), and suppose that $X \underset{p}{\sim} L_p^{2n_1+1} \times \cdots \times L_p^{2n_s+1}$. Suppose that $G = (\mathbb{Z}/(p))^r$ acts freely on X and

trivially on $H^*(X; \mathbb{F}_p)$. Then $r \le s$.

Proof. Let $u_1, \ldots, u_s$ generate $H^1(X; \mathbb{F}_p)$. In the spectral sequence of $X_G \longrightarrow BG$ let u_i transgress to $\lambda_i + \phi_i$, where λ_i is a linear polynomial and ϕ_i is exterior. If $r > s$, let $K \subseteq G$ be the subgroup with $PK = (\lambda_1, \ldots, \lambda_s)$. So $\operatorname{rk} K \ge r - s$. In the spectral sequence for $X_K \longrightarrow BK$, therefore, $u_1, \ldots, u_s$ have purely exterior transgressions. Hence any cyclic subgroup of K has a fixed point. $\square$

Now it is natural to conjecture the following.

(5.2) Conjecture. Let X be a space satisfying condition (LT), and suppose that $X \underset{p}{\sim} L_p^{2n_1+1} \times \cdots \times L_p^{2n_s+1}$. Then $\operatorname{frk}_p(X) \le s$.

Suppose that $X = (L_p^{2n+1})^s := L_p^{2n+1} \times \cdots \times L_p^{2n+1}$ s times; and suppose that $G = (\mathbb{Z}/(p))^r$ is acting freely on X. Let $\widetilde{X} = (S^{2n+1})^s$ be the universal covering space. By [Bre], Chapter I, Theorem 9.3, there is a group extension $1 \longrightarrow K \longrightarrow \widetilde{G} \longrightarrow G \longrightarrow 1$ with $K = (\mathbb{Z}/(p))^s$, such that $\widetilde{G}$ acts freely on $\widetilde{X}$. By a theorem of Adem and Browder ([A, B]), $\operatorname{rk}_p(\widetilde{G}) = s$. Hence one is led to ask the following question: does there exist such a group extension with $r > s$ and $\operatorname{rk}_p(\widetilde{G}) = s$? The answer is clearly "no" if K is central in $\widetilde{G}$ by an argument very similar to the proof of Lemma (5.1). The answer in general, however, is "yes", as is shown by the following example owing to A. Caranti.

(5.3) Example (A. Caranti). Let H be a free group on generators $x_1, \ldots, x_d$ in the variety of groups of exponent p and nilpotency class 2. Then the derived group H' is generated by the commutators $[x_i, x_j]$ for $1 \le i < j \le d$; and so $H' \cong (\mathbb{Z}/(p))^e$ where $e = \binom{d}{2}$. Let $B = (\mathbb{Z}/(p^2))^e$ generated by $\{y_{ij} : 1 \le i < j \le d\}$. Let $\widetilde{G}$ be the quotient of $H \times B$ obtained by identifying y_{ij}^p with $[x_i, x_j]$ for $1 \le i < j \le d$. So $|\widetilde{G}| = p^{d+2e}$. Let $K \subseteq \widetilde{G}$ be the subgroup generated by H' and x_d. Then $K \cong (\mathbb{Z}/(p))^{e+1}$; and it is not hard to show that K is normal in $\widetilde{G}$, and $\operatorname{rk}_p(\widetilde{G}) = e + 1$. And it is clear that $G := \widetilde{G}/K \cong (\mathbb{Z}/(p))^{d+e-1}$. Thus $r = d + e - 1$ and $s = e + 1$; and so $r > s$ if $d > 2$.

REFERENCES

[A]   A. Adem, *Torsion in equivariant cohomology*, Comment. Math. Helv. **64** (1989), 401–411.
[A, B] A. Adem and W. Browder, *The free rank of symmetry of* $(S^n)^k$, Invent. Math. **92** (1988), 431–440.
[A, P] C. Allday and V. Puppe, *Cohomological Methods in Transformation Groups*, Cambridge University Press (to appear).
[Bre] G. Bredon, *Introduction to Compact Transformation Groups*, Academic Press, New York, 1972.
[Bro] W. Browder, *Actions of elementary abelian p-groups*, Topology **27** (1988), 459–472.
[G]   D. Gottlieb, *The trace of an action and the degree of a map*, Trans. Amer. Math. Soc. **293** (1986), 381–410.
[Hs]   W.-Y. Hsiang, *Cohomology Theory of Topological Transformation Groups*, Ergebnisse der Math. und ihrer Grenzgebiete Vol. 85, Springer-Verlag, Berlin, 1975.

[McL] S. MacLane, *Homology*, Grundlehren der Math. Wissenschaften Vol. 114, Springer-Verlag, Berlin, 1963.

[M]   H. Matsumura, *Commutative Ring Theory*, Cambridge studies in advanced mathematics Vol. 8, Cambridge University Press, Cambridge , 1986.

[McC] J. McCleary, *User's Guide to Spectral Sequences*, Math. Lecture Series Vol. 12, Publish or Perish, Wilmington, 1985.

[Q1]   D. Quillen, *The mod.2 cohomology rings of extra-special 2-groups and the spinor groups*, Math. Ann. **194** (1971), 197–212.

[Q2]   D. Quillen, *The spectrum of an equivariant cohomology ring : I*, Ann. of Math. **94** (1971), 549–572.

[S]   J.-P. Serre, *Sur la dimension cohomologique des groupes profinis*, Topology **3** (1965), 413–420.

[Sk]   T. Skjelbred, *Actions of p-tori on projective space*, University of Oslo, Institute of Mathematics, Preprint Series No. 4, 1975.

[T, Y] M. Tezuka and N. Yagita, *The varieties of the mod.p cohomology rings of extra special p-groups for an odd prime p*, Math. Proc. Camb. Phil. Soc. **94** (1983), 449–459.

DEPARTMENT OF MATHEMATICS, UNIVERSITY OF HAWAII, HONOLULU, HI 96822
E-mail address: allday@ uhunix.bitnet

EQUIVARIANT SURGERY AND APPLICATIONS

ANTHONY BAK AND MASAHARU MORIMOTO

Universität Bielefeld and Okayama University

ABSTRACT. Strong vanishing theorems for $(4k+3)$-dimensional $(k \geq 0)$ equivariant surgery obstruction groups are proved and applied to problems concerning transformation groups on spheres and lens spaces. New results concerning 3–dimensional equivariant surgery theory are announced and used in the applications above.

0. INTRODUCTION

This report describes some recent advances in surgery and transformation groups. It focuses on 3–dimensional surgery and new vanishing theorems for $4k+3$–dimensional $(k \geq 0)$ G–equivariant surgery obstruction groups and applies the results obtained here to prove the existence of one fixed point smooth actions of the alternating group A_5 on S^7 and to prove the existence of smooth conjugations on homotopy lens spaces. The vanishing theorems are for the class of finite groups G whose 2–hyperelementary subgroups are abelian or dihedral, and they show that all $(4k+3)$–dimensional $(k \geq 0)$ G–equivariant surgery obstruction groups $W_{4k+3}(\mathbb{Z}[G], \Lambda(G(X)),$ trivial $\omega_X)$ vanish providing the set $G(X)$ of involutions in G is sufficiently large. The vanishing results are proved using K–theory exact sequences and by applying number theory, in particular the vanishing of certain congruence kernels, to compute terms and maps in the exact sequences. The vanishing theorems imply that G–equivariant surgery operations can always be performed. The existence of these operations is used in proving the geometric results above.

The rest of the report is organized as follows. In section 1, notation and basic definitions are recalled and the 3–dimensional surgery theorem is stated. The precise definition of the 3–dimensional equivariant surgery obstruction groups is deferred till section 3. In section 2, n–dimensional $(n \geq 5)$ surgery theorems are recalled. In section 3, the 3–dimensional and n–dimensional $(n \geq 5)$ equivariant surgery obstruction groups are defined. In section 4, the vanishing theorems are proved. In section 5, the application to S^7 is stated and proved, and the application to lens spaces is stated, but the proof is referenced to another article of the authors.

13

1. 3–DIMENSIONAL EQUIVARIANT SURGERY

Three dimensional equivariant surgery is concerned with the problem of converting a smooth G–map $f : X \to Y$, satisfying certain conditions, between 3–dimensional smooth G–manifolds to a homology equivalence $f' : X' \to Y$.

We begin by recalling standard notation and conventions in the n–dimensional ($n \geq 1$) situation, which will be used also in later sections of this report.

Let G denote a finite group. Let $\mathfrak{M}^n(G)$ denote the category of all compact, connected, oriented, smooth, n–dimensional manifolds with a smooth action of G and all smooth G–equivariant maps between such manifolds. For a subgroup H of G and $X \in \mathfrak{M}^n(G)$, let $X^H = \{x | x \in X, x^h = x \ \forall h \in H\}$. The connected components of X^H are submanifolds of X and by definition,

$$\dim(X^H) = \sup\{\dim(M) \mid M \text{ is a connected component of } X^H\}.$$

Let $X_s = \cup_{H \subseteq G, H \neq 1} X^H$. X_s is called the *singular set* of X. A map $f : X \to Y$ in $\mathfrak{M}^n(G)$ is called a *singularity* equivalence if one of the following three conditions is satisfied:

(1) The restriction $f|_{X_s} : X_s \to Y_s$ of f is an integral homology equivalence.
(2) The reduced projective class group $\tilde{K}_0(\mathbb{Z}[G]) = 0$ and $f|_{X^P} : X^P \to Y^P$ is a mod p homology equivalence for each nontrivial p–subgroup P of G.
(3) $f|_{X^H} : X^H \to Y^H$ is an integral homology equivalence for each nontrivial hyperelementary subgroup H of G.

Let ∂X denote the boundary of X. We allow the case where ∂X is empty. A map $f : X \to Y$ is called a *boundary equivalence* if $f(\partial X) \subseteq \partial Y$ and $f|_{\partial X} : \partial X \to \partial Y$ is an integral homology equivalence. A map $f : X \to Y$ is called a *degree* 1 map if the induced map $f_* : H_n(X, \mathbb{Z}) \to H_n(Y, \mathbb{Z})$ takes the distinguished generator of $H_n(X, \mathbb{Z})$ ($\cong \mathbb{Z}$) given by the orientation of X to the distinguished generator of $H_n(Y, \mathbb{Z})$. Let $T(X)$ denote the tangent bundle of X. Since the action of G on X is smooth, there is a canonical action of G on $T(X)$ making it a G–vector bundle (of dimension n) over X. A G–framed map $(f, b) : (X, T(X) \oplus f^*\xi_-) \to (Y, f^*\xi_+)$ is a pair (f, b) such that $f : X \to Y \in \mathfrak{M}^n(G), \xi_{\mp}$ are G–vector bundles over Y such that $\dim(T(X)) = \dim(f^*\xi_+) - \dim(f^*\xi_-)$, and $b : T(X) \oplus f^*\xi_- \xrightarrow{\cong} f^*\xi_+$ is a *stable* isomorphism of G–vector bundles where $f^*\xi$ is the pullback bundle on X of ξ on Y.

For $X \in \mathfrak{M}^n(G)$, let

$$\omega_X : G \to \{\pm 1\}$$

$$g \mapsto \begin{cases} +1, & \text{if the operation of } g \text{ on } X \text{ is orientation preserving} \\ -1, & \text{otherwise.} \end{cases}$$

ω_X is called the *orientation homomorphism*.

An element of order 2 of G is called an *involution*. Let

$$G(2) = \{g \mid g \in G, \ g \text{ an involution}\}$$
$$G(> 2) = \{g \mid g \in G, \ \text{order}(g) > 2\}.$$

In [M1], an interesting and important relation between the involutions in G and the operation of surgery on G–manifolds is described. If $g \in G$, let $< g >$ denote the subgroup of G generated by g. If r is a real number, let $[r]$ denote the largest integer $\leq r$ Following [M1], we define for $X \in \mathfrak{M}^n(G)$,

$$G(X) = \left\{ g \mid g \in G, \ g \text{ an involution, } \dim(X^{<g>}) = \left[\frac{n-1}{2} \right] \right\}.$$

The role the set $G(X)$ plays in 3–dimensional equivariant surgery is explained in Theorem 1 below and in n–dimensional equivariant surgery, $n \geq 5$, in Theorem 2 of the next section.

Theorem 1 [BM1]. *Let $f : X \to Y \in \mathfrak{M}^3(G)$. Suppose that Y is simply connected, that f has degree 1 and is a boundary and singularity equivalence, and that*

$$\dim X^{<g>} \leq 1 \quad \forall g \in G - \{1\},$$
$$\dim X^{<g>} = 1 \quad \forall g \in G(2), \ i.e., \ G(X) = G(2).$$

Then:

a) *There is an equivariant surgery obstruction group $W_3\big(\mathbb{Z}[G], \Lambda(G(X) \cup \{1\})$, $\omega_X\big)$ (defined in section 3) and an obstruction element $\sigma(f) \in W_3\big(\mathbb{Z}[G]$, $\Lambda(G(X) \cup \{1\}), \omega_X\big)$ such that if $\sigma(f) = 0$ then one can alter $f : X \to Y$, by a G–surgery operation leaving $\partial X \cup X_s$ fixed, to an integral homology equivalence $f' : X' \to Y \in \mathfrak{M}^3(G)$.*

b) *If $(f, b) : (X, T(X) \oplus f^*\xi_-) \to (Y, f^*\xi_+)$ is a G–framed map then there is an equivariant surgery obstruction group $W_3(\mathbb{Z}[G], \Lambda(G(X)), \omega_X)$ (defined in section 3) and an obstruction element $\sigma(f) \in W_3(\mathbb{Z}[G], \Lambda(G(X)), \omega_X)$ such that if $\sigma(f) = 0$ then one can alter (f, b) by a G–surgery operation leaving $\partial X \cup X_s$ fixed, to a G–framed integral homology equivalence $(f', b') : (X', T(X') \oplus f'^*\xi_-) \to (Y, f'^*\xi_+)$ where b' is an 'extension' of b to $T(X')$.*

One of the fundamental problems of equivariant surgery theory is to establish when equivariant surgery groups are trivial. This problem will be addressed in section 4 and the results applied in section 5.

2. n–Dimensional Equivariant Surgery, $n \geq 5$

We recall in a uniform way equivariant surgery theorems in T. Petrie [P1] (see also [PR, §3 (12.4)] and [DP]) and [M1], [M2]. The result which we state is found originally in [P1] under the assumption that $G(X) = \emptyset$ and in [M1], [M2] in general. Note that $G(X) = \emptyset$ whenever $|G|$ is odd.

Adopt the notation and conventions of the previous section.

Theorem 2. *Let $(f, b) : (X, T(X) \oplus f^*\xi_-) \to (Y, f^*\xi_+)$ be a degree 1, G–framed map satisfying the following conditions:*

(1) $f : X \to Y \in \mathfrak{M}^n(G), n \geq 5$.

(2) Y is simply connected.

(3) f is a boundary equivalence.

(4) f is a singularity equivalence.

(5) $\dim X_s \leq \left[\frac{n-1}{2}\right]$ (gap hypothesis).

Then there is an equivariant surgery obstruction group $W_n(\mathbb{Z}[G], \Lambda(G(X)), \omega_X)$ and a surgery obstruction $\sigma(f, b) \in W_n(\mathbb{Z}[G], \Lambda(G(X)), \omega_X)$ such that $\sigma(f, b) = 0$ if and only if one can alter (f, b) by a G–surgery operation leaving $\partial X \cup X_s$ fixed to a G–framed homotopy equivalence $(f', b') : (X', T(X') \oplus f'^* \xi_-) \to (Y, f'^* \xi_+)$ where b' is an 'extension' of b to $T(X')$.

3. Equivariant Surgery Obstruction Groups

We begin by describing the general K–theoretic setting where surgery obstruction groups live. Then we define the equivariant surgery obstruction groups required in Theorems 1 and 2 above. In the next section, we shall use K–theory exact sequences to show that a large class of $(4k + 3)$–dimensional equivariant surgery obstruction groups vanish. A general reference for the current section is [B3].

Let A denote a ring with involution $a \mapsto \bar{a}$ and let $\lambda \in \text{center}(A)$ such that $\lambda\bar{\lambda} = 1$. We have in mind the case $A = \mathbb{Z}[G]$ and $\lambda = \pm 1$ where the involution on A sends each element $g \in G$ to $\omega(g)g^{-1}$ and ω is some homomorphism $\omega : G \to \{\pm 1\}$.

A λ–form parameter Λ on A is an additive subgroup of A such that

1) $\{a - \lambda\bar{a} \mid a \in A\} \subseteq \Lambda \subseteq \{a \mid a \in A, a = -\lambda\bar{a}\}$

2) $a\Lambda\bar{a} \subseteq \Lambda \ \forall a \in A$.

It is easy to check that the left and right hand extremes in 1) satisfy the closure condition in 2) and that they coincide whenever there is an element $c \in \text{center}(A)$ such that $c + \bar{c} = 1$, e.g. $\frac{1}{2} \in A$. The extremes are denoted by $\min^\lambda(A)$ and $\max^\lambda(A)$, respectively.

If $S \subseteq \max^\lambda(A)$ is a subset, let $\Lambda(S)$ denote the smallest λ–form parameter containing S. If $S = \emptyset$ then $\Lambda(S) = \min^\lambda(A)$.

Let M be a right A–module. A sesquilinear form ϕ on M is a biadditive pairing $\phi : M \times M \to A$ such that $\phi(ma, nb) = \bar{a}\phi(m, n)b$ for all $m, n \in M$ and $a, b \in A$. One associates to any sesquilinear form ϕ on M a Λ–quadratic form

$$q_\phi : M \to A/\Lambda, \ m \mapsto [\phi(m, m)],$$

and a λ–Hermitian form

$$\langle \, , \, \rangle_\phi : M \times M \to A, \ (m, n) \mapsto \phi(m, n) + \lambda\overline{\phi(n, m)}.$$

A Λ–quadratic module is a triple $(M, q_\phi, \langle \, , \, \rangle_\phi)$. Such a module is called *non-singular* if M is finitely generated and projective over A and if the map $M \to \text{Hom}_A(M, A), m \mapsto \langle m, \, \rangle_\phi$ is bijective. The *orthogonal sum* of two quadratic modules is defined by $(M, q_\phi, \langle \, , \, \rangle_\phi) \perp (N, q_\psi, \langle \, , \, \rangle_\psi) = (M \oplus N, q_{\phi\oplus\psi}, \langle \, , \, \rangle_{\phi\oplus\psi})$. A *morphism* $f : (M, q_\phi, \langle \ , \ \rangle_\phi) \to (N, q_\psi, \langle \ , \ \rangle_\psi)$ of Λ–quadratic modules is an A–linear map $f : M \to N$ which preserves the Λ–quadratic and λ–Hermitian forms. Let

$$\underline{Q}^\lambda(A, \Lambda)$$

denote the category with product (symmetric monoidal category) of nonsingular Λ–quadratic modules. If the underlying modules M above are assumed to be free of even dimension then we write

$$\underline{Q}^\lambda(A, \Lambda)_{\text{even-free}} \text{ instead of } \underline{Q}^\lambda(A, \Lambda).$$

An important example of a free nonsingular Λ–quadratic module is the *hyperbolic plane* $\mathbb{H}(A) = (A \oplus \text{Hom}_A(A, A), q_\phi, \langle \ , \ \rangle_\phi)$ where $\phi((a, f), (b, g)) = f(b)$.

Define á la Quillen [Q] the K–theory groups

$$KQ_i^\lambda(A, \Lambda) = K_i(\underline{Q}^\lambda(A, \Lambda)), \ i \geq 0,$$

$$KQ_i^\lambda(A, \Lambda)_{\text{even-free}} = K_i(\underline{Q}^\lambda(A, \Lambda)_{\text{even-free}}), \ i \geq 0.$$

One can show that

$$KQ_i^\lambda(A, \Lambda) = KQ_i^\lambda(A, \Lambda)_{\text{even-free}} \text{ for } i \geq 1.$$

The group $KQ_0^\lambda(A, \Lambda)_{\text{even-free}}$ will suffice to define the even dimensional equivariant surgery obstruction groups. To define the odd dimensional surgery obstruction groups, it is convenient to use a matrix interpretation of $KQ_1^\lambda(A, \Lambda)$.

If n is a natural number, let $\mathbb{M}_n(A)$ denote the ring of $n \times n$–matrices with coefficients in A and involution $(a_{ij}) \mapsto (a'_{ij})$ where $a'_{ij} = \bar{a}_{ji}$. This involution is called the *conjugate transpose* involution and will be denoted by $\alpha \mapsto \bar{\alpha}$. Let $GQ_{2n}^\lambda(A, \Lambda)$ denote the group of all invertible $2n \times 2n$ matrices $\begin{pmatrix} \alpha & \beta \\ \gamma & \delta \end{pmatrix}$ such that

1) $\begin{pmatrix} \alpha & \beta \\ \gamma & \delta \end{pmatrix}^{-1} = \begin{pmatrix} \bar{\delta} & \lambda\bar{\beta} \\ \overline{\lambda\gamma} & \bar{\alpha} \end{pmatrix}$,

2) the diagonal coefficients of $\bar{\gamma}\alpha$ and $\bar{\delta}\beta$ lie in Λ.

Frequently, $GQ_{2n}^\lambda(A, \Lambda)$ is denoted by $SU_n^\lambda(A, \Lambda)$ as in [M1]. Let $EQ_{2n}^\lambda(A, \Lambda)$ denote the subgroup generated by all matrices of the kind

$$\begin{pmatrix} I & \beta \\ 0 & I \end{pmatrix} \text{ and } \begin{pmatrix} I & 0 \\ \gamma & I \end{pmatrix} \in GQ_{2n}^\lambda(A, \Lambda)$$

where I denotes the $n \times n$ identity matrix. A necessary and sufficient condition for such matrices to be in $GQ_{2n}^\lambda(A, \Lambda)$ is that $\beta = -\lambda\bar{\beta}$, $\gamma = -\bar{\lambda}\bar{\gamma}$, and the diagonal entries of β and $\bar{\gamma}$ lie in Λ. $EQ_{2n}^\lambda(A, \Lambda)$ is called the elementary subgroup of

$GQ_{2n}^\lambda(A, \Lambda)$. If $\alpha \in GL_n(A)$ then the $2n \times 2n$ matrix $\begin{pmatrix} \alpha & 0 \\ 0 & \overline{\alpha}^{-1} \end{pmatrix} \in GQ_{2n}^\lambda(A, \Lambda)$. This matrix is called the *hyperbolic matrix* associated to α. The $2n \times 2n$ matrix

$$\begin{pmatrix} 0 & 0 & \lambda & 0 \\ 0 & I_{n-1} & 0 & 0 \\ 1 & 0 & 0 & 0 \\ 0 & 0 & 0 & I_{n-1} \end{pmatrix}$$

is also in $GQ_{2n}^\lambda(A, \Lambda)$. Let $RQ_{2n}^\lambda(A, \Lambda)$ denote the subgroup of $GQ_{2n}^\lambda(A, \Lambda)$ generated by $EQ_{2n}^\lambda(A, \Lambda)$, the hyperbolic matrices, and the matrix above. The *stabilization homomorphism* $GQ_{2n}^\lambda(A, \Lambda) \to GQ_{2(n+1)}^\lambda(A, \Lambda)$,

$$\begin{pmatrix} \alpha & \beta \\ \gamma & \delta \end{pmatrix} \mapsto \begin{pmatrix} \alpha & 0 & \beta & 0 \\ 0 & 1 & 0 & 0 \\ \gamma & 0 & \delta & 0 \\ 0 & 0 & 0 & 1 \end{pmatrix},$$

induces homomorphisms $EQ_{2n}^\lambda(A, \Lambda) \longrightarrow EQ_{2(n+1)}^\lambda(A, \Lambda)$ and $RQ_{2n}^\lambda(A, \Lambda) \longrightarrow RQ_{2(n+1)}^\lambda(A, \Lambda)$. Define $GQ^\lambda(A, \Lambda)$, $EQ^\lambda(A, \Lambda)$, and $RQ^\lambda(A, \Lambda)$, respectively, as the direct limit over n of the systems $GQ_{2n}^\lambda(A, \Lambda)$, $EQ_{2n}^\lambda(A, \Lambda)$, and $RQ_{2n}^\lambda(A, \Lambda)$, respectively. Then

$$KQ_1^\lambda(A, \Lambda) = GQ^\lambda(A, \Lambda)/EQ^\lambda(A, \Lambda).$$

We are now ready to define the equivariant surgery obstruction groups.

Let G denote a group and $\omega : G \to \{\pm 1\}$ a group homomorphism called the *orientation homomorphism*. We use the orientation homomorphism to put an involution on the integral group ring $\mathbb{Z}[G]$ by defining $\overline{g} = \omega(g)g^{-1}$ for each $g \in G$. Let $S \subset \max^{(-1)^{\left[\frac{n}{2}\right]}}(\mathbb{Z}[G])$. Recall that $\Lambda(S)$ is the smallest $(-1)^{\left[\frac{n}{2}\right]}$–form parameter containing S. Define

$W_n(\mathbb{Z}[G], \Lambda(S), \omega)$

$$= \begin{cases} KQ_0^{(-1)^{\left[\frac{n}{2}\right]}}(\mathbb{Z}[G], \Lambda(S))_{\text{even-free}} \Big/ [\mathbb{H}(A)], & \text{if } n \text{ even} \\[2ex] KQ_1^{(-1)^{\left[\frac{n}{2}\right]}}(\mathbb{Z}[G], \Lambda(S)) \Big/ \left[RQ^{(-1)^{\left[\frac{n}{2}\right]}}(\mathbb{Z}[G], \Lambda(S)) \right], & \text{if } n \text{ odd.} \end{cases}$$

If ω is trivial or is assumed to be fixed, one often writes $W_n(\mathbb{Z}[G], \Lambda(S))$ in place of $W_n(\mathbb{Z}[G], \Lambda(S), \omega)$.

If $X \in \mathfrak{M}^n(G)$ then the set $G(X)$ defined in section 1 is contained in the maximal form parameter $\max^{(-1)^{\left[\frac{n}{2}\right]}}(\mathbb{Z}[G])$. We shall show this below. Assuming this has been done, one gets the equivariant surgery obstruction groups required in Theorem

1 b) and Theorem 2 by setting above $S = G(X)$. One gets the 3–dimensional equivariant surgery obstruction group required in Theorem 1 a) by setting above $S = G(X) \cup \{1\}$.

Suppose $g \in G(X)$. Let x denote a point in a component of $X^{<g>}$ of dimension $\left[\frac{n-1}{2}\right]$. The group $< g >$ acts on the tangent space $T_x(X)$ at x and we have $T_x(X) \cong T_x(X)^{<g>} \oplus T_x(X)_{<g>}$. Clearly, $\omega_X(g) = 1$ (resp. -1) if and only if $\dim T_x(X)_{<g>}$ is even (resp. odd). But $\dim T_x(X)_{<g>} = \dim T_x(X) - \dim T_x(X)^{<g>} = n - \left[\frac{n-1}{2}\right] = \left[\frac{n}{2}\right] + 1$. Thus, $\omega_X(g) = 1$ (resp. -1) if and only if $n \equiv 2, 3$ (resp. $n \equiv 0, 1$) mod 4. Thus, $g = (-1)^{\left[\frac{n}{2}\right]-1}\omega_X(g)g^{-1}$, i.e., $g \in \max^{(-1)^{\left[\frac{n}{2}\right]}}(\mathbb{Z}[G])$.

In the next section, we shall show that the size of $G(X)$ plays an important role in vanishing theorems for equivariant surgery obstruction groups.

4. Vanishing Theorems

There is a general rule [B3, (11.2) and (11.4)] which says that the size of an equivariant surgery obstruction group of a finite group G is inversely proportional to the size of its form parameter. Since the size of the latter depends directly on that of the set $G(X)$ of involutions, it follows that the bigger $G(X)$ is, the more likely it is that the corresponding surgery obstruction group vanishes. The results of this section will show that if $G(X)$ is big and if the 2–hyperelementary subgroups of G are abelian or dihedral then $W_{4k+3}(\mathbb{Z}[G], G(X), \text{trivial}) = 0$. In particular, it follows for such G's that the 3–dimensional equivariant surgery obstruction groups required in Theorem 1 vanish. This means that one can do a lot of surgery.

We begin by recalling the known general vanishing theorems [B1], [B2] for equivariant surgery obstruction groups.

Let G denote a finite group. Assume that the orientation homomorphism $\omega : G \to \{\pm 1\}$ is trivial. Note that this is automatically the case if $|G|$ is odd. Note also that this implies $G(X) = \emptyset$ if $n \equiv 0, 1 \mod 4$ and $G(X) = G(2)$ if $n \equiv 2, 3 \mod 4$. Following our convention in section 3, we shall drop, henceforth, ω from our notation for equivariant surgery obstruction groups.

Theorem 3.

a) [B1] *If n and $|G|$ are odd then $W_n(\mathbb{Z}[G], \Lambda) = 0$ for any Λ. (If $n \equiv 1 \mod 4$ then there is only one Λ and if $n \equiv 3 \mod 4$ then there are precisely two Λ's, namely $\min^{-1}(\mathbb{Z}[G])$ and $\max^{-1}(\mathbb{Z}[G])$.)*

b) *[B2, Theorem 15] If the 2–hyperelementary subgroups of G are abelian, e.g. G is abelian itself, then $W_3(\mathbb{Z}[G], \Lambda(_2G)) = 0$ where $_2G$ denotes the set of elements of exponent 2 in G. ($\Lambda(_2G) = \max^{-1}(\mathbb{Z}[G])$).*

If G is as in Theorem 3 b) and $G(X) = \emptyset$ then $W_3(\mathbb{Z}[G], \Lambda(G(X)))$ is in general highly nontrivial, by [B2, Theorem 6]. Thus, we are encouraged to believe that a big $G(X)$ can influence significantly the vanishing of W_3. The next result bears this out.

Theorem 4 [BM1]. *If G is as in Theorem 3 b) then $W_3(\mathbb{Z}[G], \Lambda(F)) = 0$ for any subset F of $_2G$ which contains all but one element of $_2G$. In particular, F can be the set $G(2)$ of involutions defined in section 1.*

Proof. By induction [B3, §12], we can reduce to the case where G is abelian. Let $SKQ_1^{-1}(\mathbb{Z}[G], \Lambda(F)) = \text{Ker}(\text{Determinant} : KQ^{-1}(\mathbb{Z}[G], \Lambda(F)) \to \text{units}(\mathbb{Z}[G]))$. The proof of [B2, Theorems 6 and 7], when extended routinely to an arbitrary form parameter, shows that the canonical homomorphism $SKQ_1^{-1}(\mathbb{Z}[G], \Lambda(F)) \to W_3(\mathbb{Z}[G], \Lambda(F))$ is surjective. (The notation KQ_i and SKQ_i of the current article is replaced in [B2] by the notation KU_i and SKU_i.) Thus, it suffices to show that $SKQ_1^{-1}(\mathbb{Z}[G], \Lambda(F))$ is generated by $\begin{pmatrix} 0 & -1 \\ 1 & 0 \end{pmatrix}$ and matrices $\begin{pmatrix} g & 0 \\ 0 & g \end{pmatrix}$ where $g \in {}_2G$.

The proof of [B2, Theorem 10], when extended routinely to an arbitrary form parameter, shows that the canonical homomorphism $SKQ_1^{-1}(\mathbb{Z}[G], \Lambda(F)) \xrightarrow{\cong} SKQ_1^{-1}(\hat{\mathbb{Z}}_2[G_2], \hat{\mathbb{Z}}_2\Lambda_2(F))$ is an isomorphism where G_2 is the 2–Sylow subgroup of G and $\Lambda_2(F)$ is the image of $\Lambda(F)$ in $\mathbb{Z}[G_2]$ under the canonical homomorphism $\mathbb{Z}[G] \to \mathbb{Z}[G_2]$. By reduction theory [B3, §10], one shows that $SKQ_1^{-1}(\hat{\mathbb{Z}}_2[G_2], \hat{\mathbb{Z}}_2\Lambda_2(F))$ is generated by $\begin{pmatrix} 0 & -1 \\ 1 & 0 \end{pmatrix}$ and all hyperbolic matrices $\begin{pmatrix} a & 0 \\ 0 & a^{-1} \end{pmatrix}$ such that $a \in \text{units}(\hat{\mathbb{Z}}[G_2])$ and $a = \bar{a}$. The definition of F guarantees that either $a \in \Lambda(F)$ or $ag \in \Lambda(F)$ for some $g \in {}_2G$. But for any $b \in \Lambda(F) \cap \text{units}(\hat{\mathbb{Z}}_2[G_2])$, we have $\begin{pmatrix} b & 0 \\ 0 & b^{-1} \end{pmatrix} = \begin{pmatrix} 1 & b \\ 0 & 1 \end{pmatrix} \begin{pmatrix} 1 & 0 \\ -b^{-1} & 1 \end{pmatrix} \begin{pmatrix} 1 & b \\ 0 & 1 \end{pmatrix} \begin{pmatrix} 0 & -1 \\ 1 & 1 \end{pmatrix}$. This completes the proof.

The next theorem shows that $G(X)$ can be much smaller than $G(2)$ and still one has vanishing theorems.

Theorem 5 [BM1]. *Let G be the semidirect product $G = \pi \rtimes <\tau>$ where π is abelian and τ generates a group $<\tau>$ of order 2 and acts on π by inverting each element, e.g., G is a dihedral group. Then $W_3(\mathbb{Z}[G], \Lambda(F)) = 0$ for any $F \subseteq {}_2G$ such that $\pi\tau \subseteq F$ and either $F \cap \pi \neq \emptyset$ or the 2–Sylow group π_2 of π is trivial.*

Proof. The proof combines number theory, in particular the vanishing of certain congruence kernels [B4], and K–theory exact sequences to reduce the problem to one over the 2–adic integers $\hat{\mathbb{Z}}_2$. This is solved using a computational trick. The current proof differs in some details from that in [BM1].

Unfortunately, the induction machinery in the literature has not been set up to handle the problem we face. (The difficulty lies with the form parameters.) So we tackle it directly. By [B3, (7.36)], there is a K–theory exact sequence

$$KQ_2^{-1}(\mathbb{Q}[G], \mathbb{Q}\Lambda(F)) \xrightarrow{\theta} \coprod_p \text{coker}(KQ_2^{-1}(\hat{\mathbb{Z}}_p[G], \hat{\mathbb{Z}}_p\Lambda(F)) \to$$

$$KQ_2^{-1}(\hat{\mathbb{Q}}_p[G], \hat{\mathbb{Q}}_p\Lambda(F))) \to KQ_1^{-1}(\mathbb{Z}[G], \Lambda(F)) \to$$

$$\prod_p KQ_1^{-1}(\hat{\mathbb{Z}}_p[G], \hat{\mathbb{Z}}_p\Lambda(F)) \oplus KQ_1^{-1}(\mathbb{Q}[G], \mathbb{Q}\Lambda(F)) \to$$

$$\coprod_p (KQ_1^{-1}(\hat{\mathbb{Q}}_p[G], \hat{\mathbb{Q}}_p\Lambda(F)), KQ_1^{-1}(\hat{\mathbb{Z}}_p[G], \hat{\mathbb{Z}}_p\Lambda(F)))$$

where $\coprod_p$ denotes the restricted direct product. The cokernel of θ is a quotient of the direct sum of the congruence kernels corresponding to the simple factors of $\mathbb{Q}[G]$. However, each simple factor of $\mathbb{Q}[G]$ is a matrix ring over a real field with trivial involution. Thus, the congruence kernels above vanish by [B4]. Using classical results concerning involutions on matrix rings over fields, one adjusts the involution on each simple factor of $\mathbb{Q}[G]$ so that it is the conjugate transpose involution. By Morita theory [B3, §9], the computation of $KQ_1^{-1}(\mathbb{Q}[G], \mathbb{Q}\Lambda(F))$ reduces to that of $KQ_1^{-1}(k, \max^{-1}(k))$ where k runs over the centers of the simple factors of $\mathbb{Q}[G]$. But, $KQ_1^{-1}(k, \max^{-1}(k))$ is K_1 of the symplectic group which is well known to be trivial. In a similar vein, one shows that $KQ_1^{-1}(\hat{\mathbb{Q}}_p[G], \hat{\mathbb{Q}}_p\Lambda(F)) = 0$ and that $KQ_1^{-1}(\hat{\mathbb{Z}}_p[G], \hat{\mathbb{Z}}_p\Lambda(F)) = 0$ for $p \neq 2$. For the latter computation, one uses in a routine way reduction theory [B3, §10] when $p \mid |G|$. Thus, $KQ_1^{-1}(\mathbb{Z}[G], \Lambda(F)) \xrightarrow{\cong} KQ_1^{-1}(\hat{\mathbb{Z}}_2[G], \hat{\mathbb{Z}}_2\Lambda(F))$.

One writes $\pi = \pi_2' \times \pi_2$ and factors $\hat{\mathbb{Z}}_2[\pi_2']$ as a product of 2–adic cyclotomic integers. The involution leaves a factor invariant or exchanges it with another one. $\hat{\mathbb{Z}}_2[G]$ has a corresponding decomposition. By [B3, (11.6)], KQ_1^{-1} respects this decomposition and one shows that the only factor which survives is $KQ_1^{-1}(\hat{\mathbb{Z}}_2[\pi_2 \rtimes < \tau >], \hat{\mathbb{Z}}_2\Lambda(F))$ where F is viewed now as subset of $\pi_2 \rtimes < \tau >$. Thus, we can assume that $\pi = \pi_2$.

Next, we show that $\begin{pmatrix} 0 & -1 \\ 1 & 0 \end{pmatrix}$ generates $KQ_1^{-1}(\hat{\mathbb{Z}}_2[G], \hat{\mathbb{Z}}_2\Lambda(F))$. This will complete the proof of the theorem.

The group ring $\hat{\mathbb{Z}}_2[G]$ can be viewed as the twisted group ring $(\hat{\mathbb{Z}}_2[\pi]) < \tau >$ of $< \tau >$ over $\hat{\mathbb{Z}}_2[\pi]$ such that the action of τ on $\hat{\mathbb{Z}}_2[\pi]$ is given by the involution on $\hat{\mathbb{Z}}_2[\pi]$. Let $1 - \pi$ denote the set $\{1 - g \mid g \in \pi\}$. Let $\mathfrak{q}$ denote the ideal of $\hat{\mathbb{Z}}_2[G]$ generated by 2 and $1 - \pi$. Let $\mathbb{F}_2$ denote the field ot two elements. The quotient ring $\hat{\mathbb{Z}}_2[G]/\mathfrak{q}$ is canonically identified with the group ring $\mathbb{F}_2 < \tau >$ and the latter has only 4 elements. Let Γ denote the image of $\hat{\mathbb{Z}}_2\Lambda(F)$ in $\mathbb{F}_2 < \tau >$. Using the fact that $\tau \in \Gamma$, one shows easily that $KQ_1^{-1}(\mathbb{F}_2 < \tau >, \Gamma)$ is generated by $\begin{pmatrix} 0 & -1 \\ 1 & 0 \end{pmatrix}$. Let $\mathfrak{p}$ denote the ideal of $\hat{\mathbb{Z}}_2[\pi]$ generated by 2 and $1 - \pi$. Thus, $\mathfrak{q} = \mathfrak{p} + \mathfrak{p}\tau$. Since $\hat{\mathbb{Z}}_2[G]$ is complete with respect to $\mathfrak{q}$, it follows from [B3, (10.6) a)] that $KQ_1^{-1}(\hat{\mathbb{Z}}_2[G], \hat{\mathbb{Z}}_2\Lambda(F))$ is generated by $\begin{pmatrix} 0 & -1 \\ 1 & 0 \end{pmatrix}$ and by all 2×2 matrices $\begin{pmatrix} x & 0 \\ 0 & \bar{x}^{-1} \end{pmatrix}$ where $x = 1 + a + b\tau$ for some $a, b \in \mathfrak{p}$. If we can show that x can be expressed as a product of elements in $\hat{\mathbb{Z}}_2\Lambda(F)$ then it will follow as at the end of the proof of Theorem 4 that $\begin{pmatrix} 0 & -1 \\ 1 & 0 \end{pmatrix}$ generates $KQ_1^{-1}(\hat{\mathbb{Z}}_2[G], \hat{\mathbb{Z}}_2\Lambda(F))$ and we shall be finished.

Suppose $\pi_2 = 1$. (Under the assumption $\pi = \pi_2$ made three paragraphs above, this means that $\pi = 1$.) Then $x = ((1 + a)\tau + b)\tau$. The conditions on F guarantee that τ and $(1 + a)\tau \in \hat{\mathbb{Z}}_2\Lambda(F)$. Since $b \in 2\hat{\mathbb{Z}}_2$, $b \in \hat{\mathbb{Z}}_2\Lambda(F)$. Thus, $(1 + a)\tau + b \in \hat{\mathbb{Z}}_2\Lambda(F)$.

Suppose $F \cap \pi \neq \emptyset$ and let $g \in F \cap \pi$. Then $x = ((1 + a)g)(g + g(1 + a)^{-1}b\tau)$. One checks easily that $(1 + a)g$ and $(g + g(1 + a)^{-1}b\tau) \in \hat{\mathbb{Z}}_2\Lambda(F)$. This completes the proof.

Corollary 6. *If the 2–hyperelementary subgroups or G are abelian or dihedral then $W_3(\mathbb{Z}[G], \Lambda(_2G)) = W_3(\mathbb{Z}[G], \Lambda(G(2))) = 0$. In particular, all the 3–dimensional equivariant surgery obstruction groups in Theorem 1 vanish for G as above.*

Proof. One reduces by induction [B3, §12] to the case G is abelian or dihedral and finishes by Theorems 4 and 5.

5. Applications to Transformation Groups on Spheres and Lens Spaces

We apply equivariant surgery operations and vanishing theorems for equivariant surgery obstruction groups to problems concerning transformation groups on spheres and lens spaces. We shall show that the standard sphere S^7 has a smooth one fixed point action of the alternating group A_5 and that certain homotopy lens spaces have conjugations.

Let us recall some background concerning smooth one fixed point actions of finite groups on spheres. A half century ago, P.A. Smith showed that if G is a p–group and X a $(\mathbb{Z}/p)$–homology sphere then the fixed point set X^G of the action of G on X is a $(\mathbb{Z}/p)$–homology sphere. It follows that if X^G is finite and nonempty then X^G has precisely two points. In the 1950's, Montgomery asked whether some finite groups can act with one fixed point on a homology sphere. In the late 1950's, the work of E. Floyd and R. Richardson showed that A_5 could act topologically with one fixed point on spheres of undetermined dimensions. In 1977, E. Stein [S] showed that there is a smooth one fixed point action of $SL(2,5)$ on S^7 and in the period 1978–82, T. Petrie [P1], [P2], [P3] showed using equivariant surgery techniques that various finite groups can act smoothly with one fixed point on higher dimensional spheres. More recently, it was shown in [M3], [M4], using equivariant surgery techniques that A_5 acts smoothly with one fixed point on all spheres A^n where $n = 6$, or ≥ 9. In addition, Buchdahl–Kwasik–Schultz [BKS] constructed locally linear (but not necessarily smooth), one fixed point actions of A_5 on all spheres S^n where $n \geq 6$. We shall show using 3–dimensional and 7–dimensional equivariant surgery operations the following.

Theorem 7 [BM1]. *A_5 acts smoothly with one fixed point on the standard sphere S^7.*

Proof. The proof follows the pattern of those in [M3] and [M4]. One uses Petrie's transversality method to construct a smooth equivariant map $f : X \to S^7$ for some space $X \in \mathfrak{M}^7(A_5)$ such that $\partial X = \emptyset$ and X^{A_5} has exactly one point. The map f is automatically a boundary equivalence. One hopes now to be able to apply equivariant surgery to convert f to a homotopy equivalence $f' : X' \to S^7$, knowing fully well that if he is successful, i.e., if all the intervening surgery obstructions vanish, then the A_5–manifold X' constructed via equivariant surgery operations will have precisely one fixed point. This done, one completes the proof as follows.

For a 7–dimensional homotopy sphere Z with A_5–action, denote by $[Z]$ the class of Z in the group $\Theta(7)$ [KM] of 7–dimensional homotopy spheres. By [KM], $\Theta(7)$ is a finite group. Consider the equivariant connected sum X'' of X' with $A_5 \times_H \operatorname{res}_H^{A_5} X'$ at points of isotropy type H where $H = A_4$ or D_{10}. For details see [M3] section 3. X'' is a homotopy sphere having exactly one fixed point. On the other hand, $[X''] = (1 + k)[X']$ in $\Theta(7)$ for $k = 5$ or 6 corresponding to $H = A_4$ or D_{10}, respectively. Since 5 and 6 are relatively prime, we can obtain the standard sphere as the underlying space of an equivariant connected sum X''' of X' with several $(A_5 \times_{A_4} \operatorname{res}_{A_4}^{A_5} X')$'s and $(A_5 \times_{D_{10}} \operatorname{res}_{D_{10}}^{A_5} X')$'s. By construction, X''' has one fixed point and we are finished.

It remains to solve the surgery problem. The map f is got as follows. Up to conjugation, the subgroups of A_5 are A_5, A_4, D_{10}, D_4, C_5, C_3, C_2 and $\{1\}$. Using a character table of A_5 and the formula $\dim V^H = \frac{1}{|H|} \sum_{g \in H} \chi_V(g)$ where V is any representation of G, H any subgroup of G, and $\chi_V(g) = \operatorname{trace}(g : V \to V)$, one can show that there is a 7–dimensional orthogonal representation V of A_5 such that $V^H = 0$ if $H \cong A_5$ or D_{10}, $\dim V^H = 1$ if $H \cong A_4$, D_6, D_4 or C_5 and $\dim V^H = 3$ if $H \cong C_3$ or C_2. Let A_5 act trivially on $\mathbb{R}$, give $\mathbb{R} \oplus V$ the diagonal action of A_5, and let $Y = S(\mathbb{R} \oplus V)$. The action of A_5 on Y has precisely two fixed points. Call them p_+ and p_-.

Let $\Omega(A_5)$ denote the Burnside ring of A_5. Define $\omega \in \Omega(A_5)$ by $\omega = 1 - ([A_5/A_4] + [A_5/A_{10}] + [A_5/D_6] - [A_5/C_3] - 2[A_5/C_2] + [A_5/\{1\}])$. The element ω is an idempotent, $\chi_{A_5}(\omega) = 1$, and $\operatorname{res}_H^{A_5}(\omega) = 0 \in \Omega(H)$ for any proper subgroup H of A_5. By T. Petrie's transversality construction [P2], one can obtain the following: There is an A_5–normal map $w = (f, b, c) : (X, T(X), \nu(X)) \to (Y, f^*T(Y), f^*\nu(Y))$ and for all proper subgroups H of A_5, there are H–normal cobordisms $W_H = (F_H, B_H, C_H)$ between $\operatorname{res}_H^{A_5} w$ and $\operatorname{res}_H^{A_5} 1_{A_5}$, where 1_{A_5} denotes the identity A_5–normal map on Y and $F_H : W_H \to [0, 1] \times Y$, such that the following conditions are satisfied:

(i) $X^{A_5} = (f^{-1}(p_-))^{A_5}$ and $|X^{A_5}| = 1$. (We shall denote the one fixed point of X by p_- again.)

(ii) For any subgroup H of A_5 such that $H \cong D_{10}$, $X^H = \{p_-\} \cup (f^{-1}(p_+))^H$ and $|(f^{-1}(p_+))^H| = 1$. (Thus, X^H has precisely two fixed points.)

We fix orientations of X^H and Y^H as follows. Y^H is orientable for any H such that $\dim Y^H \geq 1$. For each such H, pick an orientation of Y^H such that the orientations of Y^H and $Y^{H'}$ coincide whenever $Y^H = Y^{H'}$. For each pair (H, K) of subgroups $K \subseteq H \subseteq A_5$, the manifold W_H^K is orientable. Choose an orientation of W_H^K such that $F_H^K : W_H^K \to I \times Y^K$ has degree one. The boundary $\partial W_H^K = X^K \cup (-Y^K)$ and we give X^K the orientation induced from that on ∂W_H^K. The map $f^K : X^K \to Y^K$ has then degree one. Thus, the orientation of X^K has been chosen independently of the choice of H containing K.

We want to operate via equivariant surgery on the map $w = (f, b, c)$. We begin as usual by performing operations which adjust w so that for every proper subgroup H of A_5, $f^H : X^H \to Y^H$ becomes a homology equivalence, beginning with $\dim X^H$

as small as possible and then going up a dimension at a time. We can invoke the singularity equivalence criteria in Theorems 1 and 2 since by [R], $\tilde{K}_0(\mathbb{Z}[H]) = 0$ for any subgroup or subquotient H of G.

The $\dim X^H = \dim Y^H$ and by construction the dimension of the fixed spaces Y^H of Y are 0, 1, 3, and 7. If $\dim X^H = 0$ then $H \cong A_5$ or D_{10}, and by (i) and (ii), f^H is already a homeomorphism. So there is nothing to do. If $\dim X^H = 1$ then $H \cong A_4$, D_6, D_4, or C_5 and X^H is a disjoint union of circles. By equivariant surgery of isotropy type H, we kill the circles which do not contain p_- and modify w rel $\cup_K X^K (K \cong D_{10})$ so that f^H becomes a homotopy equivalence for $H \cong A_4$, D_6, D_4, or C_5. If $\dim X^H = 3$ then $H \cong C_3$ or C_2. The equivariant surgery obstruction to converting w keeping $\cup_K X^K (K \cong A_4$, D_6, D_4, C_5, and $D_{10})$ fixed to an A_5–normal map such that $f^H : X^H \to Y^H$ is a homology equivalence lies in the 3–dimensional equivariant surgery obstruction group $W_3(\mathbb{Z}[N/H]$, $\Lambda((N/H)(2))$, trivial) of Theorem 1 b), where N is the normalizer of H in A_5. Since $N/H \cong C_2$, it follows by Theorem 4 that $W_3(\mathbb{Z}[N/H], \Lambda((N/H)(2))$, trivial)$= 0$. Thus, by the conclusion of Theorem 1 b), we can perform the desired equivariant surgery operation. It remains to show we can perform equivariant surgery in the case $X = X^H$, $H = \{1\}$. Our surgery operations until now guarantee that f is a singularity equivalence. The equivariant surgery obstruction to converting w keeping X_s fixed to an A_5–normal map w' such that $f' : X \to Y$ is a homotopy equivalence lies in the equivariant surgery obstruction group $W_3(\mathbb{Z}[A_5], \Lambda(A_5(2))$, trivial) of Theorem 2. Since the 2–hyperelementary subgroups of A_5 are abelian or dihedral, $W_3(\mathbb{Z}[A_5], \Lambda(A_5(2))$, trivial) $= 0$ by Corollary 6 of Theorems 4 and 5. Thus, by the conclusion of Theorem 2, we can perform the desired equivariant surgery operation. This completes the proof.

We record without proof another application of equivariant surgery theory and the vanishing theorems.

Recall that a homotopy lens space is a smooth manifold whose fundamental group is finite cyclic and whose universal covering space is a homotopy sphere. Let τ be the generator of a group $< \tau >$ of order 2. A *conjugation* on a homotopy lens space L of dimension $2m + 1$, with fundamental group π, is a smooth action of $< \tau >$ on L such that

$$L^{<\tau>} \cong_{\text{diffeo}} \begin{cases} S^m & \text{if } |\pi| \text{ is odd} \\ \mathbb{R}P^m \amalg \mathbb{R}P^m & \text{if } |\pi| \text{ is even.} \end{cases}$$

Theorem 8. *Let m be a natural number ≥ 2. For each natural number $q \geq 7$ or $q = 5$ there is a conjugation on some homotopy lens space L of dimension $2m + 1$ such that the fundamental group of L is cyclic of order q and L is not diffeomorphic to any standard lens space.*

Details of the proof are given in [BM2].

Problem. Does any homotopy lens space have a conjugation?

REFERENCES

[B1]   A. Bak, *Odd dimension surgery groups of odd torsion groups vanish*, Topology **14** (1975), 367–374.

[B2]   A. Bak, *The computation of surgery groups of finite groups with abelian 2-hyperelementary subgroups*, Lec. Notes in Math. **551** (1976), 384–409.

[B3]   A. Bak, *K-theory of Forms*, Ann. Math. Stud. 98, Princeton Univ. Press, 1981.

[B4]   A. Bak, *Le problème des sous-groups de congruence et le problème métaplectique de rang > 1*, C. R. Acad. Sc. Paris **t. 292, Sér. 1** (1981), 307–309.

[BKS] N. P. Buchdahl, S. Kwasik, and R. Schultz, *One fixed point actions on low-dimensional spheres*, Invent. Math. **102** (1990), 633–662.

[BM1] A. Bak and M. Morimoto, *3-dimensional equivariant surgery and applications*, preprint.

[BM2] A. Bak and M. Morimoto, *Conjugations and homotopy lens spaces*, preprint.

[DP]   K. Dovermann and T. Petrie, Memoirs of the Amer. Math. Soc. **260** (1982).

[KM]   M. Kervaire and J. Milnor, *Groups of homotopy spheres I*, Ann. Math. **77** (1963), 504–537.

[M1]   M. Morimoto, *Bak Groups and Equivariant Surgery*, K-Theory **2** (1989), 465–483.

[M2]   M. Morimoto, *Bak Groups and Equivariant Surgery II*, K-Theory **3** (1990), 505–521.

[M3]   M. Morimoto, *Most of the standard spheres have one fixed point action of A_5*, Lec. Notes in Math. **1375** (1989), 240–258.

[M4]   M. Morimoto, *Most standard spheres have smooth one fixed point actions of A_5 II*, K-Theory **4** (1991), 289–302.

[P1]   T. Petrie, *Pseudoequivalences of G-manifolds*, Proc. Symp. Pure Math. **XXXII** (1978), 169–210.

[P2]   T. Petrie, *One fixed point actions on spheres, I*, Adv. in Math. **46** (1982), 3–14.

[P3]   T. Petrie, *One fixed point actions on spheres, II*, Adv. in Math. **46** (1982), 15–70.

[P4]   T. Petrie, *Smith equivalence of representations*, Proc. Camb. Phil. Soc. **94** (1983), 61–99.

[PR]   T. Petrie and J. Randall, *Transformation Groups on Manifolds*, Pure and Appl. Math., Marcel Dekker Inc., New York, 1984.

[Q]   D. Quillen, *Higher algebraic K-theory I*, Lec. Notes in Math. **341** (1973), 85–147.

[R]   I. Reiner, *Class groups and Picard groups of group rings and orders*, A.M.S. Regional Conf. Ser. in Math. **26** (1976).

[S]   E. Stein, *Surgery on products with finite fundamental group*, Topology **16** (1977), 473–493.

DEPT. OF MATHEMATICS, UNIVERSITY OF BIELEFELD, 4800 BIELEFELD, GERMANY
E-mail address: bak@math1.mathematik.uni-bielefeld.de

DEPT. OF MATHEMATICS, COLLEGE OF LIBERAL ARTS AND SCIENCES, OKAYAMA UNIVERSITY, OKAYAMA 700, JAPAN
E-mail address: cticq111@jpnoucc.bitnet

GENERALIZED VERSION OF STABLE K-THEORY

STANISLAW BETLEY

Uniwersytet Warszawski

ABSTRACT. We introduce a new tool for computing $H_*(Gl(R); \lim_{n \to \infty} s(R^n, R^n))$, where $s : (f.g.R - mod)^{op} \times (f.g.R - mod) \longrightarrow Ab$ is a bifunctor and R is any commutative ring with unit. The idea is to generalize the notion of stable algebraic K-theory to this new situation – stable K-theory was introduced to deal with $s = \mathrm{Hom}(\cdot, \cdot)$.

INTRODUCTION

Let R_{Mod} denote the category of R–modules, where R is any ring and let R_{fMod} stand for the category of finitely generated free R–modules. As usual Ab will denote the category of abelian groups. The aim of this paper is to generalize the definition of stable algebraic K–theory to the case when we have a given commutative ring R with unit and a functor $t : R_{fMod} \to Ab$ or a bifunctor $s : (R_{fMod})^{op} \times R_{fMod} \to Ab$. It is known that if M denotes the full group of matrices over R with $Gl(R)$–action given by conjugation then $H_*(Gl(R); M) = h_*(BGl(R))$, where $h_*(\cdot)$ denotes the homology theory defined by the stable K–theory spectrum (see section 1, remark 1.5 for the definition). We shall denote this spectrum by K. Using the fact that K is a module–spectrum over an Eilenberg–MacLane spectrum Bökstedt obtained the formula:

(A) $$h_*(BGl(R)) = \bigoplus_p H_{*-p}(BGl(R); \pi_p(K))$$

where $GL(R)$ acts trivially on $\pi_*(K)$. We would like to see how close we can approach this result in the more general case.

Let $T = \lim_{n \to \infty} t(R^n)$ and $S = \lim_{n \to \infty} s(R^n, R^n)$, where t and s are as before and the limits are taken with respect to the standard inclusion $R^n \to R^{n+1}$ taking $(x_1, \cdots, x_n)$ to $(x_1, \cdots, x_n, 0)$ and the standard projection $R^{n+1} \to R^n$ taking $(x_1, \cdots, x_n, x_{n+1})$ to $(x_1, \cdots, x_n)$. The action of $Gl_n(R)$ on $s(R^n, R^n)$ is given by conjugation. Let $h_*(\cdot; L)$ denote the homology theory defined by a given spectrum L. We will construct spectra K^t and K^s respectively such that

$$H_*(Gl(R);T) = h_*(BGl(R);K^t) \text{ and } H_*(Gl(R);S) = h_*(BGl(R);K^s)$$

In section 3 we shall explain how we can approach Bökstedt's formula (A) in the general case.

Of course there is a question whether the groups $\pi_*(K^t)$ or $\pi_*(K^s)$ are easy to compute or at least computable. The main theorem of [SSW] (unpublished yet) states that $K = THH(R)$, where $THH(R)$ denotes topological Hochschild homology spectrum defined in [B1]. In [B] Bökstedt calculated homotopy groups of $THH(R)$ for $R = Z$ or $R = Z/p$. In [P-W] there appears a definition of topological Hochschild homology with coefficients in a functor or a bifunctor. It is highly possible (and expected) that the proof $K = THH(R)$ can be generalized to $K^s = THH(R;s)$ (i.e., to the case of coefficients in a bifunctor rather than just in a bimodule). This would give us a strong computational tool towards calculating $\pi_*(K^s)$ and hence $H_*(Gl(R);S)$.

Throughout the paper we shall use the convention space = simplicial set. Moreover we shall treat mostly only the case of a functor and we shall explain the necessary modifications when the case of a functor differs from the case of a bifunctor.

Section 1

Let X be a connected simplicial set and let M be a $\pi_1(X)$–module. We have a map of spaces $f : X \longrightarrow K(\pi_1(X),1)$ inducing an isomorphism on π_1. Let $M[S^n]$ denote the abelian simplicial group spanned freely by M on $[S^n]$, where $[S^n]$ is any simplicial model for S^n, and let $\tilde{M}[S^n] = M[S^n]/M[*]$. Consider $\pi_1(X)$ as a simplicial group in a trivial way and let G^n denote the simplicial group obtained as the semi–direct product of $\tilde{M}[S^n]$ and $\pi_1(X)$. We have the short exact sequence of groups

$$(*) \qquad \tilde{M}[S^n] \to G^n \to \pi_1(X)$$

Let $Y^n \longrightarrow X$ be the space over X obtained as a pull–back along f of the map $BG^n \longrightarrow K(\pi_1(X),1)$. Then we have the fibration:

$$(**) \qquad B\tilde{M}[S^n] \to Y^n \to X$$

The sequence $(*)$ has the obvious section and hence $(**)$ is also equipped with a section.

1.1. Lemma. *There is an isomorphism* $H_i(X;M) = \pi^s_{i+n+1}(Y^n,X)$ *for* n *sufficiently large with respect to* i.

Proof. Observe first that $B\tilde{M}[S^n]$ is homotopy equivalent to $K(M,n+1)$. We obtain the following formula for stable homotopy by Freudenthal's theorem:

$$\pi^s_i(B\tilde{M}[S^n]) = \begin{cases} M \text{ for } i = n+1 \\ 0 \text{ for } i \neq n+1 \text{ and } i < 2n \end{cases}.$$

We have an Atiyah–Hirzebruch type spectral sequence with

$$E^2_{p,q} = H_p(X; \tilde{\pi}^s_q(B\tilde{M}[S^n]))$$

converging to $\pi^s_{p+q}(Y^n, X)$. Hence $\pi^s_k(Y^n, X) = H_{k-n-1}(X; M)$ for $k < 2n$.

1.2. Remark. The maps $f_n : B\tilde{M}[S^n] \longrightarrow \Omega B\tilde{M}[S^{n+1}]$ between vertical fibers in the diagram

$$
\begin{array}{ccc}
Y^n & \longrightarrow & X \\
\downarrow & & \downarrow \\
X & \longrightarrow & Y^{n+1}
\end{array}
$$

give us a spectrum L, where the map $X \longrightarrow Y^{n+1}$ is given by the section of $(**)$. Obviously in that case L is equivalent to the Eilenberg–MacLane spectrum $K(M, n)$. Moreover if Y^n is an H–space (which can happen only if M is a trivial $\pi_1(X)$–module) then $Y^n = X \times B\tilde{M}[S^n]$ and $\pi^s_i(Y^n, X) = \pi^s_i(X_+ \wedge B\tilde{M}[S^n])$, and these groups are equal to the $(i - n - 1)$-th homology group of X in the theory defined by L.

Usually Y^n does not have a product structure but sometimes (and that is an interesting case for us) Y^{n+} has such a structure ("+" denotes here the Quillen's plus-construction with respect to the maximal perfect subgroup of π_1). For example that is the case when Y^n is of the type $BGl(R)$. On the other hand we are interested in finding homology groups so we can change our space by the "+–construction" and we still preserve the interesting information. Let F_n denote the homotopy fiber of the map $Y^{n+} \to X^+$. We have the following diagram obtained from the diagram in 1.2 by performing +–construction:

$$
\begin{array}{ccc}
F_n & \xrightarrow{h_n} & \Omega F_{n+1} \\
\downarrow & & \downarrow \\
Y^{n+} & \longrightarrow & X^+ \\
\downarrow & & \downarrow \\
X^+ & \longrightarrow & Y^{(n+1)+}
\end{array}
$$

1.3. Lemma. *The map h_n defined above is $(2n - 2)$–connected.*

Proof. We will proceed as in [W; Prop. 1.3]. Let $Y^n \to X'$ denote the cofibration obtained in the standard way from $Y^n \to X$. Then the map between homotopy fibers of the vertical maps in the diagram:

$$
\begin{array}{ccc}
Y^n & \longrightarrow & X' \\
\downarrow & & \downarrow \\
X' & \longrightarrow & X' \cup_{Y^n} X'
\end{array}
$$

is $(2n-2)$–connected by the homotopy excision theorem. Hence the canonical map $X' \cup_{Y^n} X' \to Y^{n+1}$ is $(2n-2)$–connected. Observe that in our case $(X' \cup_{Y^n} X')^+ = (X')^+ \cup_{Y^{n+}} (X')^+$ because both spaces have the same π_1 and are homologically equivalent. This immediately implies that the standard map $(X')^+ \cup_{Y^{n+}} (X')^+ \to (Y^{n+1})^+$ is $(2n-2)$–connected and that is equivalent to the statement that h_n is $(2n-2)$–connected.

Now let L denote the spectrum obtained from the maps h_n. Moreover assume that Y^{n+} has a product structure for any n. Then

$$Y^{n+} = X^+ \times F_n$$

and the groups $\pi_i^s(Y^{n+}, X^+) = \pi_i^s(F_n \wedge (X^+)_+)$ are equal to the $(i-n-1)$-th homology group of the pair (Y^{n+}, X^+) in the theory defined by L. On the other hand the "+–construction" does not change any homology so the stable homotopy also remains unchanged after applying the "+–construction". Hence as a direct consequence of Lemma 1.1 we can state:

1.4. Corollary. *If X is a space such that the spaces Y^{n+} have a product structure then for any $\pi_1(X)$-module M:*

$$H_*(X; M) = h_*(X; L)$$

1.5. Remark. If $X = BGl(R)$ for some ring R and $M = M(R)$ (the full group of matrices over R), and if $Gl(R)$ acts on $M(R)$ by conjugation, then the spectrum L obtained from the maps h_n is equal to the stable K–theory spectrum defined for example in [K].

Section 2

Assume now that R is any commutative ring with unit and $t : R_{fMod} \to Ab$ is a functor (= covariant functor). We shall assume always that $t(0) = 0$. If this is not true then $t = t(0) \oplus t'$, where $t'(0) = 0$. We shall write T for $\lim\limits_{n \to \infty} t(R^n)$ and T_n for $t(R^n)$. We would like to calculate $H_*(Gl(R); T)$ in the spirit of section 1. This means we want to take $X = BGl(R)$ and $M = T$, and then apply the machinery from section 1 to produce some spectrum (denoted by L in 1.4), which we will denote by K^t. This spectrum will have the property that

$$H_*(Gl(R); T) = h_*(BGl(R); K^t)$$

if we can prove that the spaces Y^{n+} have some product structure, where Y^n is a space which fits into the fibration:

$$B\tilde{T}[S^n] \to Y^n \to BGl(R)$$

2.1. Lemma. *For any n, Y^{n+} is an infinite loop space.*

Proof. We shall use the Γ–space machinery from [S]. Observe that $Y^n = BG^n$, where G^n is the simplicial group obtained as the semi–direct product of simplicial groups:

$$\tilde{T}[S^n] \to G^n \to Gl(R)$$

and $G^n = \lim_{k \to \infty} G^n_k$, where the groups G^n_k fit (as semi–direct products) into:

$$\tilde{T}_k[S^n] \to G^n_k \to Gl_k(R).$$

The action of $Gl_k(R)$ on $\tilde{T}_k[S^n]$ is given by the functor t. We can define a pairing $G^n_k \times G^n_l \to G^n_{l+k}$ as follows: it takes $A \in Gl_k(R)$ and $B \in Gl_l(R)$ to $\begin{pmatrix} A & 0 \\ 0 & B \end{pmatrix} \in Gl_{k+l}(R)$ and the map $\tilde{T}_k[S^n] \times \tilde{T}_l[S^n] \to \tilde{T}_{k+l}[S^n]$ is induced by the embedding (as a direct summand) $T_k \oplus T_l \hookrightarrow T_{k+l}$ coming from the identification $R^{l+k} = R^k \oplus R^l$. An easy check shows that this pairing is well defined and has sufficiently good properties in order to use it for constructing a Γ–space. Hence following [S; §2] we obtain a Γ–space A with

$$A(1) = \bigsqcup_{k \geq 0} BG^n_k$$

Moreover by [S; §4] the 0–space of the Ω–spectrum associated to A is equivalent to $BG^{n+} = Y^{n+}$.

2.2. Corollary. *In the notation from the beginning of this section we have*

$$H_*(Gl(R); T) = h_*(BGl(R); K^t)$$

2.3. Remark. We can obtain the Γ–space A through a categorical approach (see [S; §2]). Let $\bar{R}$ denote the following category:

    i) $\mathrm{Ob}(\bar{R}) = \mathrm{Ob}(R_{fMod})$;

    ii) $\mathrm{Hom}_{\bar{R}}(R^m, R^k)$ is the simplicial set given by the product $\mathrm{Hom}_{R_{fMod}}(R^m; R^k) \times \tilde{T}_k[S^n]$, where $\mathrm{Hom}_{R_{fMod}}(R^m; R^k)$ is considered as a simplicial set in a trivial way.

The composition of morphisms is given degreewise by

$$(x, a) \circ (y, b) = (x \cdot y, a + t(x)_*(b))$$

Usually we do not have sums in this category, never the less, we can get a Γ–category out of $\bar{R}$ (just like in [S; §2]). Let s be a finite set and $P(s)$ denote the category of subsets of s with inclusions as morphisms. Denote by $R(s)$ the category of functors $F : P(s) \to \bar{R}$ such that:

    i) $F(x \sqcup y) = F(x) \oplus F(y)$ in the category R_{fMod};

    ii) $F(x \hookrightarrow y)$ is induced by $Gl_x \hookrightarrow Gl_y$ and $T_x \hookrightarrow T_y$ as a direct summand. The morphisms in $R(s)$ are given by isomorphisms of functors.

Note. The case of bifunctors differs a little bit from the case of functors at this point (actually, the bifunctor case is more general). The correct $\bar{R}$ is given in that case by the extension of R_{fMod} by a given bifunctor (see [B-W, §2] with all constructions to be done simplicially).

2.4. Remark. Let P be an R–bimodule. Let K^P denote the stable K–theory with coefficients in P, like in [K]. If we set $s = \mathrm{Hom}(\cdot, \cdot \otimes P)$, then obviously $K^s = K^P$.

2.5. Remark. Let t be a functor of finite degree. Then by [Be] we know that the map $BG^n \longrightarrow BGl(R)$ is an isomorphism on homology. Hence, after the application of the "+–construction" we obtain a homotopy equivalence $BG^{n+} \simeq BGl(R)^+$. This means that the spaces F_n are contractible and the spectrum K^t is trivial. Hence our calculations from [Be] and results from this paper agree for functors of finite degree.

2.6. Remark. The groups $\pi_*(BG^{0+})$ can be considered as a version of K–theory for the category R_{fMod} extended in the sense of [B-W] by s or t. Then K^s or K^t is a stable version of this K–theory.

The question is now: how can we compute the homology groups defined by the spectra K^t?

SECTION 3

In this section we shall write everything for bifunctors rather than functors because we want to indicate again that both cases can be treated in the same way. So let s denote a bifunctor $R^{op}_{fMod} \times R_{fMod} \to R_{Mod}$. Let P denote the Eilenberg–MacLane spectrum defined by a given ring R (commutative with unit). We make the following remark:

3.1. Remark. We can obtain P by the same stabilization process as K^t. We must only apply 1.2 to $X = BGl_1(R)$ and $M = R$. The space P_n is obtained as the fiber of the map $BGl_1(R \oplus \tilde{R}[S^n]) \longrightarrow BGl_1(R)$, where $R \oplus \tilde{R}[S^n]$ is the "suspended" version of the dual numbers over R.

Now observe that the space K^s_m is obtained as the fiber of the map $BG^{m+} \longrightarrow BGl(R)^+$. We can define a simplicial group homomorphism

$$G^m_k \times Gl_1(R \oplus \tilde{R}[S^n]) \to G^{m+n}_k$$

given essentially by

$$\tilde{S}_j[S^m] \otimes \tilde{R}[S^n] \simeq \tilde{S}_j[S^m \times S^n] \to \tilde{S}_j[S^{n+m}]$$

which covers the tensor product map $Gl_1(R) \times Gl_k(R) \to Gl_k(R)$, where of course $S_j = s(R^j, R^j)$. By improving this homomorphism as in [L] on the level of classifying spaces, passing to fibers and to the limit with respect to k we should get a P–module structure on the spectrum K^s. That is certainly true for $s = \mathrm{Hom}(\cdot, \cdot)$ and was used in [B]. But so far we are not able to check whether this structure is well defined in the general case so we state it only as a conjecture:

3.2. Conjecture. *The spectra K^s are module spectra over P.*

Let $S = \lim\limits_{n \to \infty} s(R^n, R^n)$.

3.3. Corollary. *If Conjecture 3.3 holds for $s(\cdot, \cdot)$ then*

$$H_n(Gl(R); S) = \bigoplus_{i+j=n} H_i(Gl(R); \pi_j(K^s))$$

Let us conclude this paper with some remarks concerning computations and relations to classical results in stable K–theory. We shall assume now that s goes to abelian groups. Let $F(R)$ denote the homotopy fiber of the plus–construction map $BGl(R) \to BGl(R)^+$. Then we have the Hochschild–Serre spectral sequence

$$E_{p,q}^2 = H_p(BGl(R)^+; H_q(F(R); S)) \Rightarrow H_{p+q}(Gl(R); S)$$

where the action of $\pi_1(BGl(R)^+)$ on $H_q(F(R); S)$ is trivial. The purpose for triviality of this action is that the action of $Gl(R)$ on S factors through elementary matrices as in [K1, proof of 2.16]. The same considerations as in [K, pages 387–388] show that for any i

$$\pi_i(K^s) = H_i(F(R); S)$$

Observe that $H_0(F(R); S) = H_0(Gl(R); S) = S_{Gl(R)}$. Hence the groups $\pi_i(K^s)$ can be viewed as obstructions for the map $S \to S_{Gl(R)}$ to induce a homology isomorphism $H_*(Gl(R); S) \longrightarrow H_*(Gl(R); S_{Gl(R)})$. This observation is a generalization of one of the origins of stable K–theory groups which were introduced (for example) as obstructions for the trace map $M(R) \to R$ to induce a homology isomorphism for $Gl(R)$.

REFERENCES

[B-W] H.-J. Baues and G. Wirsching, *Cohomology of small categories*, J. Pure and Applied Algebra **38** (1985), 187–211.

[Be] S. Betley, *Homology of $Gl(R)$ with coefficients in a functor of finite degree*, to appear, J. Algebra.

[B] M. Bökstedt, *The topological Hochschild homology of Z and Z/p*, to appear, Annals of Math..

[B1] M. Bökstedt, *Topological Hochschild homology*, to appear, Topology.

[K] C. Kassel, *La K-théorie stable*, Bull. Soc. Math. France **110** (1982), 381–416.

[K1] C. Kassel, *Calcul Algébrique de l'Homologie de Certains Groupes de Matrices*, J. Algebra **80** (1983), 235–260.

[L] J-L. Loday, *K-théorie algébrique et représentations de groupes*, Ann. Scient. Éc. Norm. Sup. **9** (1976), 309–377.

[P-W] T. Pirashvili and F. Waldhausen, *MacLane homology and topological Hochschild homology*, to appear, J. Pure and Applied Algebra.

[S] G. Segal, *Categories and cohomology theories*, Topology **13** (1974), 293–312.

[SSW] R. Schwänzl, R. Staffeldt and F. Waldhausen, in preparation.

[W] F. Waldhausen, *Algebraic K-theory of topological spaces I*, Proc. Symp. Pure Math. **32** (1978), 35–60.

INSTYTUT MATEMATYKI, UNIWERSYTET WARSZAWSKI, UL. BANACHA 2, 00-913 WARSZAWA 59, POLAND

A NOTE ON TWO–ORBIT VARIETIES

MICHEL BRION

Institut Fourier

ABSTRACT.

We consider a complete, normal algebraic variety X on which a connected affine algebraic group acts with a dense orbit Ω; we assume that $X \setminus \Omega$ is another orbit, and has codimension one in X. We show that the automorphism group of X is transitive if and only if there exist non–constant regular functions on Ω.

1. INTRODUCTION

Define a *two–orbit variety* as a normal, complete algebraic variety X with an action of a connected affine algebraic group G, such that G has a dense orbit in X whose complement is only one (closed) orbit; all varieties are defined over the field of complex numbers. Two–orbit varieties were introduced and classified by D. N. Ahiezer, in the case where the codimension of the closed orbit is one; see [A], and also [HS] where Kähler varieties are considered. In particular, there is a complete, and rather small, list of two–orbit varieties whose open orbit is *affine*. Moreover, it can be checked on this list, that such a variety is always homogeneous under some affine algebraic overgroup of G; see [A, p.67] and [HS, p.769]. The aim of the present note is to give a classification–free proof of this fact. More generally, we prove the following

Theorem. *For a two–orbit G–variety X whose closed orbit Y has codimension one, the following conditions are equivalent:*

   (i) *There exist non–constant regular functions on the open orbit of X.*
   (ii) *The normal bundle of Y in X has non–zero global sections.*
   (iii) *The automorphism group of X is transitive.*

The following example will illustrate our result. Let q be a non–degenerate quadratic form on $V = \mathbb{C}^{2n+1}$, where n is any positive integer. Denote by X the variety of pairs (D, E) where D is a line in V, and E is a n–dimensional subspace of V such that $q|E = 0$ and that D is orthogonal to E. The special orthogonal group $SO(2n + 1)$ acts on X with two orbits (according as $q|D$ is zero or not). Moreover, via the map $p_2 : (D, E) \to E$, we can see X as a $\mathbf{P}^n$–bundle over a projective homogeneous variety, hence X is projective and smooth. The open orbit Ω is mapped via $p_1 : (D, E) \to D$ onto the complement of the quadric $(q = 0)$ in $\mathbf{P}^{2n}$, hence Ω has non–constant regular functions, and its complement has codimension one in X. By our theorem, X must be homogeneous; in fact, it can be checked that the action of $SO(2n+1)$ on X extends to a transitive action of $SO(2n+2)$.

The (easy) equivalence of (i) and (ii) is proved in §2. The proof of (ii)$\Longrightarrow$(iii) fills §§3 to 6 and the beginning of §7. Finally, we prove (iii)$\Longrightarrow$(ii).

Our methods could give some insight into the automorphism groups of spherical varieties; see [Br1], [Br2] for some connections between two–orbit varieties and embeddings of spherical homogeneous spaces.

The author thanks the University of Bochum for its hospitality while this work was done, and Dmitri Ahiezer for very useful discussions.

2. Proof of (i) $\Longleftrightarrow$ (ii)

From now on, X denotes a two–orbit G–variety, Ω its open orbit, and Y its closed orbit. We assume that Y is a divisor of X, and we denote by $\mathcal{N}_Y$ its normal sheaf.

Proof of (i)$\Longrightarrow$(ii). Let f be a non–constant regular function on Ω. It is a rational function on X, with a pole of some order $n > 0$ along Y. Then f defines a global section of $\mathcal{O}_X(nY)$. Moreover, the set of points of X where f generates $\mathcal{O}_X(nY)$ contains a dense open subset of Y. Hence, because Y is homogeneous, $\mathcal{O}_Y(nY)$ is generated by its global sections. So the same holds for $\mathcal{O}_Y(nY) = \mathcal{N}_Y^n$. From the homogeneity of Y, it follows that $\mathcal{N}_Y$ is generated by its global sections, too.

Proof of (ii)$\Longrightarrow$(i). From the exact sequence

$$0 \to \mathcal{O}_X \to \mathcal{O}_X(Y) \to \mathcal{N}_Y \to 0$$

we get an exact sequence

$$0 \to \mathbb{C} = H^0(X, \mathcal{O}_X) \to H^0(X, \mathcal{O}_X(Y)) \to H^0(Y, \mathcal{N}_Y) \to H^1(X, \mathcal{O}_X) .$$

But $H^1(X, \mathcal{O}_X) = 0$ because X is projective, smooth and unirational (see [S, Lemma 1]). Hence every non–zero global section of $\mathcal{N}_Y$ lifts to a non–constant global section of $\mathcal{O}_X(Y)$, which corresponds to a non–constant regular function on Ω.

3. The Tangent Bundle

In this section, we analyze the automorphism group and the tangent bundle of any two–orbit variety X with a one–codimensional closed orbit. Because X is complete, smooth and unirational, its automorphism group $\mathrm{Aut}(X)$ is affine algebraic; moreover, the Lie algebra of $\mathrm{Aut}(X)$ consists in all global vector fields on X (see [MO, Theorems 3.6 and 3.7] and [R, Corollary 1]). Under assumption (ii) of the theorem, we will produce such vector fields, which move Y; this will imply that X is homogeneous.

Denote by $\mathcal{T}_X$ the sheaf of vector fields on X, i.e., of derivations of $\mathcal{O}_X$ into itself. Denote by $\mathcal{S}_X$ its subsheaf consisting of all derivations which preserve the ideal $\mathcal{O}_X(-Y)$ of Y in X.

Proposition.

(i) *The sheaf $\mathcal{S}_X$ is locally free.*

(ii) *There is an exact sequence*

$$(*) \qquad\qquad 0 \longrightarrow \mathcal{S}_X \longrightarrow \mathcal{T}_X \longrightarrow \mathcal{N}_Y \longrightarrow 0 .$$

(iii) *The restriction of $\mathcal{S}_X$ to Y is isomorphic to its tangent sheaf $\mathcal{T}_Y$.*

Proof. Let $p \in X$. Because Y is a smooth divisor of X, we can choose a system of parameters $x_1, \ldots, x_n$ in $\mathcal{O}_{X,p}$ (the local ring of X at p) such that x_n generates the ideal of Y. Then $(dx_1, \ldots, dx_n)$ is a basis of the module of differentials of $\mathcal{O}_{X,p}$ over $\mathbb{C}$. Denote by $(\partial_1, \ldots, \partial_n)$ its dual basis: it is a basis of $\mathcal{T}_{X,p}$. Moreover, $\mathcal{S}_{X,p}$ is the $\mathcal{O}_{X,p}$–submodule of $\mathcal{T}_{X,p}$ generated by $\partial_1, \ldots, \partial_{n-1}, x_n\partial_n$, hence (i).

For (ii), define a map $\pi : \mathcal{T}_X \to \mathcal{N}_Y$ as the composition

$$\mathcal{T}_X = \mathrm{Der}(\mathcal{O}_X, \mathcal{O}_X) \to \mathrm{Der}(\mathcal{O}_X, \mathcal{O}_Y) \to \mathrm{Hom}(\mathcal{O}_X(-Y), \mathcal{O}_Y) = \mathcal{O}_Y(Y) = \mathcal{N}_Y .$$

Then the kernel of π is $\mathcal{S}_X$. Moreover, if $p \in X$ is as before, then $\pi(\partial_n)$ sends $x_n \in \mathcal{O}_{X,p}(-Y)$ to 1, hence $\pi(\partial_n)$ generates $\mathrm{Hom}(\mathcal{O}_X(-Y), \mathcal{O}_Y)_p = \mathcal{N}_{Y,p}$. So π is surjective.

In order to prove (iii), observe that the restriction to Y of $\pi : \mathcal{T}_X \to \mathcal{N}_Y$ coincides at p with the canonical map $\mathcal{T}_X|Y \to \mathcal{N}_Y$. Hence its kernel $\mathcal{S}_X|Y$ is the tangent sheaf of Y.

4. A First Characterization of Homogeneity

From the exact sequence $(*)$ in the proposition in the previous section, we get an exact sequence

$$(**) \quad 0 \to H^0(X, \mathcal{S}_X) \to H^0(X, \mathcal{T}_X) \to H^0(Y, \mathcal{N}_Y) \to H^1(X, \mathcal{S}_X) \to H^1(X, \mathcal{T}_X) .$$

Moreover, $H^0(X, \mathcal{S}_X)$ contains the image of the natural map from the Lie algebra of G, to $H^0(X, \mathcal{T}_X)$.

Set $V = H^0(Y, \mathcal{N}_Y)$; it is a simple G–module, because $\mathcal{N}_Y$ is a line bundle on the G–homogeneous projective variety Y. For any G–module M, denote by M_V the space of all G–module morphisms from V to M.

Proposition. *Under assumption (ii) of the theorem, the group $\mathrm{Aut}(X)$ acts transitively on X if and only if $H^1(X, \mathcal{S}_X)_V = 0$.*

Proof. If $H^1(X, \mathcal{S}_X)_V = 0$, then any non–zero global section of $\mathcal{N}_Y$ lifts to a global vector field on X, which does not preserve the ideal of Y. Hence $\mathrm{Aut}(X)$ does leave Y stable, so it acts transitively on X.

Conversely, if X is homogeneous, then $H^1(X, \mathcal{T}_X) = 0$ (cf. [Bo, Theorem VII] or [D, Théorème 2]). Let us show that $H^1(X, \mathcal{S}_X) = 0$. If not, then the map $V \to H^1(X, \mathcal{S}_X)$ is an isomorphism because V is simple. Hence the map $H^0(X, \mathcal{S}_X) \to H^0(X, \mathcal{T}_X)$ is an isomorphism, too, which contradicts the homogeneity of X.

5. A Vanishing Lemma

In this section and the next one, we prove two technical results which will imply the vanishing of $H^1(X, \mathcal{S}_X)_V$.

Lemma. *For any integer $n \geq 1$, we have*

$$H^0(Y, \mathcal{T}_Y \otimes_{\mathcal{O}_Y} \mathcal{N}_Y^n)_V = H^1(Y, \mathcal{T}_Y \otimes_{\mathcal{O}_Y} \mathcal{N}_Y^n)_V = 0 \ .$$

Proof of the lemma. The unipotent radical of G acts trivially on Y and on V, hence we may assume that G is reductive. Choose a Borel subgroup B of G, and a maximal torus T of B. Denote by λ the highest weight of V. There is a filtration of the homogeneous vector bundle $\mathcal{T}_Y$ on Y, such that every subquotient is the homogeneous vector bundle associated with a simple P–module, whose highest weight is a positive root (compare [D, p. 184]). Hence by Bott's theorem (see [Bo]), the highest weight of any simple submodule of $H^0(Y, \mathcal{T}_Y \otimes_{\mathcal{O}_Y} \mathcal{N}_Y^n)$ can be written as $\beta + n\lambda$, where β is a positive root. If $H^0(Y, \mathcal{T}_Y \otimes_{\mathcal{O}_Y} \mathcal{N}_Y^n)_V \neq 0$, then $\beta + n\lambda = \lambda$, hence $-(n-1)\lambda$ is a positive root. This is impossible, because $(n-1)\lambda$ is a dominant weight.

Similarly, the highest weights of any simple submodule of $H^1(Y, \mathcal{T}_Y \otimes_{\mathcal{O}_Y} \mathcal{N}_Y^n)$ can be written as

$$s_\alpha(\beta + n\lambda + \rho) - \rho$$

where α is a simple root, s_α is its associated simple reflection, and ρ is the half–sum of all positive roots. If $H^1(Y, \mathcal{T}_Y \otimes_{\mathcal{O}_Y} \mathcal{N}_Y^n)_V \neq 0$, then $s_\alpha(\beta + n\lambda + \rho) = \rho + \lambda$, hence

$$\beta + n\lambda + \rho = s_\alpha(\rho + \lambda) = \rho - \alpha + \lambda - m\alpha$$

where m is a non–negative integer. So

$$\beta + (m+1)\alpha = (1-n)\lambda \ .$$

The left–hand side is a non–zero sum of positive roots, and the right–hand side is the opposite of a dominant weight: a contradiction.

6. Another Vanishing Lemma

Because $\mathcal{N}_Y$ has non–zero global sections and Y is homogeneous, $\mathcal{N}_Y$ is generated by its global sections. As in the proof of (i)$\Longleftrightarrow$(ii), it follows that $\mathcal{O}_X(Y)$ is generated by its global sections, too. Let $\pi : X \to \mathbf{P}^N$ be the associated morphism to a projective space; then $\mathcal{O}_X(Y)$ is equal to $\pi^*\mathcal{O}(1)$.

Lemma. *For every $i \geq 1$, we have $R^i\pi_*\mathcal{S}_X = 0$.*

Proof. There is an action of G on $\mathbf{P}^N$ such that π is G–equivariant. Moreover, $X' := \pi(X)$ is normal and positive–dimensional (because $\mathcal{O}_X(Y)$ is non–trivial). Hence X' is a two–orbit G–variety; moreover, its closed orbit is a hyperplane section of X', so its open orbit is affine.

By Stein factorization, we can write $\pi : X \to X'$ as the composition of a morphism with connected fibers $p : X \to X''$ and a finite morphism $q : X'' \to X'$. Moerover, X'' is a normal G–variety, and p, q are G–equivariant. Because the morphism q is affine, the open orbit in X'' is affine, too; and its suffices to show that $R^i p_* S_X = 0$ for every $i \geq 1$.

Observe that p preserves the codimension of the closed orbit, hence p is equidimensional. Because X and X'' are smooth, p is flat. Hence it is enough to show that $H^i(F, S_X|F) = 0$ for every $i \geq 1$.

If $F = p^{-1}(x'')$ where x'' is in the open G–orbit of X'', then $F \simeq H''/H$ where H'' (resp. H) denotes the isotropy group of x'' (resp. of $x \in F$) in G. Moreover, the restriction of S_X (i.e., of T_X) to F, is the H''–linearized sheaf on H''/H associated to the H–module $\mathrm{Lie}(G)/\mathrm{Lie}(H)$. From the exact sequence of H–modules

$$0 \to \mathrm{Lie}(H'')/\mathrm{Lie}(H) \to \mathrm{Lie}(G)/\mathrm{Lie}(H) \to \mathrm{Lie}(G)/\mathrm{Lie}(H'') \to 0$$

we get an exact sequence

$$0 \to T_F \to S_X|F \to \mathcal{O}_F \otimes_{\mathbb{C}} \mathrm{Lie}(G)/\mathrm{Lie}(H'') \to 0$$

because the H–module $\mathrm{Lie}(G)/\mathrm{Lie}(H'')$ is the restriction of an H''–module. But $H^i(F, T_F) = H^i(F, \mathcal{O}_F) = 0$ for every $i \geq 1$, because F is projective and homogeneous under an affine algebraic group; see [Bo, Theorem VII]. Hence $H^i(F, S_X|F) = 0$ for every $i \geq 1$.

If F is the fiber of a point in the closed orbit of X'', then F is contained in Y and the proof is the same, because $S_X|Y = T_Y$ by (iii) of the proposition in Section 3.

7. CONCLUSION OF THE PROOF

End of the proof of (ii)$\Longrightarrow$(iii). Choose an integer $n \geq 1$. From the exact sequence

$$0 \to \mathcal{O}_X((n-1)Y) \to \mathcal{O}_X(nY) \to \mathcal{O}_Y(nY) = \mathcal{N}_Y^n \to 0$$

and the fact that S_X is locally free ((i) of the proposition in Section 3), it follows that the sequence

$$0 \to S_X((n-1)Y) \to S_X(nY) \to S_X \otimes_{\mathcal{O}_X} \mathcal{N}_Y^n \to 0$$

is exact. Moreover, $S_X \otimes_{\mathcal{O}_X} \mathcal{N}_Y^n$ is isomorphic to $T_Y \otimes_{\mathcal{O}_Y} \mathcal{N}_Y^n$ by (iii) of the proposition in Section 3. Hence we have a long exact sequence of G–modules

$$H^0(Y, T_Y \otimes_{\mathcal{O}_Y} \mathcal{N}_Y^n) \to H^1(X, S_X((n-1)Y))$$
$$\to H^1(X, S_X(nY)) \to H^1(Y, T_Y \otimes_{\mathcal{O}_Y} \mathcal{N}_Y^n) .$$

So, by the lemma in Section 5, the map

$$H^1(X, \mathcal{S}_X((n-1)Y))_V \to H^1(Y, \mathcal{S}_X(nY))_V$$

is an isomorphism. Hence

$$H^1(X, \mathcal{S}_X)_V \simeq H^1(X, \mathcal{S}_X(nY))_V \ .$$

But $H^1(X, \mathcal{S}_X(nY)) = H^1(X, \mathcal{S}_X \otimes \pi^* \mathcal{O}(n))$ is isomorphic to $H^1(\mathbf{P}^N, (\pi_* \mathcal{S}_X)(n))$ by the Lemma in Section 6 and the projection formula. The latter group vanishes for n large enough, so $H^1(X, \mathcal{S}_X)_V = 0$.

Proof of (iii)$\Longrightarrow$(ii). There exist global vector fields on X which do not preserve Y. Hence the map $H^0(X, \mathcal{S}_X) \to H^0(X, \mathcal{T}_X)$ in the exact sequence (∗∗) is not onto; so $H^0(Y, \mathcal{N}_Y)$ is non–zero.

REFERENCES

[A]   D. N. Ahiezer, *Equivariant completions of homogeneous varieties by homogeneous divisors*, Ann. Glob. Analysis and Geometry **1** (1983), 49–78.

[Bo] R. Bott, *Homogeneous vector bundles*, Ann. of Math. **66** (1957), 203–248.

[Br1] M. Brion, *On spherical varieties of rank one*, Canadian Math. Soc. Conference Proceedings **10** (1989), 31–41.

[Br2] M. Brion, *Vers une généralisation des espaces symétriques*, J. of Alg. **134** (1990), 115–143.

[D]   M. Demazure, *Automorphismes et déformations des variétés de Borel*, Invent. math. **39** (1977), 179–186.

[HS] A. Huckleberry, D. Snow, *Almost homogeneous Kähler manifolds with hypersurface orbits*, Osaka J. of Math. **19** (1982), 763–786.

[MO] H. Matsumura, F. Oort, *Representability of group functors, and automorphisms of algebraic schemes*, Invent. math. **4** (1997), 1–25.

[R]   C. P. Ramanujam, *A note on automorphism groups of algebraic varieties*, Math. Ann. **156** (1964), 25–33.

[S]   J. P. Serre, *On the fundamental group of a unirational variety*, J. London Math. Soc. **34** (1959), 481–484.

INSTITUT FOURIER, UNIVERSITÉ DE GRENOBLE, B. P. 74, F-38402 SAINT-MARTIN D'HÈRES CEDEX, FRANCE

E-mail address: mbrion@ frgren81.bitnet

HOCHSCHILD HOMOLOGY FOR A POLYNOMIAL COCHAIN ALGEBRA OF A NILMANIFOLD

BOHUMIL CENKL

Northeastern University

ABSTRACT. With the fundamental group G of a k–dimensional nilmanifold there is associated a free differential graded algebra $P(G)$ over $\mathbb{Z}$ whose cohomology is isomorphic to the integer cohomology of G, [2]. In this note we define a spectral sequence converging to the Hochschild homology $HH_*(P(G))$ of the free differential graded algebra (DGA) $P(G)$. The E^1-term is isomorphic to the Hochschild homology of $P(H)$, where H is a free abelian group on k generators. The E^1-term is computed explicitly. We also outline an algorithm for the computation of $HH_*(P(G))$ in terms of the group structure of G. This result should facilitate a computation of the homology of the free loop space of a nilmanifold.

1. POLYNOMIAL COCHAIN ALGEBRA $P(G)$

The study of nilmanifolds is simplified and many computations can be explicitly carried out by replacing the standard cochains by rational integer valued polynomial functions.

A *nilmanifold* is a compact manifold N which is the space of cosets of a connected simply connected nilpotent Lie group by a discrete uniform subgroup G. The fundamental group G of N is a finitely generated torsion free nilpotent group. Thus a nilmanifold N can be identified with the Eilenberg-Mac Lane space $K(G, 1)$. Therefore the cohomology of N is the same as the cohomology of G. In this part we introduce a free algebra $P(G)$ over $\mathbb{Z}$ in place of the standard cochains for the group G.

Suppose that the nilmanifold N is k-dimensional. Then, using the Malcev basis $\{g_1, \ldots, g_k\}$, any element g of G can be written in the form $g = g_1^{x_1} \cdots g_k^{x_k}$ where x_j is an integer. Thus G can be identified with $\mathbb{Z}^k$ by sending g to the sequence $x = (x_1, \ldots, x_k)$. Then $\mathbb{Z}^k$ inherits a group structure from G with the group operation on the elements $x = (x_1, \ldots, x_k)$ and $y = (y_1, \ldots, y_k)$ given by the formula

$$x \cdot y = x + y + \tau(x, y) = (x_1 + y_1 + \tau_1(x, y), \ldots, x_k + y_k + \tau_k(x, y)),$$

1991 *Mathematics Subject Classification.* Primary: 57T15,16A61 Secondary: 18G35,55N25.
Key words and phrases. Hochschild homology, nilmanifold, noncommutative algebra.
I wish to thank to M.Vigué-Poirrier for useful comments concerning this paper

where

$$\tau_j = \tau_j(x, y) = \tau(x_1, \ldots, x_{j-1}, y_1, \ldots, y_{j-1})$$

are polynomials with rational coefficients.

Let $P^1(G)$ be the $\mathbb{Z}$-module generated by the $k \times 1$ matrices

$$a = \begin{pmatrix} a_k \\ \vdots \\ a_1 \end{pmatrix}$$

of nonnegative integers such that $a_1 + \cdots + a_k > 0$. A matrix a can be identified with a rational polynomial function from $\mathbb{Z}^k$ to $\mathbb{Z}$ sending a to the polynomial

$$\langle a, x \rangle = \prod_{r=1}^{k} \binom{x_r}{a_r},$$

where

$$x = \begin{pmatrix} x_k \\ \vdots \\ x_1 \end{pmatrix}, \quad \binom{x_.}{a_.} = \frac{x_.(x_. - 1) \ldots (x_. - a_. + 1)}{a_.!}, \quad \binom{x_.}{0} = 1.$$

We use the symbol $P^1(G)$ for the $\mathbb{Z}$-module generated by such polynomials. Let $P(G) = T(P^1(G))$ be the free associative algebra with unit generated by $P^1(G)$. The product of two elements a, b is denoted by ab. Let $d : P^1(G) \longrightarrow P^2(G) = P^1(G) \otimes P^1(G)$ be the operation defined by the formulas

$$d\langle a, x \rangle = \langle a, y \rangle + \langle a, z \rangle + \langle a, y + z + \tau(y, z) \rangle,$$

where the right hand side, as an element of $P^2(G)$, is a polynomial in $2k$ variables. Then d is extended to $P(G)$ as a derivation, and $(P(G), d)$ becomes a DGA over $\mathbb{Z}$. $P(G)$ is called the *polynomial cochain algebra* of G.

Consider $P(G)$ as a subcomplex of the cochain complex $C^*(G) = \operatorname{Hom}(C_*(G); \mathbb{Z})$. The differential d is the dual of the differential ∂ on the chain complex. The inclusion of $P(G)$ into the cochain algebra $C^*(G)$ for the group G induces an isomorphism between the cohomology ring $H^*(P(G))$ and the standard cohomology ring $H^*(G; \mathbb{Z})$ of the group G, [2].

Let

$$G = G_1 > G_2 > \cdots > G_s > G_{s+1} = 1$$

be the shortest central series such that
   (i) G_i/G_{i+1} is a free abelian group and for which
   (ii) the commutator induces the maps

$$[G_i, G_j] \leq G_{i+j} \quad \text{for all} \quad i, j.$$

Such a central series can be constructed as follows: The group G_i is generated by all the commutators $[a, b]$, $a \in G$, $b \in G_{i-1}$ and by all elements $c \in G_{i-1}$, some nonzero power of which is a product of such commutators. That such a series satisfies the condition (i) is obvious. The property (ii) follows from the Jacobi identity, when applied to the sequence of ideals of the Lie algebra associated to G (see [3]).

Let $\{g_1, g_2, \ldots, g_k\}$ be the Malcev canonical basis for the group G. Suppose that $\{g_{i_{j-1}+1}, g_{i_{j-1}+2}, \ldots, g_{i_j}\}$ is the subset of such a basis which projects to a basis for the abelian group G_j/G_{j+1}. In this way the Malcev canonical basis is divided into s mutually exclusive subsets

$$\{g_1, \ldots, g_{i_1}; \ldots; g_{i_{s-1}+1}, \ldots, g_k\}.$$

This decomposition induces an equivalent partition

$$\{1, \ldots, i_1; \ldots; i_{s-1}+1, \ldots, k\}$$

of the indices $1, 2, \ldots, k$. The resulting subsets are referred to as levels. For example the indices $i_{j-1}+1, \ldots, i_j$ are of level j for $j = 1, 2, \ldots, s$; $i_0 = 0$; $i_s = k$.

Using this partition of the sequence $\{1, 2, \ldots, k\}$ we define a graded module structure on $P(G)$. To each element a in $P^1(G)$ we assign a nonnegative integer $\|a\|$ called the norm according to the rule

$$\|a\| = (a_{i_1+1} + \cdots + a_{i_2}) + \cdots + j \left(a_{i_j+1} + \cdots + a_{i_{j+1}}\right) + \cdots$$
$$+ (s-1) \left(a_{i_{s-1}+1} + \cdots + a_k\right).$$

And for any nonnegative integer m, $\|ma\| = \|a\|$. Furthermore

$$\|b + c\| = \max\{\|b\|, \|c\|\}$$

and

$$\|bc\| = \|b\| + \|c\|$$

for any b and c in $P(G)$.

As was mentioned earlier, an element $a \in P^1(G)$ can be identified with the polynomial function $\prod \begin{pmatrix} x_r \\ a_r \end{pmatrix}$. It follows that this identification preserves norms if the norm of the binomial is the integer

$$\left\| \begin{pmatrix} x_r \\ a_r \end{pmatrix} \right\| = (j-1)a_r,$$

where j is the level of r. In particular $\left\| \begin{pmatrix} x_r \\ 1 \end{pmatrix} \right\| = \|x_r\| = j - 1$. The definition of the norm of a binomial can be extended to rational polynomials. If $p(x_1, \ldots, x_k)$ is such a polynomial then we set

$$\|p(x_1, \ldots, x_k)\| = \left\| \begin{pmatrix} p(x_1, \ldots, x_k) \\ 1 \end{pmatrix} \right\|,$$

where $\|qx_r\| = \|x_r\|$ for any rational number q. Then

$$\left\| \binom{p(x_1, \ldots, x_k)}{a_r} \right\| = a_r \|p(x_1, \ldots, x_k)\|.$$

Using the norm $\| \ \|$ we define a filtration on $P(G)$ by setting

$$\mathcal{G}^i = \mathcal{G}^i P(G) = \{a \in P(G) \mid \|a\| \leq i\}, \quad i = 0, 1, 2, \ldots,$$

$$\mathcal{G}^0 \subset \mathcal{G}^1 \subset \cdots \subset \mathcal{G}^{i-1} \subset \mathcal{G}^i \subset \cdots \subset P(G).$$

Lemma 1 [3]. *The product of cochains in $P(G)$ induces the map*

$$\mathcal{G}^i \times \mathcal{G}^j \longrightarrow \mathcal{G}^{i+j}$$

and the filtration is preserved by the differential d.

Proof. The first part of the statement is obvious. The proof of the second part follows from the bound

$$\|\tau_r(x, y)\| \leq j - 2$$

which holds for any index r of level j, i.e., $i_{j-i} + 1 \leq r \leq i_j$. This crucial inequality follows from the analysis of the functions τ_r in [4]. It is also proved in [3] by using the Lie algebra associated with the above central series for G.

2. HOCHSCHILD HOMOLOGY

With the algebra $(P(G), d)$ we associate a negatively graded algebra (P_*, d_*) over $\mathbb{Z}$, with the differential d_* of degree -1 by setting $P_* = P^{-*}(G)$. In order to establish a notation we review the definition of the *Hochschild homology* for a negatively graded DGA (A_*, d_*) over a ring with a unity. Let $(A_*^{\mathrm{op}}, d_*^{\mathrm{op}})$ be the opposite DGA with $A^{\mathrm{op}} \simeq A$, $a^{\mathrm{op}} \cdot b^{\mathrm{op}} = (-1)^{|a||b|}(ba)^{\mathrm{op}}, d^{\mathrm{op}}(a^{\mathrm{op}}) = (da)^{\mathrm{op}}$. The tensor product $A^{\mathrm{e}} = A \otimes A^{\mathrm{op}}$ is the associated enveloping algebra. The operations

$$m_\ell : A^{\mathrm{e}} \otimes A \longrightarrow A, \quad m_\ell(a \otimes b^{\mathrm{op}}, c) = (-1)^{|b||c|}acb$$

and

$$m_r : A \otimes A^{\mathrm{e}} \longrightarrow A, \quad m_r(c, a \otimes b^{\mathrm{op}}) = (-1)^{|b|(|a|+|c|)}bca$$

give A the structure of a left and right A^{e}-module respectively. Then the Hochschild homology of (A_*, d_*) is the graded module

$$HH_*(P_*, d_*) = \mathrm{Tor}_*^{A^{\mathrm{e}}}(A_*, A_*).$$

Our goal is to compute the Hochschild homology of (P_*, d_*). In order to emphasize the original positively graded algebra $P(G)$ we write $HH_*(P(G))$ instead of $HH_*(P_*, d_*)$. In [4] it was demonstrated that this homology, for a free algebra P_*

can be computed as the homology of a certain "small" complex associated with P_*. In what follows we study this "small" complex for the free DGA (P_*, d_*).

Let $V = V_{-1} = P^1(G)$, $\overline{V} = \overline{V}_0 = P^1(G)$ and let $S : P \otimes P \longrightarrow P \otimes \overline{V}$ be a $\mathbb{Z}$-linear map defined by

$$
S(a, v_1 \cdots v_p) = (-1)^{|a|} \sum_{i=1}^{p-1} (-1)^{\epsilon_i} v_{i+1} \cdots v_p a v_1 \cdots v_{i-1} \otimes \overline{v}_i
$$
$$
+ (-1)^{|a|} a v_1 \cdots v_{p-1} \otimes \overline{v}_p,
$$

where $\epsilon_i = (p - i)(|a| + i - 1)$, for $a \in P$, $v_i \in V$. Then

Lemma 2 [6]. *The Hochschild homology of the free DGA (P_*, d_*) is isomorphic to the homology of the complex $\{P \oplus (P \otimes \overline{V}), \delta\}$ where*
$$
\delta|_P = d,
$$
$$
\delta(a \otimes \overline{v}) = da \otimes \overline{v} + (-1)^{|a| + |\overline{v}|}(av - (-1)^{|a||v|} va) - S(a, dv).
$$

Using the norm $\| \ \|$ we define a filtration on the complex $\{P \oplus (P \otimes \overline{V}), \delta\}$ by setting

$$
F_i = \{a \oplus (b \otimes \overline{v}) | \ \|a \oplus (b \otimes \overline{v})\| = \max(\|a\|, \|b\| + \|\overline{v}\|) \le i\}.
$$

It follows that

$$
F_0 \subset F_1 \subset \cdots \subset F_{i-1} \subset F_i \subset \cdots \subset P \oplus (P \otimes \overline{V}).
$$

Furthermore, by Lemma 1,
$$
\delta F_i \subset F_i.
$$

Let $\{E^r, d^r\}$, $r \ge 0$, be the spectral sequence corresponding to the filtration $\{F_i\}$.

3. Computation of E^1

Since $F_i = F_i' \oplus F_i''$, where

$$
F_i' = \{a \in P | \ \|a\| \le i\},
$$
$$
F_i'' = \{a \otimes \overline{v} \in P \otimes \overline{V} | \|a \otimes \overline{v}\| \le i\},
$$

the projection of F_i onto $E_i^0 = F_i / F_{i-1}$ has the form

$$
p = p' \oplus p'' : F_i = F_i' \oplus F_i'' \longrightarrow E_i^0.
$$

The differential $\delta : F_i \longrightarrow F_i$ can be written as

$$
\delta = d + \delta' + \delta'',
$$

where

$$d = \delta|_P,$$
$$\delta'(a \otimes \bar{v}) = (-1)^{|a|+|\bar{v}|}(av - (-1)^{|a||v|}va),$$
$$\delta''(a \otimes \bar{v}) = da \otimes \bar{v} - S(a, dv)$$

and δ', δ'' are both zero maps on P.

Next we analyze the images of d, δ', and δ'' on an element of norm equal to i. Since d is a derivation on $P(G)$ it suffices to consider the image of d on an element $v = \begin{pmatrix} v_k \\ \vdots \\ v_1 \end{pmatrix}$ of $P^1(G)$ with the norm $\|v\| = \sum_{r=1}^{k} \|x_r\| v_r = i$. Let

$$(x, y) = \begin{pmatrix} x_k & y_k \\ \vdots & \vdots \\ x_1 & y_1 \end{pmatrix}$$

be an element of $C_2(G)$. Then

$$\langle dv, (x, y) \rangle = \langle v, \partial(x, y) \rangle$$
$$= \langle v, x \rangle + \langle v, y \rangle - \langle v, x + y + \tau(x, y) \rangle$$
$$= \langle v, x \rangle + \langle v, y \rangle - \prod_{r=1}^{k} \sum_{l=0}^{v_r} \binom{x_r + y_r}{v_r - l} \binom{\tau_r}{l}$$
$$= -\prod_{r=1}^{k} \sum_{t=1}^{v_r-1} \binom{x_r}{v_r - t} \binom{y_r}{t} - \prod_{r=1}^{k} \sum_{l=1}^{v_r} \sum_{t=0}^{v_r-l} \binom{x_r}{v_r - l - t} \binom{y_r}{t} \binom{\tau_r}{l}.$$

The first term can be written in the form

$$\sum_{t_r=1}^{v_r-1} \langle \begin{pmatrix} v_k - t_k, & t_k \\ \vdots & \vdots \\ v_1 - t_1, & t_1 \end{pmatrix} \begin{pmatrix} x_r & y_r \\ \vdots & \vdots \\ x_1 & y_1 \end{pmatrix} \rangle,$$

with $t = 1, 2, \ldots, k$. Its norm is equal to $i = \|v\|$, because

$$\left\| \prod_{r=1}^{k} \sum_{l=1}^{v_r-1} \binom{x_r}{v_r - t} \binom{y_r}{t} \right\| = \sum_{r=1}^{k} \max \left\{ \left\| \binom{x_r}{v_r - t} \right\| + \left\| \binom{y_r}{t} \right\| \right\}$$
$$= \sum_{r=1}^{k} v_r \|x_r\| = i \quad \text{as} \quad \|x_r\| = \|y_r\|.$$

On the other hand the norm of the second term is strictly smaller than i because

$$\left\| \prod_{r=1}^{k} \sum_{l=1}^{v_r} \sum_{t=1}^{v_r-l} \binom{x_r}{v_r-l-t} \binom{y_r}{t} \binom{\tau_r}{l} \right\|$$

$$= \sum_{r=1}^{k} \sum_{l=1}^{v_r} \{ \|x_r\|(v_r-l) + \|\tau_r\|l \} < \sum_{r=1}^{k} \|x_r\| v_r = i.$$

Therefore we can conclude that $v \in \overline{V}$, $\|v\| = i$,

$$dv = - \sum_{t_r=1}^{v_r-1} \begin{pmatrix} v_k-t_k, & t_k \\ \vdots & \vdots \\ v_1-t_1, & t_1 \end{pmatrix} + \cdots,$$

where the first term has norm equal to i while the terms $\cdots$ have norm strictly smaller than i. If we use the notation

$$v^t = \begin{pmatrix} v_k-t_k \\ \vdots \\ v_1-t_1 \end{pmatrix}, \qquad t = \begin{pmatrix} t_k \\ \vdots \\ t_1 \end{pmatrix}, \quad \text{and} \quad a = \begin{pmatrix} a_k \\ \vdots \\ a_1 \end{pmatrix}$$

we can write

$$dv = - \sum v^t \cdot t + \cdots .$$

It is clear that δ' preserves not only the filtration but also the norm. And it is sufficient to analyze the operator δ'' only on the elements $a \otimes \overline{v} \in P^1(G) \otimes \overline{V}, \|a \otimes \overline{v}\| = i$, because d is a derivation on $P(G)$. On such an element $a \otimes \overline{v}$ we have

$$da \otimes \overline{v} = - \sum a^t \cdot t \otimes \overline{v} + \text{terms of lower filtration}$$

and

$$S(a, dv) = - \sum S(a, v^t \cdot t) + \text{terms of lower filtration}.$$

Summarizing we get

Lemma 3. Let $a \oplus (b \otimes \overline{v}) \in P^1 \oplus (P^1 \otimes \overline{V})$ be an element of norm i. Then

$$p\delta(a \oplus (b \otimes \overline{v})) = -p' \sum a^t \cdot t + p'(bv + vb) + p'' \sum S(b, v^t \cdot t) - p'' \sum b^4 \cdot t \otimes \overline{v}.$$

Then we can conclude that

Lemma 4. *The E^1-term of the spectral sequence $\{E^r, d^r\}$ is isomorphic to the graded Hochschild homology*

$$HH_*(P(H))$$

for the free DGA $P(H)$ of a free abelian group H on k generators.

Proof. For a free abelian group H on k generators, with Malcev basis $\{h_1, \ldots, h_k\}$, the group structure is given by the formula

$$x \cdot y = x + y = (x_1 + y_1, \cdots, x_n + y_n),$$

i.e., all the polynomial functions τ_j (see Section 1) are identically equal to zero.

Let $(\mathcal{P}_*, \partial_*)$ be the negatively graded differential algebra associated with the free DGA $P(H)$. Then the Hochschild homology of the free DGA $(\mathcal{P}_*, \partial_*)$ is according to Lemma 2, isomorphic to the homology of the complex $\{\mathcal{P} \oplus (\mathcal{P} \otimes \overline{\mathcal{V}}), {}^{\backprime}\delta\}$, where $\overline{\mathcal{V}} = P^1(H)$. Let

$$\mathcal{F}_0 \subset \mathcal{F}_1 \subset \cdots \subset \mathcal{F}_{i-1} \subset \mathcal{F}_i \subset \cdots \subset \mathcal{P} \oplus (\mathcal{P} \otimes \overline{\mathcal{V}})$$

be the filtration, defined by using the norm $\| \ \|$ in the same way as for the group G, and let $\{\mathcal{E}^r, \partial^r\}, r \geq 0$, be the corresponding spectral sequence. The formulas for the differential ${}^{\backprime}\delta$ are given by the formulas for δ when all the polynomial functions τ_j are equal to zero for all j. Therefore all the terms of lower filtration in the formulas for $dv, da \otimes \overline{v}$ and $S(a, dv)$, when $v \in P^1(H)$, $a \otimes \overline{v} \in P^1(H) \otimes \overline{\mathcal{V}}$, as written in the proof of Lemma 3, are equal to zero.

An identification of the Malcev bases for H and G, $h_j \longmapsto g_j$, $j = 1, 2, \ldots, k$, induces an isomorphism of graded algebras $\iota : P(H) \longrightarrow P(G)$ such that $\iota(\mathcal{F}_i) \subset F_i$. The map ι does not commute with the differentials. However, the induced map $\iota : \mathcal{E}^0 \longrightarrow E^0$ is a morphism of differential graded modules. The differential ∂^0 is induced by d through the formulas in Lemma 3, when applied to $P^1(H)$.

Since the ${}^{\backprime}\delta$-image of an element of the norm equal to i is of the same norm (not smaller), it follows that $\partial^1 = 0$. Therefore the spectral sequence $\{\mathcal{E}^r, \partial^r\}$ collapses and $\mathcal{E}_i^1 \simeq \mathcal{E}_i^\infty$ for all $i = 0, 1, \ldots$. Hence we have

$$\mathcal{E}_i^\infty \simeq F_i HH_*(P(H))/F_{i-1}HH_*(P(H))$$

where $F_i HH_*(P(H)) = HH_*(\mathcal{F}_i)$. Therefore the statement follows from the isomorphism $\mathcal{E}_i^1 \simeq \mathcal{E}_i^\infty$.

The computation of the $\mathcal{E}^1$ - term of the spectral sequence can be greatly simplified by replacing the algebra $P(H)$ by its "model" $M(H)$. For any nilmanifold N with fundamental group G, $(M(G), D)$ is a graded algebra whose cohomology is isomorphic to the cohomology of the free DGA $(P(G), d)$; $M(G) = M \otimes \cdots \otimes M$, ($k$-times, $k = \dim N$, $M = H^*(S^1, \mathbb{Z})$), as a $\mathbb{Z}$-module; D and the product $\cdot$ are determine by the group structure of G. More specifically, there exist degree preserving chain maps I, P and a homotopy H,

$$
\begin{array}{ccc}
P(G) & d & H \\
I\uparrow \ \downarrow P & & \\
M(G) & D &
\end{array}
$$

such that $PI = $ identity, $IP = dH + Hd + $ identity. Once the maps I, P, H and the differential D have been constructed then the product $\cdot$ on $M(G)$ is induced by the cup product on $P(G)$ by the formula

$$
a \cdot b = P((Ia) \cup (Ib)).
$$

The construction of I, P, H and D proceeds as follows. Let

$$
1 = G^k < G^{k-1} < \cdots < G^1 < G^0 = G
$$

be a central series such that the successive quotients G^{i-1}/G^i are isomorphic to $\mathbb{Z}$. Note that this series is a refinement of the shortest central series from Section 1. Let $G_i = G/G^i$. Then

$$
0 \leftarrow G_{s-1} \leftarrow G_s \leftarrow G_1 = \mathbb{Z} \leftarrow 0
$$

is the central extension and

$$
K(G_1, 1) \rightarrow K(G_s, 1) \rightarrow K(G_{s-1}, 1)
$$

is the corresponding fibration. This fibration gives a homotopy equivalence

$$
\begin{array}{ccc}
P(G_s) & d_s & h_s \\
i_s\uparrow \ \downarrow P_s & & \\
P(G_1) \otimes P(G_{s-1}) & d'_s &
\end{array}
$$

for $s = 1, 2, \ldots, k$. The differential d_s is the differential d on the algebra $P(G_s)$ which is associated with the nilmanifold $K(G_s, 1)$. All the other maps i_s, p_s, h_s and d'_s have to be constructed.

Since $M(G_1)$ is the cohomology of S^1 , there is a chain homotopy equivalence

$$
\begin{array}{ccc}
P(G_1) & d & h \\
i\uparrow \ \downarrow p & & \\
M(G_1) & d' = 0 &
\end{array}
$$

with d given on $P(G_1)$.

These constructions lead to two families of chain homotopy equivalences. Namely

$$M \otimes \cdots \otimes M \otimes P(G_s) \qquad\qquad \mathcal{D}_{2s} \qquad \mathcal{H}_{2s-1}$$

$$\mathcal{I}_{2s-1} \uparrow \quad \downarrow \mathcal{P}_{2s-1}$$

$$M \otimes \cdots \otimes M \otimes P(G_1) \otimes P(G_{s-1}) \quad \mathcal{D}_{2s-1} \qquad \mathcal{H}_{2s-2}$$

$$\mathcal{I}_{2s-2} \uparrow \quad \downarrow \mathcal{P}_{2s-2}$$

$$M \otimes \cdots \otimes M \otimes M \otimes P(G_{s-1}) \qquad \mathcal{D}_{2s-2}$$

for $s = 2, 3, \ldots, k$ with $\mathcal{D}_{2k} = d$, and

$$M \otimes \cdots \otimes M \otimes P(G_1) \qquad \mathcal{D}_2 \qquad \mathcal{H}_1$$

$$\mathcal{I}_1 \uparrow \quad \downarrow \mathcal{P}_1$$

$$M(G) = M \otimes \cdots \otimes M \otimes M \qquad \mathcal{D}_1.$$

Thus the differential d on $P(G)$, $G = G_k$, is "pushed down" so that finally we have the differential D on $M(G)$, and

$$I = \mathcal{I}_{2k-1} \mathcal{I}_{2k-2} \cdots \mathcal{I}_2 \mathcal{I}_1,$$
$$P = \mathcal{P}_1 \mathcal{P}_2 \cdots \mathcal{P}_{2k-1},$$
$$H = \mathcal{H}_{2k-1} + \mathcal{I}_{2k-1} \mathcal{H}_{2k-2} \mathcal{P}_{2k-1} + \cdots$$
$$+ \mathcal{I}_{2k-1} \cdots \mathcal{I}_2 \mathcal{H}_1 \mathcal{P}_2 \cdots \mathcal{P}_{2k-1}.$$

The details can be found in [2]. Then the homologies $HH_*(P(G))$ and $HH_*(M(G))$ are isomorphic. If H is a free abelian group on k generators then the DGA $(M(H), D)$ is isomorphic to the exterior algebra $(E = \Lambda(f_1, \ldots, f_k), d = 0)$ on k one dimensional generators with zero differential, [2]. Therefore the computation of $HH_*(P(H))$ reduces to the computation of $HH_*(E)$ over an arbitrary commutative ring R.

Let W be the R–module on one dimensional generators $f_1, \ldots, f_k$, and let $E = \Lambda W$ be the exterior algebra on W. Denote by $\overline{W}$ the R–module with generators of dimension zero and let $\overline{E} = \Lambda \overline{W}$. Consider the resolution of E by graded E^e–modules

$$\cdots \longrightarrow E \otimes \overset{2}{\Lambda} \overline{W} \otimes E \overset{b_2}{\longrightarrow} E \otimes \overline{W} \otimes E \overset{b_1}{\longrightarrow} E \otimes E \overset{m}{\longrightarrow} E \longrightarrow 0,$$

where $b_1(a \otimes \overline{w} \otimes b) = (-1)^{|b|}(aw \otimes b - a \otimes wb), \cdots$ and $m(a \otimes b) = a \wedge$. By tensoring the resolution by the E^e– module E on the left we obtain the complex whose homology is isomorphic to the homology of the standard complex of Hochschild $HH_*(E)$ of E. Hence we have

Lemma 5. $HH_*(E) = \overline{E} \otimes E^{-*}$.

On the homology $HH_*(E)$ there is a filtration which is compatible with the filtrations on $P(H)$ and on $P(G)$. This fact permits us to describe more precisely the structure of E^1. First of all we denote by $\{e_k^{\epsilon_k} \otimes \cdots \otimes e_1^{\epsilon_1}\}, \epsilon_j = 0$ or 1, the basis elements of $M(G)$. Then the norm $\|\ \|$ on $M(G)$ is defined by setting

$$\|e_k^{\epsilon_k} \otimes \cdots \otimes e_1^{\epsilon_1}\| = (\epsilon_{i_1+1} + \cdots + \epsilon_{i_2}) + \cdots$$
$$+ j(\epsilon_{i_j+1} + \cdots + \epsilon_{i_{j+1}}) + \cdots + (s-1)(\epsilon_{i_{s-2}+1} + \cdots + \epsilon_{i_k}),$$

where the grouping of the indices is identical with that in Section 1. The filtration associated with this norm together with the inclusion map

$$i : M(H) \longrightarrow P(H), \qquad i(e_j) = \begin{pmatrix} 0 \\ \vdots \\ 1 \\ \vdots \\ 0 \end{pmatrix} \begin{matrix} (k) \\ \\ (j) \\ \\ (1) \end{matrix}$$

determine the graded structure on $HH_*(E)$ and therefore on $HH_*(G)$. Thus we can summarize.

Theorem. *There is a spectral sequence $\{E^r, d^r\}$; associated with the norm $\|\ \|$; such that*

$$E^1 \Longrightarrow HH_*(P(G)).$$

The E^1-term of this spectral sequence is isomorphic to the graded Hochschild homology $HH_*(E)$ of the exterior algebra over a field of characteristic zero.

With the central series

$$G = G_1 > G_2 > \cdots > G_s > G_{s+1} = 1$$

there are associated the central extensions

$$0 \longrightarrow G_{j+1}/G_j \longrightarrow G/G_{j+1} \longrightarrow G/G_j \longrightarrow 1,$$

$j = 2, 3, \ldots, s+1$. By using the Malcev basis, we can give to the set G/G_{j+1} the direct product structure $G_{j+1}/G_j \times G/G_j$. And more generally the direct product

$$G/G_{j+1} \times G_{j+1}/G_{j+2} \times \cdots \times G_s/G_{s+1}$$

induces a group structure on G. We denote this group by G^j, $j = 1, \ldots, s$, with $G^s = G$. On the algebra $P(G/G_{j+1}) \otimes P(G_{j+1}/G_{j+2}) \otimes \cdots \otimes P(G_s/G_{s+1})$, there is a differential induced by the individual groups. The canonical inclusion of this algebra into $P(G^j)$ determines a differential d_j on $P(G^j)$. Note that $P((G^1), d_1)$ is exactly the DGA $(P(H), d)$ defined above and that $(P(G^s), d_s)$ is just the algebra $(P(G), d)$.

The differentials d_s and d_1 can be connected in an obvious way by writing simply $d_s = d_1 + (d_2 - d_1) + \cdots + (d_s - d_{s-1})$, where each component reflects the group structure induced by an appropriate level. The higher differentials in the spectral sequence $\{E^r, d^r\}$ can be derived from the following formula for $d_j v$, where $v \in P^1(G^j)$:

$$\langle d_j v, (x, y) \rangle = - \sum_{z_1=0}^{v_1} \cdots \sum_{z_k=0}^{v_k} \prod_{r=1}^{k} \binom{x_r}{v_r - z_r} \binom{y_r}{v_r}$$

$$- \sum_{\ell_1=0}^{v_1} \cdots \sum_{\ell_{i_j}=0}^{v_{i_j}} \prod_{r=1}^{i_j} \binom{x_r + y_r}{x_r - \ell_r} \prod_{r=i_j+1}^{k} \binom{x_r + y_r}{v_r} \prod_{r=1}^{i_j} \binom{\tau_r^j}{\ell_r},$$

where $z_1 + \cdots + z_k \neq 0$ and $\neq v_1 + \cdots + v_k$; $\ell_1 + \cdots + \ell_{i_j} \neq 0$, and the τ^j's are the polynomials τ corresponding to the group G^j. It should be pointed out that the map induced by the operator $d_{j+1} - d_j$ does not lower the filtration by j for $j \geq 3$. However there is a component of $d_{j+1} - d_j$ which does just that. This part plays an important role in the computation of the higher differentials in the spectral sequence $\{E^r, d^r\}$.

REFERENCES

1. D. Burghelea, *Cyclic homology and the algebraic K-theory of spaces I*, Contemp. Math. **55** (1986), 89–115.
2. B.Cenkl and R.Porter, *Polynomial cochains on nilmanifolds*, preprint.
3. B.Cenkl and R.Porter, *Spectral sequence for polynomial cochains*, preprint.
4. P.Hall, *Nilpotent Groups*, Mathematics Notes, Queens Mary College, 1969.
5. D.S.Jones, *Cyclic homology and equivariant homology*, Invent.Math. **87** (1987), 403–423.
6. M.Vigué-Poirrier, *Homologie de Hochschild et homologie cyclique des algèbres différentielles graduées*, PUB.IRMA,Lille **17** (1989), no. 8, 1–15.

DEPARTMENT OF MATHEMATICS, NORTHEASTERN UNIVERSITY, 360 HUNTINGTON AVENUE, BOSTON, MA 02115
E-mail address: cenkl@northeastern.edu

AN ANALYTIC MODEL FOR S³-EQUIVARIANT COHOMOLOGY

HUA CHEN

The Ohio State University at Lima

ABSTRACT. There is an Atiyah-Bott type model, constructed in terms of invariant differential forms, for the S^3-equivariant cohomology of a smooth manifold. Using this model, one can calculate the S^3-equivariant cohomology of the singular point set of an S^3-manifold in terms of the cohomology of the set.

0. INTRODUCTION

If M is a smooth S^1-manifold, it is proved in [AB] that

$$H^*_{S^1}(M, \mathbb{R}) = H^*(\mathbb{R}[u] \otimes \Omega^*_{S^1}(M), d + u \cdot i_X),$$

where $deg(u) = 2$ and i_X is the contraction operator in the direction of the left invariant vector field on M induced by the action of S^1. In this note, I will propose a similar analytic model for the S^3-equivariant cohomology of a smooth S^3-manifold. I will also use this model to calculate the S^3-equivariant cohomology of the singular point set of an S^3-action.

This paper originated from a part of my Ph.D. dissertation completed under the direction of Professor Dan Burghelea. I wish to express my gratitude to Professor Burghelea for his invaluable help.

1. G–EQUIVARIANT COHOMOLOGY

In this section, I will establish basic terminologies and review the analytic model described in [AB] for equivariant cohomology. Proofs of some results can be found in [AB] and [GHV].

Let G be a compact connected Lie group and $g = <e_1, \cdots, e_n | [e_\alpha, e_\beta] = C^\gamma_{\alpha\beta} e_\gamma >$ the Lie algebra of G. Then the *Weil algebra* $W(g)$ is defined to be the tensor product of the exterior algebra $\wedge g^*$ and the symmetric algebra Sg^* on the dual algebra of g. $W(g)$ is graded by assigning degree 1 to an element $\theta \in g^*$ in $\wedge g^*$, and degree 2 to the corresponding element $u \in g^*$ in Sg^*. If $\{\theta^1, \cdots, \theta^n\}$ and $\{u_1, \cdots, u_n\}$ are the sets of generators in $\wedge g^*$ and Sg^* dual to $\{e_1, \cdots, e_n\}$, then

$$W(g) = \wedge \{\theta^1, \cdots, \theta^n, u_1, \cdots, u_n\}.$$

In addition, $W(g)$ is endowed with a differential δ, which on the generators is defined by

$$\delta\theta^\alpha = -\frac{1}{2}\sum_{\beta \geq \gamma} C^\alpha_{\beta\gamma}\theta^\beta\theta^\gamma - u_\alpha$$

$$\delta u_\alpha = \sum C^\alpha_{\beta\gamma} u_\beta \theta^\gamma.$$

For any $X \in g$, one can define the *contraction* i_X and the *Lie derivative* L_X on $W(g)$ by the following formulas

$$i_X(\theta) = \theta(X) \qquad\qquad \theta \in g^* \subset \wedge g^*$$
$$i_X(u) = 0 \qquad\qquad u \in g^* \subset Sg^*$$
$$L_X = \delta \circ i_X + i_X \circ \delta.$$

Let $M = (M, \mu)$ denote a smooth manifold M with a smooth action $\mu : G \times M \to M$. Each element $X \in g$ induces a left invariant vector field on M, which is defined by

$$Z_X(m) = d\mu_{(e,m)}(X_e, 0),$$

where e is the unit element in G. Z_X is called the *fundamental vector field* induced by X. We simply denote the contraction $i_{Z_X} : \Omega^n(M) \to \Omega^{n-1}(M)$ by i_X and the Lie derivative $L_{Z_X} : \Omega^n(M) \to \Omega^n(M)$ by L_X. Again these two operators are related by the equation $L_X = d \circ i_X + i_X \circ d$.

Proposition 1.1. (1) *Let $\Omega^*_G(M)$ be the subalgebra of all G-invariant forms on M. Then*

$$\Omega^*_G(M) = \{\omega \in \Omega^*(M) | L_X \omega = 0, \ \forall X \in g\}.$$

(2) $H^*(\Omega^*_G(M), d) = H^*(M; \mathbb{R})$. ([GHV, Vol. 2, 146-153])

We consider the free G-space $EG \times M$ as a smooth manifold modelled on a Banach space. Let V be a principal connection of the principal G-bundle $p : EG \times M \to EG \times_G M$. Let $\omega_p \in \Omega^1(EG \times M; g)$ and $\Omega_p \in \Omega^2(EG \times M; g)$ denote the connection form and the curvature form of V. We can define a homomorphism of differential graded algebras

$$I_0 : (W(g) \otimes \Omega^*(M), \delta \otimes 1 + 1 \otimes d) \to (\Omega^*(EG \times M), d)$$

by the formulas

$$I_0(\theta) = \theta(\omega_p) \in \Omega^1(EG \times M), \quad \theta \in g^* \subset \wedge g^*;$$
$$I_0(u) = u(\Omega_p) \in \Omega^2(EG \times M), \quad u \in g^* \subset Sg^*;$$
$$I_0(a) = p_2^*(a), \qquad\qquad a \in \Omega^*(M), \ p_2 : EG \times M \to M.$$

Then the induced homomorphism

$$I_0^* : H^*(W(g) \otimes \Omega^*_G(M), \delta \otimes 1 + 1 \otimes d) \longrightarrow H^*(EG \times M; \mathbb{R}) \cong H^*(M; \mathbb{R})$$

is an isomorphism.

Observation 1.2. (1) *Let*

$$\text{Basic}(\Omega^*(EG \times M)) = \{\omega \in \Omega^*(EG \times M) \mid L_X\omega = 0,\ i_X\omega = 0,\ \forall X \in g\}.$$

Then $p^* : \Omega^*(EG \times_G M) \to \text{Basic}(\Omega^*(EG \times M))$ *is an isomorphism* ([GHV, Vol. 2]).
(2) *Let* $B_G^*(M) = \{\omega \in W(g) \otimes \Omega^*(M) \mid L_X\omega = 0, i_X\omega = 0, \forall X \in g\}$. *Naturally, we have* $I_0(B_G^*(M)) \subset \text{Basic}(\Omega^*(EG \times M))$.

Let $I : B_G^*(M) \to \Omega^*(EG \times_G M)$ *be the composition* $(p^*)^{-1} \circ I_0|_{B_G^*(M)}$.

Theorem 1.3. *I induces an isomorphism*

$$I^* : H^*(B_G^*(M), \delta \otimes 1 + 1 \otimes d) \xrightarrow{\cong} H^*(\Omega^*(EG \times_G M), d) \cong H_G^*(M; \mathbb{R}).$$

([AB]).

2. S^3-Equivariant Cohomology

Now we consider the case when $G = S^3$. Since $S^3 \to SO(3)$ is a 2-fold covering, the Lie algebra g of S^3 is equal to that of $SO(3)$, which is the Lie algebra of all 3×3 skew-symmetric matrices with real entries. In particular, if one chooses as its generators

$$e_2 = \begin{pmatrix} 0 & 0 & 1 \\ 0 & 0 & 0 \\ -1 & 0 & 0 \end{pmatrix} \quad e_1 = \begin{pmatrix} 0 & 1 & 0 \\ -1 & 0 & 0 \\ 0 & 0 & 0 \end{pmatrix} \quad e_3 = \begin{pmatrix} 0 & 0 & 0 \\ 0 & 0 & 1 \\ 0 & -1 & 0 \end{pmatrix}$$

then $g = <e_1, e_2, e_3 \mid [e_1, e_2] = e_3,\ [e_3, e_1] = e_2,\ [e_2, e_3] = e_1 >$. By choosing generators in g^* dual to $\{e_\alpha\}$, the Weil algebra of g is completely described as follows:

(1) $W(g) = \wedge\{\theta^1, \theta^2, \theta^3, u_1, u_2, u_3\}$, $\deg \theta^\alpha = 1$, $\deg u_\alpha = 2$.
(2) $\delta\theta^1 = -\theta^2\theta^3 - u_1$, $\delta\theta^2 = -\theta^3\theta^1 - u_2$, $\delta\theta^3 = -\theta^1\theta^2 - u_3$,
$\delta u_1 = u_2\theta^3 - u_3\theta^2$, $\delta u_2 = u_3\theta^1 - u_1\theta^3$, $\delta u_3 = u_1\theta^2 - u_2\theta^1$.
(3) $i_{e_\alpha}(\theta^\beta) = \delta_\alpha^\beta$ and $i_{e_\alpha}(u_\beta) = 0$.
(4) Since $L_{e_\alpha}(\theta^\beta) = \sum_\gamma C_{\gamma\alpha}^\beta \theta^\gamma$ and $L_{e_\alpha}(u_\beta) = \sum_\gamma C_{\gamma\alpha}^\beta u_\gamma$, we have $L_{e_\alpha}(\theta^\beta) = \theta^{[\alpha,\beta]}$ and $L_{e_\alpha}(u_\beta) = u_{[\alpha,\beta]}$ where

$$\begin{array}{lll} [1,1] = 0 & [1,2] = 3 & [1,3] = -2 \\ [2,1] = -3 & [2,2] = 0 & [2,3] = 1 \\ [3,1] = 2 & [3,2] = -1 & [3,3] = 0 \end{array}$$

and by the formulas $\theta^0 = 0$, $u_0 = 0$ and $\theta^{-\alpha} = -\theta^\alpha$, $u_{-\alpha} = -u_\alpha$.

Let M be a smooth S^3-manifold. From a straightforward computation we conclude that

$$B^*_{S^3}(M) = \{\omega \in W(g) \otimes \Omega^*(M) \mid L_{e_\alpha}\omega = 0, i_{e_\alpha}\omega = 0\}$$
$$= \{F_\theta(f(u_1, u_2, u_3)) \mid f(u_1, u_2, u_3) \in \Omega^*(M) \otimes \mathbb{R}[u_1, u_2, u_3], L_{e_\alpha}f = 0\}$$

where F_θ is the zero-degree operator on $W(g) \otimes \Omega^*(M)$ defined by

$$F_\theta = id - [\theta^1 i_{e_1} + \theta^2 i_{e_2} + \theta^3 i_{e_3}]$$
$$+ [\theta^2\theta^3 i_{e_3} \circ i_{e_2} + \theta^3\theta^1 i_{e_1} \circ i_{e_3} + \theta^1\theta^2 i_{e_2} \circ i_{e_1}] - \theta^1\theta^2\theta^3 i_{e_3} \circ i_{e_2} \circ i_{e_1}.$$

Proposition 2.1. $B^*_{S^3}(\overline{pt}) = \mathbb{R}[u]$, where $u = u_1^2 + u_2^2 + u_3^2$.

Proof. If $\omega \in B^{2N}_{S^3}(pt)$, we already know from the above computation that

$$\omega = f(u_1, u_2, u_3) = \sum_{N=l+m+n} a_{l,m,n} u_1^l u_2^m u_3^n \in \mathbb{R}[u_1, u_2, u_3],$$

where $L_{e_\alpha}f(u_1, u_2, u_3) = 0$ for $\alpha = 1, 2, 3$.

First of all, the equation

$$L_{e_1}f = \sum u_1^l u_2^m u_3^n[(m+1)a_{l,m+1,n-1} - (n+1)a_{l,m-1,n+1}] = 0$$

implies that $(m+1)a_{l,m+1,n-1} = (n+1)a_{l,m-1,n+1}$. Then

(1) if either m or n is odd, $a_{l,m,n} = 0$;

(2) if $m = 2m', n = 2n'$ then $a_{l,2m',2n'} = a_{l,2m'+2n',0}\begin{pmatrix} m'+n' \\ m' \end{pmatrix}$. Therefore

$$f(u_1, u_2, u_3) = \sum_{l+2k=N} a_{l,2k,0} u_1^l (u_2^2 + u_3^2)^k.$$

Similarly, the equation

$$L_{e_3}f = \sum u_1^{l-1}(u_2^2 + u_3^2)^k u_2[la_{l,2k,0} - 2(k+1)a_{l-2,2k+2,0}] = 0$$

implies that $la_{l,2k,0} = 2(k+1)a_{l-1,2k+2,0}$. Then:

(1) if l is odd, $a_{l,2k,0} = 0$;

(2) if $l = 2l'$, $a_{2l',2k,0} = a_{2l'+2k,0,0}\begin{pmatrix} l'+k \\ l' \end{pmatrix}$. Therefore

$$f(u_1, u_2, u_3) = a_{2N',0,0}(u_1^2 + u_2^2 + u_3^2)^{N'}, \quad 2N' = N.$$

Q.E.D.

Since $H_{S^3}^*(\overline{pt}; \mathbb{R}) = H^*(\mathbb{R}[u], 0) = \mathbb{R}[u]$, the homomorphism

$$p^n : \mathbb{R}[u] = H_{S^3}^n(\overline{pt}; \mathbb{R}) \to H_{S^3}^n(M; \mathbb{R})$$

induced by the canonical projection $p : M \to pt$ introduces a graded $\mathbb{R}[u]$-algebra structure on $H_{S^3}^n(M; \mathbb{R})$.

Furthermore, the S^3-bundle $S^3 \xrightarrow{i} ES^3 \times M \xrightarrow{P} ES^3 \times_{S^3} M$ induces the following long exact sequence in cohomologies, usually called the Gysin sequence

$$\cdots \to H_{S^3}^{n-4}(M; \mathbb{R}) \xrightarrow{\cdot \chi_P} H_{S^3}^n(M; \mathbb{R}) \xrightarrow{P^*} H^n(M; i^*\mathbb{R}) \xrightarrow{\int_{S^3}} H_{S^3}^{n-3}(M; \mathbb{R}) \to \cdots$$

where χ_P is the Euler class of the bundle P and $\int_{S^3}$ is the integration over fibre S^3. In terms of the analytic model constructed in this section, we can describe the Gysin sequence by the following commutative diagram

$$
\begin{array}{ccccccc}
H_{S^3}^{n-4}(M; \mathbb{R}) & \xrightarrow{\cdot \chi_P} & H_{S^3}^n(M; \mathbb{R}) & \xrightarrow{P^n} & H^n(M; \mathbb{R}) & \xrightarrow{\int_{S^3}} & H_{S^3}^{n-3}(M; \mathbb{R}) \\
\cong \uparrow {\scriptstyle I^{n-4}} & & \cong \uparrow {\scriptstyle I^n} & & \cong \uparrow & & \cong \uparrow {\scriptstyle I^{n-3}} \\
H^{n-4}(B_{S^3}^*(M)) & \xrightarrow{\cdot u} & H^n(B_{S^3}^*(M)) & \xrightarrow{Q^n} & H^n(\Omega_{S^3}^*(M)) & \xrightarrow{\beta^n} & H^{n-3}(B_{S^3}^*(M))
\end{array}
$$

The last two squares are induced by the following commutative diagram of de Rham cochain complexes

$$
\begin{array}{ccccc}
\Omega^n(ES^3 \times_{S^3} M) & \xrightarrow{P^n} & \Omega_{S^3}^n(ES^3 \times M) & \xrightarrow{\int_{S^3}} & \Omega^{n-3}(ES^3 \times_{S^3} M) \\
\uparrow {\scriptstyle I} & & \uparrow & & \uparrow {\scriptstyle I} \\
B_{S^3}^n(M) & \xrightarrow{Q} & \Omega_{S^3}^n(M) & \xrightarrow{\beta} & B_{S^3}^{n-3}(M)
\end{array}
$$

where $\beta = i_{e_3} \circ i_{e_2} \circ i_{e_1} : \Omega_{S^3}^n(M) \to B_{S^3}^{n-3}(M)$ [GHV, Vol. 2, 242-244], and $Q : B_{S^3}^*(M) \to \Omega_{S^3}^*(M)$ is the chain homomorphism defined by the formula

$$Q(F_\theta(f(u_1, u_2, u_3))) = a_0 \in \Omega_{S^3}^*(M)$$

with $f(u_1, u_2, u_3) = a_0 + a_1 u_1 + a_2 u_2 + a_3 u_3 + \cdots$.

3. AN ANALYTIC MODEL FOR S^3-EQUIVARIANT COHOMOLOGY

We now consider the S^3-equivariant cohomology of a manifold M in the context of the Borel fibration $M \to ES^3 \times_{S^3} M \to BS^3$. Since

$$H^*(M; \mathbb{R}) = H^*(\Omega_{S^3}^*(M), d)$$
$$H^*(BS^3; \mathbb{R}) = H^*(\mathbb{R}[u], 0),$$

we expect that the tensor product $\mathbb{R}[u] \otimes \Omega^*_{S^3}(M)$ will provide the underlying graded module of a model which can be used to calculate $H^*_{S^3}(M; \mathbb{R})$. We also expect that there is an injective homomorphism of graded modules

$$\lambda : \mathbb{R}[u] \otimes \Omega^*_{S^3}(M) \to B^*_{S^3}(M)$$

of degree zero, such that the following diagram commutes.

$$
\begin{array}{ccccc}
\mathbb{R}[u] & \longrightarrow & B^*_{S^3}(M) & \xrightarrow{\ Q\ } & \Omega^*_{S^3}(M) \\
\uparrow{\scriptstyle id} & & \uparrow{\scriptstyle \lambda} & & \uparrow{\scriptstyle id} \\
\mathbb{R}[u] & \longrightarrow & \mathbb{R}[u] \otimes \Omega^*_{S^3}(M) & \longrightarrow & \Omega^*_{S^3}(M)
\end{array}
$$

Then the differential D on $\mathbb{R}[u] \otimes \Omega^*_{S^3}(M)$ is defined so that

$$\lambda \circ D = (\delta \otimes id + id \otimes d) \circ \lambda.$$

One candidate for the desired homomorphism is defined as follows

$$\lambda|_{\mathbb{R}[u] \otimes 1} = id_{\mathbb{R}[u] \otimes 1} \quad \text{and} \quad \lambda|_{1 \otimes \Omega^*_{S^3}(M)} = \lambda_M = F_\theta \circ P_u,$$

where F_θ is defined as in the previous section and

$$P_u = id + u_1 i_{e_3} \circ i_{e_2} + u_2 i_{e_1} \circ i_{e_3} + u_3 i_{e_2} \circ i_{e_1}.$$

Since $L_{e_\alpha} P_u(a) = 0$ for any $a \in \Omega^*_{S^3}(M)$, therefore $F_\theta \circ P_u(\mathbb{R}[u] \otimes \Omega^*_{S^3}(M)) \subset B^*_{S^3}(M)$. On the other hand, since

$$
\begin{aligned}
[\delta \otimes id + id \otimes d](\lambda(a)) &= P_u(da) + (u_1^2 + u_2^2 + u_3^2) i_{e_3} \circ i_{e_2} \circ i_{e_1} a \\
&\quad - (\theta^1 i_{e_1} + \theta^2 i_{e_2} + \theta^3 i_{e_3}) P_u(da) \\
&\quad + (\theta^2 \theta^3 i_{e_3} i_{e_2} + \theta^3 \theta^1 i_{e_1} i_{e_3} + \theta^1 \theta^2 i_{e_2} i_{e_1}) P_u(da) \\
&\quad - \theta^1 \theta^2 \theta^3 i_{e_3} i_{e_2} i_{e_1} P_u(da) \\
&= F_\theta \circ P_u(da) + u\beta a.
\end{aligned}
$$

and since $F_\theta \circ P_u(u\beta a) = u\beta a$, we have

$$[\delta \otimes id + id \otimes d](\lambda(a)) = F_\theta \circ P_u(da + u\beta a).$$

Hence the differential on $\Omega^*_{S^3}(M)[u]$ is defined to be

$$D(\omega) = d\omega + u\beta\omega \quad \text{for } \omega \in \mathbb{R}[u] \otimes \Omega^*_{S^3}(M).$$

Theorem A. $H^*(\mathbb{R}[u] \otimes \Omega^*_{S^3}(M); D) \xrightarrow{I^* \circ \lambda_*} H^*_{S^3}(M; \mathbb{R})$ *is an isomorphism.*

Proof. There is a short exact sequence of differential cochain complexes

$$0 \to [\Omega^*_{S^3}(M) \otimes \mathbb{R}[u]]_n \xrightarrow{\cdot u} [\Omega^*_{S^3}(M) \otimes \mathbb{R}[u]]_{n+4} \xrightarrow{P_0} \Omega^{n+4}_{S^3}(M) \to 0$$

where $P_0(u) = 0$ and $P_0|_{\Omega^*_{S^3}(M)} = id$. It induces a long exact sequence in cohomology

$$\cdots \to H^n(\mathbb{R}[u] \otimes \Omega^*_{S^3}(M)) \xrightarrow{\cdot u} H^{n+4}(\mathbb{R}[u] \otimes \Omega^*_{S^3}(M)) \xrightarrow{P_0^{n+4}} H^{n+4}(\Omega^*_{S^3}(M)) \xrightarrow{\beta^{n+4}}$$

$$\to H^{n+1}(\mathbb{R}[u] \otimes \Omega^*_{S^3}(M)) \to \cdots$$

such that the following diagram is commutative. We abbreviate $\Omega^*_{S^3}$ as Ω.

$$
\begin{array}{ccccccc}
H^{n-4}(\Omega(M)[u]) & \xrightarrow{\cdot u} & H^n(\Omega(M)[u]) & \xrightarrow{P_0^n} & H^n(\Omega(M)) & \xrightarrow{\beta^n} & H^{n-3}(\Omega(M)[u]) \\
\downarrow{\scriptstyle I^{n-4} \circ \lambda^{n-4}} & & \downarrow{\scriptstyle I^n \circ \lambda^n} & & \downarrow{\scriptstyle =} & & \downarrow{\scriptstyle I^{n-3} \circ \lambda^{n-3}} \\
H^{n-4}_{S^3}(M; \mathbb{R}) & \xrightarrow{\cdot \chi_P} & H^n_{S^3}(M; \mathbb{R}) & \xrightarrow{P^n} & H^n(M; \mathbb{R}) & \xrightarrow{\int_{S^3}} & H^{n-3}_{S^3}(M; \mathbb{R})
\end{array}
$$

By induction on n one can easily prove that $I^n \circ \lambda^n$ is an isomorphism for all n. Q.E.D.

Remark 3.1. *Generally, the model* $(\mathbb{R}[u] \otimes \Omega^*_{S^3}(M), D)$ *is not a differential graded algebra, because β does not satisfy the product rule.*

4. LOCALIZATION THEOREM

Let k be a field of characteristic zero and let $\{V\}$ represent an S^3-local coefficient system of k-vector spaces. Then $H^*_{S^3}(\overline{pt}; k) = k[u]$ is a principal ideal domain and consequently, if $\mathcal{M} = H^*_{S^3}(X; \{V\})$ is a finitely generated $k[u]$-module, then $\mathcal{M} = \mathcal{M}_f \oplus \mathcal{M}_t$, where $\mathcal{M}_f$ is a free $k[u]$-module and $\mathcal{M}_t$ is a torsion $k[u]$-module.

Let X be an S^3-space. A point in X is called a singular point of the S^3-action if its isotropy group has positive dimension. We use ΣX to denote the set of all singular points on X. Then the following localization theorem is well known:

Theorem 4.1. *Let X be a finite dimensional S^3-space. Then $\Sigma X \hookrightarrow X$ induces an isomorphism modulo $k[u]$-torsion in S^3-equivariant cohomology over $\{V\}$.*

Using the model constructed in this note, we can compute the S^3-equivariant cohomology of the space ΣM over the constant real coefficient.

Theorem B. *Let M be any S^3-manifold. Then*

$$H^n_{S^3}(\Sigma M; \mathbb{R}) \cong (H^*(\Sigma M; \mathbb{R}) \otimes \mathbb{R}[u])^n.$$

Proof. First we assume that ΣM is a smooth manifold. We will prove that

$$\beta|_{\Sigma M} = i_{e_3} \circ i_{e_2} \circ i_{e_1}|_{\Sigma M} = 0.$$

Since β is linear with respect to smooth functions on M, it is sufficient to show that

$$\beta_m : \wedge^* T_m^*(M) \to \wedge^{*-3} T_m^*(M)$$

equals to zero for any $m \in \Sigma M$.

Since $\dim G_m \geq 1$ for any $m \in \Sigma M$, there exists a fundamental vector field W on M such that $W(m) = 0$. Therefore $i_W \omega(m) = 0$. Without loss of generality one may assume that $W = a_1 Z_{e_1} + a_2 Z_{e_2} + a_3 Z_{e_3}$ with $a_1 \neq 0$. Hence

$$\beta_m \omega(m) = \frac{1}{a_1} i_{e_3} \circ i_{e_2} \circ i_W \omega(m) = 0.$$

In order to prove the general case of Theorem B, we consider the following observations:

Observation 4.2. *Theorem A remains true if one replaces the smooth S^3-manifold by an abstract stratified S^3-space as defined in [V1], and S^3-invariant differential forms by the controlled S^3-invariant forms on the stratified space.*

Observation 4.3. *Let G be any compact Lie group and M a smooth G-manifold. Then the singular point set ΣM has a structure of G-abstract stratification:*

$$\Sigma M = \bigsqcup_\alpha M_\alpha$$

where M_α is the set of all points $x \in M$ with the isotropy group G_x conjugate to a subgroup G_α with $\dim(G_\alpha) \geq 1$, [V2].

Repeating the argument used in part 1 of the proof, we can show that β is a zero operator on the controlled S^3-invariant forms on ΣM. The general case of Theorem B follows from Observation 4.2.   Q.E.D.

Remark 4.4. *The isomorphism in Theorem B is not multiplicative in general.*

Corollary 4.5. $\operatorname{rank}(H_{S^3}^*(M;\mathbb{R})) = \dim(H^*(\Sigma M;\mathbb{R}))$

Corollary 4.6. $\dim(H_{S^3}^*(M;\mathbb{R})) \geq \dim(H^*(\Sigma M;\mathbb{R}))$

Proof. It suffices to show that $\dim(H^*(M;\mathbb{R})) \geq \operatorname{rank}(H_{S^3}^*(M;\mathbb{R}))$. We consider the Gysin sequence

$$\cdots \to H_{S^3}^{n-4}(M;\mathbb{R}) \xrightarrow{\cdot u} H_{S^3}^n(M;\mathbb{R}) \xrightarrow{P^*} H^n(M;\mathbb{R}) \xrightarrow{\beta} H_{S^3}^{n-3}(M;\mathbb{R}) \to \cdots$$

Let $x \in H_{S^3}^*(M;\mathbb{R})$ be any non-zero element and k the largest integer such that $u^k y = x$ for some $y \in H_{S^3}^*(M;\mathbb{R})$. It is easy to see that $P^*(y) \neq 0$ in $H^*(M;\mathbb{R})$ and $\beta \circ P^*(y) = 0$. Therefore we have

$$\operatorname{rank}(H_{S^3}^*(M;\mathbb{R})) \leq \dim(\ker(\beta : H^*(M;\mathbb{R}) \to H_{S^3}^{*-3}(M;\mathbb{R}))).$$

Hence the inequality.   Q.E.D.

REFERENCES

[AB]   M. F. Atiyah and R. Bott, *The Moment Map and Equivariant Cohomology*, Topology **23** (1984), no. 1, 1–28.

[GHV] W. Greub, S. Halperin, and R. Vanstone, *Connections, Curvatures and Cohomology*, Vol. 1, 2, and 3, Academic Press, 1973.

[V1]   A. Verona, *Théorème de de Rham Pour Préstratifications Abstraites*, C. R. Acad. Sc. Paris, Serie A **273** (1971), 886–889.

[V2]   A. Verona, *Stratified Mappings - Structure and Triangulability*, Lecture Notes in Mathematics, Springer Verlag, Vol. 1102.

DEPARTMENT OF MATHEMATICS, THE OHIO STATE UNIVERSITY AT LIMA, LIMA, OH 45804
E-mail address: chen@ function.mps.ohio-state.edu

[References]

RECENT APPLICATIONS OF UNIVERSAL CONNECTIONS

C. T. J. DODSON

University of Toronto

ABSTRACT. A connection encodes geometrical choices, and through its curvature, underlying topological information. In some situations, both in geometry and in theoretical physics, it is necessary to consider a family of connections, for example with regard to stability of certain properties. Then it is useful for the family to be presented in a convenient form that avoids introduction of infinite-dimensional spaces. An appropriate algebraic organization allows also the provision of a universal connection, of which each member of the family is a pullback. A classical success of universal connections was the elegant proof they provided for the Weil Theorem on characteristic classes. In this note, some recent uses of universal connections will be described, including applications to geometry and parametric statistical models; some speculations on further applications to physical field theory are offered.

1. INTRODUCTION

The notion of a system (or structure) of connections was introduced by Mangiarotti and Modugno [27, 28] and is valid in the more general context of connections of any order on fibred manifolds. This is important both for geometry and for field theory because a fibred manifold, that is a surjective submersion, is the least structure that can support the concept of a connection—as a section of its first jet bundle.

A **fibred manifold** is a surjective submersion $p: E \longrightarrow B$, so p has maximal rank everywhere. Every fibre bundle is a fibred manifold but not conversely. For example,

$$p: \mathbb{R}^2 - \{0\} \longrightarrow \mathbb{R} \; : (x, y) \longmapsto x$$

is a fibred manifold but not a fibre bundle.

A **connection** on a fibred manifold $p: E \longrightarrow B$ is a section Γ of the first jet bundle $j: JE \longrightarrow E$. Now, JE consists of classes of sections of p that are equivalent up to first derivative, so a typical element of JE at $X \in E$ is represented by a linear map, the derivative of σ

$$T\sigma: T_{p(X)}B \longrightarrow T_X E.$$

Since σ is a section, then $Tp \circ T\sigma = 1_{TB}$ and hence our representative $T\sigma$ corestricts to the identity on $TB \hookrightarrow TE$. Moreover, $JE \longrightarrow E$ is an affine subbundle of the vector bundle $T^*B \otimes_B TE$ and it projects onto the identity 1_{TB} when 1_{TB} is viewed as a section of TE.

Locally, the coordinate expression of a connection Γ on a fibred manifold $p\colon E \to B$ is

$$\Gamma\colon E \longrightarrow JE \hookrightarrow T^*B \otimes_B TE$$

$$(x^i, X^\alpha) \longmapsto dx^i \otimes \frac{\partial}{\partial x^i} - \Gamma_i^\alpha \, dx^i \otimes \frac{\partial}{\partial X^\alpha}$$

where $\Gamma_i^\alpha = \frac{\partial}{\partial x^i}\gamma^\alpha$ and (γ^α) represents the class of sections of E determined by Γ. The connection induces a splitting of TE into horizontal and vertical subbundles:

$$TE \longrightarrow HE \oplus_E VE$$

$$(x^i, X^\alpha, \dot{x}^i, \dot{X}^\alpha) \to (x^i, X^\alpha, \dot{x}^i, \dot{x}^i\Gamma_i^\alpha) \oplus (x^i, X^\alpha, 0, \dot{X}^\alpha - \dot{x}^i\Gamma_i^\alpha)$$

On a G–bundle the horizontal distribution would be required to be G–invariant.

A **system of connections** on a fibred manifold $p : E \to B$ consists of a fibred manifold $p_c : C \to B$ and a fibred morphism over E

$$\eta : C \times_E E \to JE$$

Then any section $\tilde{\Gamma}$ of p_c determines a unique connection $\Gamma = \eta \circ (\tilde{\Gamma} \circ p, 1_E)$ on E.

When $p : E \to B$ is a principal G-bundle then we are interested in principal or G-invariant connections and the set of all such is in 1-1 correspondence with sections of $JE/G \to B$ (cf. Garcia [18]). Hence the set of all G-invariant connections constitutes a system of connections with $C = JE/G$; further details are given in Canarutto and Dodson [4] (cf. also [5, 12]). More general theory can be found in Modugno [29] and Mangiarotti and Modugno [27].

Consider the following two ways to organise the set of all linear connections on a manifold M, this set being otherwise an infinite-dimensional space of sections.

Tangent bundle system.

The system of all (linear) connections on the fibred manifold $\pi_T\colon TM \longrightarrow M$ (which is here a vector bundle) with system space

$$C_T = (1_{T^*M} \otimes_M T\pi_T)^{\leftarrow}1_{TM} \subset T^*M \otimes_M JTM$$

where we view 1_{TM} as a section of $T^*M \otimes_M TM$ in $T^*M \otimes_{TM} TTM$.

Frame bundle system.

The system of all (G-invariant) connections on the fibred manifold $\pi_F\colon FM \longrightarrow M$ (which is here a principal G–bundle) with system space

$$C_F = JFM/G \hookrightarrow T^*M \otimes_{TM} TFM/G .$$

These two representations are equivalent because of the following (cf. Del Riego and Dodson [9]).

Theorem 1.1. *For any manifold M there are bijections among*

(a) $Con\,(FM/G) = Con\,(M) = Sec\,(JFM/FM)$;
(b) $Sec\,\big((JFM/G)/M\big) = Sec\,(C_F/M)$;
(c) $Sec\,(JTM/TM)$;
(d) $Sec\,(C_T/M)$.

This recovers the classical result of Nomizu [34] which established the equivalence of linear connections on TM with principal connections on FM.

The elegant bonus available on any system of connections is a canonical or **universal connection**: every connection in the system is a pullback of the universal connection on the system. This is a different representation of an object similar to that introduced by Narasimhan and Ramanan [32], [33] for G-bundles, also allowing a proof of Weil's theorem (cf. [25, 18, 5]):

Theorem 1.2. *Take any frame bundle $\pi_F : FM \longrightarrow M$, or one of its subbundles, with structure group G. There is a well-defined algebra homomorphism into de Rham cohomology*

$$\mathbf{w} : I(G) \longrightarrow H^*(M, \mathbb{R})$$

where $I(G) = \sum_{k=0}^{\infty} I^k(G)$ and $I^k(G)$ is the real vector space of G-invariant k-linear maps $g \times \ldots \times g \longrightarrow \mathbb{R}$ defined on the Lie algebra g of G.

1.1 Principal Bundles

We shall outline the situation for universal connections of systems of connections on principal bundles, the most common situation for applications, following the approach described in [5].

On a G–bundle $p\colon E \longrightarrow E/G$, the vertical vector valued 1–form $\lambda \in T^*JE \otimes VE$ given locally by

$$\lambda\colon (X^i, Y^\alpha, E_i^\alpha) \longmapsto (0, Y^\alpha - E_i^\alpha X^i)$$

has as kernel a G-invariant distribution and an invariant connection. For $\bar{s} \in JE$ with $j(\bar{s}) = s$,

$$\lambda_{\bar{s}} = (1_{TE} - Ts \circ Tp) \circ Tj \, ,$$

where $JE \xrightarrow{j} E \xrightarrow{p} B$. Then $\sigma \in Sec\,(JE/E)$ determines the connection 1-form of σ,

$$\sigma^*\lambda_{\bar{s}} = \lambda_{\bar{s}} \circ T\sigma = (1_{TE} - Ts \circ Tp) \circ Tj \circ T\sigma = 1_{TE} - Ts \circ Tp \, ,$$

taking values in VE. So, σ is a pullback of λ and λ is in this sense universal and we have the following.

Theorem 1.3. *Consider the first jet bundle over a principal G-bundle*

$$JE \xrightarrow{j} E \xrightarrow{\pi} E/G.$$

Then on the principal G–bundle $\pi_J: JE \longrightarrow JE/G$ there is an invariant connection Λ which has the universal property that for each (principal) connection Γ on $\pi: E \longrightarrow E/G$

$$\Gamma = \Gamma^*\Lambda \ .$$

If ω_Γ, Ω_Γ and ω_Λ, Ω_Λ are the respective connection 1–form and curvature 2–form for Γ and Λ then

$$\omega_\Gamma = \Gamma^*\omega_\Lambda \ \text{ and } \ \Omega_\Gamma = \Gamma^*\Omega_\Lambda$$

where $\Omega_\Gamma = (1/2)[\omega_\Gamma, \omega_\Gamma]$.

In terms of sections of arbitrary fibred manifolds,

$$JE \xrightarrow{j} E \xrightarrow{p} B,$$

there is a distinguished connection Λ on the fibred manifold

$$\pi_1: JE \times E \longrightarrow JE \ ,$$

and it has the universal property that $\Gamma = \Gamma^*\Lambda$ for any connection $\Gamma \in Sec\,(JE/E)$. To see this, observe the following affine subbundle inclusions:

$$JE \hookrightarrow T^*B \otimes TE, \ \ J(JE) \hookrightarrow T^*E \otimes TJE.$$

Now we define

$$\Lambda: JE \times_B E \longrightarrow J(JE \times E) \hookrightarrow T^*(JE) \otimes T(JE \times E)$$
$$(\bar{s}_x, e) \mapsto \big((X, Y, S) \mapsto (X, Y, S, Ts X)\big)$$

and it follows that Λ has the required universal properties

$$\omega_\Gamma = \Gamma^*\omega_\Lambda, \ \ \Omega_\Gamma = \Gamma_\Lambda.$$

For the frame bundle system of connections

$$JFM/G \times_M FM \xrightarrow{\eta_F} JFM \hookrightarrow T^*M \otimes_{TM} TFM/G$$
$$([\bar{s}_x], b) \mapsto \left[T_xM \xrightarrow{[\bar{s}_x]} T_bFM\right],$$

each $\widetilde{\Gamma} \in Sec\,\big((JFM/G)/M\big)$ determines the unique G-invariant connection Γ which is in $Sec\,(JFM/FM)$ with

$$\Gamma = \eta_F \circ (\widetilde{\Gamma} \circ \pi_F \circ 1_{FM}) \ .$$

On the bundle

$$\pi_1: JFM/G \times_M FM \longrightarrow JFM/G$$

there is a G-invariant connection

$$\Lambda: JFM/G \times_M FM \to J(JFM/G \times_M FM) \hookrightarrow$$
$$\hookrightarrow T^*(JFM/G) \otimes_{TM} T(JFM/G \times_M FM)$$
$$(x^i, \gamma^k_{ij}, b^i_j) \mapsto \left[(X^i, Y^k_{ij}) \mapsto (X^i, Y^k_{ij}, b^m_j \gamma^k_{mr} X^r)\right] .$$

This Γ is the universal connection for the frame bundle system. Explicitly, each $\widetilde{\Gamma}$ in $Sec((JFM/G)/M)$ gives an injection $(\widetilde{\Gamma} \circ \pi_F, 1_{FM})$, of FM into $JFM/G \times FM$. This is a section of π_1 and Γ coincides with the restriction of Λ to this section:

$$\Lambda_{|(\widetilde{\Gamma} \circ \pi_F, 1_{FM})FM} = \Gamma .$$

An analogous formulation of the universal linear connection but on the tangent bundle $\pi_T: TM \longrightarrow M$ is discussed in the context of sprays in Del Riego and Dodson [9]. Narasimhan and Ramanan [32, 33] proved that for any given Lie group G with a finite number of components and a given positive integer N, there is a principal G-bundle

$$E_G \longrightarrow E_G/G$$

carrying a connection Γ_G such that every connection Γ on any principal G-bundle $E \longrightarrow E/G$ with $\dim E \leq N$ is the pullback of Γ_G to E.

The new universal connection on systems has been used by Dodson and Modugno [13] to yield a universal calculus for geometrical field theory and in the presence of a metric on the base space there arises a universal codifferential and a Laplacian allowing gauge equations to be written canonically for any system. Equally, for Lagrangians dependent on the jets of the connection, the gauge-invariant Lagrangians are just those whose Euler-Lagrange equations admit a representation through the universal differential and codifferential. These results for fibred manifolds restrict to give the standard view on principal bundles.

Del Riego and Dodson [9] showed that certain connection-generated Lie algebras are pullbacks of the universal connection Lie algebra, and that there is a universal counterpart for sprays. Also, in [10], is studied the system of all linear connections on the second order tangent bundle, its universal connection and associated geometrical objects.

Cordero, Dodson and Parker [6, 7, 8] provided explicit characterization of the space of principal connections and the universal connections in the cases of principal circle bundles over T^2, S^2 (this latter includes the Hopf fibration) and principal r-torus bundles over homeomorphs of compact symmetric spaces. For example, from Kobayashi's theorem [23] for compact manifolds, we know that the equivalence classes of S^1 bundles over the torus T^2 are in bijective correspondence with $H^2(T^2, \mathbb{Z})$. Then it follows that up to isomorphism, all principal S^1 bundles over T^2 can be obtained as

$$K_n = H^3_n/\mathbb{Z}^3 \overset{\pi}{\twoheadrightarrow} T^2 = \mathbb{R}^2/\mathbb{Z}^2 .$$

Here H_0^3 is the 3-dimensional abelian Lie group, and for $n \neq 0$

$$H_n^3 = \left\{ \begin{pmatrix} 1 & x^1 & -\frac{x^3}{n} \\ 0 & 1 & x^2 \\ 0 & 0 & 1 \end{pmatrix} \right\}$$

defines the Heisenberg groups.

Theorem 1.4. *The space of all principal connections on H_n^3 over $\mathbb{R}^2$ is represented by the set of all $\mathbb{R}^2$-valued functions $\sigma = (\sigma_1, \sigma_2)$ on $\mathbb{R}^2$ such that σ_2 is nonvanishing.*

The space of all principal connections on $K_n = H_n^3/\mathbb{Z}^3$ over $T^2 = \mathbb{R}^2/\mathbb{Z}^2$ is represented by the set of all $\mathbb{R}^2/\mathbb{Z}^2$-valued σ on $\mathbb{R}^2/\mathbb{Z}^2$ such that σ_2 is nonvanishing. Then the explicit formulae for the possible choices of connection forms ω, and curvature forms Ω_ω reduce to

$$\omega = \omega^3 - \sigma_1 \omega^1 + (1 - \sigma_2) \omega^2$$
$$\Omega_\omega = d\omega = n \omega^1 \wedge \omega^2 - d(\sigma_1 \omega^1 + \sigma_2 \omega^2) .$$

2 UNIVERSAL CONNECTIONS AND STABILITY PROPERTIES

A convenient way to visualise the role of a system of connections $C \times_E E \to JE$ for a fibred manifold $p : E \to B$, is as follows—cf. [4, 9] for some helpful diagrams.

Each section $\tilde{\Gamma}$ of p determines a unique connection Γ as a section of $JE \to E$. Moreover, $\tilde{\Gamma}$ embeds a copy of E as a slice $\tilde{\Gamma} \circ p(E)$ in $C \times E$. The universal connection Λ is on the fibred manifold $C \times E \to C$ and coincides with Γ when restricted to the slice $\tilde{\Gamma} \circ p(E)$. So we have a family of connection geometries $\{(E, \Gamma)\}$ represented by $\{(\tilde{\Gamma} \circ p(E), \Lambda_{|\tilde{\Gamma} \circ p(E)})\}$ in $C \times E$. Across these slices we can use connection-geometric methodologies by means of the universal connection, faithfully recovering the connection-sensitive properties of each (E, Γ) on its corresponding slice in $C \times E$. For example, when E is the principle bundle of linear frames (cf. [5] for a detailed study of this bundle) then each linear connection induces a Riemannian metric on E and the slice embedding actually becomes an isometry through the universal connection. This is described in [4, 12, 5] and in [9] it is compared with the corresponding situation for linear connections represented as connections on the tangent bundle $TB \to B$.

Once a family of connections is organised in the form of a system of connections, then there is the possibility to study topological properties. One example has been the use of this approach by Canarutto and Dodson [4] to prove the stability of connection incompleteness. Problems of incomplete inextensible curves become particularly difficult in pseudo-Riemannian manifolds where the Hopf-Rinow theorem [24] is inapplicable. The difficulty can be removed by lifting the problem to the frame bundle where the connection-induced Riemannian structure can be exploited. It turns out that if a curve lifts to be inextensible and incomplete in the frame bundle with respect to one connection, then so is it with respect to nearby connections.

A consequence of this is that a singularity in general relativistic spacetime, like a black or white hole, is stable with respect to perturbations of the distribution of the matter which defines the geometry. This appears to give a useful generalisation of the result of Gotay and Isenberg [19] that a massless Klein-Gordon scalar field on a positively curved spacetime cannot escape the collapse of its state vector under geometric quantization. Now it seems that *no* quantum theory of gravity is likely to remove a classical connection singularity.

A second application has been to the differential geometric theory of parametric statistical models in mathematical statistics (cf. Amari [1, 2] and Lauritzen [22] for the background to this). Briefly, $\tilde{M}$ is the parameter space of an n-dimensional smooth family of probability distribution functions

$$M = \{p_x : \Omega \longrightarrow [0,1] | x \in \tilde{M}\}$$

for some fixed event space. For each p_x we have its log-likelihood function $l = \log p_x$. Suppose that at each point $x \in M$ the expected value

$$g_{ij} = \mathcal{E}\left(\frac{\partial l}{\partial x^i} \frac{\partial l}{\partial x^j}\right) \quad \text{(for coordinates } (x^i) \text{ about } x)$$

is a positive definite matrix. In this case it induces a statistically pertinent Riemannian metric g on M, called the **expected information metric** for the parametric model. Certain families of connections which contain the Levi-Civita connection of g turn out to have statistical importance [1]. Now the families of statistically interesting connections are revealed to be part of larger systems [11] and several stability theorems immediately follow.

3 UNIVERSAL CONNECTIONS AND FIELD THEORY

The use of universal connections in physical field theory followed some fifteen years after their invention by Narasimhan and Ramanan [32, 33]. A number of subsequent developments are relevant to their geometrical and physical roles; we summarise these, and offer some speculation on how systems of connections may be useful.

Tischler [40] used universal connections to show that every closed integral 2-form on $\mathbb{C}P^n$ is the curvature of a connection on an S^1 bundle; Frohlich [17] discussed their application to Stiefel and Grassmann manifold-valued functions involved in the construction of instantons. Schafly [38, 39] developed the geometrical theory some more and extended the methods of Narasimhan and Ramanan for orthogonal and unitary groups to the symplectic case; Kumar [26] showed that the geometric universal connection corresponds algebraically to the identity map on the Weil algebra of the Lie algebra.

Gursey and Tze [20] employed universal connections over Stiefel bundles in their comparison of 2-dimensional $\mathbb{C}P^n$ sigma models and 4-dimensional $\mathbb{H}P^n$ sigma models. Ramadas [37] used the universal connection on the Stiefel bundle of orthonormal

k-frames in a Hilbert space to show that the space of maps, from the underlying manifold to the Grassmannian of the k-frames which preserves a connection, has the homotopy type of a universal bundle for the automorphism group of that connection.

Dubois-Violette [15] replaced the Yang-Mills action on a $4k$-manifold by the integral of a conformally invariant $2k$-linear function of the curvature and used a universal connection to fuse together the Yang-Mills equations and the 2-dimensional analogue of Yang-Mills theory. Atiyah and Singer [3] used index theory and universal connections to compute obstructions relating to symmetry breaking of the classical action in gauge theory. Fischer [16] proposed a phase space for gauge field theories consisting of a quotient of the set of all connections on all G-bundles over a fixed base, studying it by means of a universal principal bundle and its universal connection. Picken [36] showed using universal connections that nontrivial solutions to the Weiss-Zumino consistency condition for gauge anomalies are related to nontrivial de Rham cohomology of the Lie group. Porta and Recht [35] obtained via universal connections a classification theorem for connections on vector bundles.

3.1 SPECULATIONS

Many of these developments will admit expression and extension through the notion of systems of connections. For the remainder of this note, we look at one particular possible development of current relevance to field theory. Donaldson's theory [14] deals with an oriented compact 4-manifold M with metric ρ, a compact Lie group G and a G-bundle $G \hookrightarrow P \to M$ of fixed topological type k. It views a connection A as an unknown, subject to the self-duality condition $\star F_A = F_A$ on its curvature F_A. The gauge group $\mathcal{G} = Aut(P)$ acts on the space of connections and interest centres on the quotient by this action, a manifold $\mathcal{M}_k(M, \rho)$. Certain invariants of $\mathcal{M}_k(M, \rho)$ are parametrised by k but are independent of the metric ρ. It would be of interest to investigate the structure and stability properties that are inherited by the self-dual connections in the appropriate system of connections, how these are parametrised by k and what role is played here by the gauge group or by some associated group action.

The fact that the condition $\star F_A = \pm F_A$ obtains at an absolute minimum of the Yang-Mills functional should admit an interpretation through the universal connection, perhaps for example via a dynamical system on the system of connections. Similarly, the stability properties of the Yang-Mills functional (cf. Laquer [21]) should be accessible by similar methods to those using systems of connections for the study of stability of connection singularities [4, 12]. Witten (cf. [41, 42], for example) devised a Lagrangian from which Donaldson's invariants can be obtained as integrals. This construction may be usefully representable through a system of connections.

The space $\mathcal{A}$ of all connections in $G \hookrightarrow P \to M$ is an infinite-dimensional affine space with a natural inner product. On it the gauge group $\mathcal{G}$ acts freely and moreover compatibly with the map

$$f : \mathcal{A} \to \Omega_+^2 : A \mapsto F_A^+$$

into the infinite-dimensional linear space of self-dual 2-forms. Suppose that we represent $\mathcal{A}$ by a finite-dimensional bundle of a system of connections, and can represent the map f in this context. Now, for each system of connections on a principal bundle we can construct a corresponding associated system for each associated bundle ([9] describes this in the case of the associated tangent bundle to the frame bundle). Hence a system representation should be constructible for the quotient $(\mathcal{A} \times \Omega_+^2)/\mathcal{G} \to \mathcal{A}/\mathcal{G}$ and the distinguished section of it induced by f. It is of any case interesting to know the extent to which finite-dimensional information from systems and their universal connections can carry that which at present is encoded by $\mathcal{A}$ as an infinite-dimensional $\mathcal{G}$-bundle. The Witten operator is an infinitesimal gauge transformation which also yields an equivariant cohomology operator and physical states are classes of it. The universal connection on a system of connections, over a manifold with a metric, generates a universal differential (cf. [13]) and codifferential whose cohomology may be pertinent to this.

REFERENCES

1. S.-I. Amari, *Differential Geometrical Methods in Statistics*, Springer Lecture Notes in Statistics 28, Springer-Verlag, Berlin, 1985.
2. S.-I. Amari, *Dual connections on the Hilbert bundles of statistical models*, Workshop on Geometrization of Statistical Theory, Lancaster 28-31 October 1987 (C. T. J. Dodson, eds.), ULDM Publications, Univ. Lancaster, 1987, pp. 123-152.
3. M. F. Atiyah and I. M. Singer, *Dirac operators coupled to vector potentials*, Proc. Nat. Acad. Sci. U.S.A., Phys. Sci. **81** (1984), no. 8, 2597-2600.
4. D. Canarutto and C. T. J. Dodson, *On the bundle of principal connections and the stability of b-incompleteness of manifolds*, Math. Proc. Camb. Phil. Soc. **98** (1985), 51-59.
5. L. A. Cordero, C. T. J. Dodson and M. deLeon, *Differential Geometry of Frame Bundles*, Kluwer, Dordrecht, 1989.
6. L. A. Cordero, C. T. J. Dodson and P. E. Parker, *Connections on principal circle bundles over the torus and the sphere*, Preprint, Department of Geometry and Topology, University of Santiago de Compostela, 1989.
7. L. A. Cordero, C. T. J. Dodson and P. E. Parker, *Connections on principal bundles over symmetric compacta*, Preprint, Department of Geometry and Topology, University of Santiago de Compostela, 1989.
8. L. A. Cordero, C. T. J. Dodson and P. E. Parker, *Examples of spaces of Connections and universal connections*, Conference on Differential Geometry, Eger, 21-23 August, 1989, Bolyai Math. Soc..
9. L. Del Riego and C. T. J. Dodson, *Sprays, universality and stability*, Math. Proc. Camb. Phil. Soc. **103** (1988), 515-534.
10. L. Del Riego and C. T. J. Dodson, *Universal reduction of second order tangent structure*, Workshop on Differential Geometry, La Laguna, 3-8 December, 1990.
11. C. T. J. Dodson, *Systems of connections for parametric models*, Proc. Workshop on Geometrization of Statistical Theory, Lancaster 28-31 October 1987 (C. T. J. Dodson, eds.), ULDM Publications, Univ. Lancaster, 1987, pp. 153-170.
12. C. T. J. Dodson, *Categories, Bundles and Spacetime Topology*, 2nd edition, Kluwer, Dordrecht, 1988.

13. C. T. J. Dodson and M. Modugno, *Connections over connections and a universal calculus*, V1 Convegno Nazionale de Relativita Generale e Fisica della Gravitazione. Florence, 10-13 October 1984, (R. Fabbri and M. Modugno,, eds.), Pitagora Editrice, Bologna, 1986, pp. 89–97.

14. S. K. Donaldson, *The geometry of 4-manifolds*, Int. Congress of Mathematicians, Berkeley, 1986 (A. M. Gleason, eds.), Amer. Math. Soc., Providence, RI, 1987, pp. 43–54.

15. M. Dubois-Violette, *Equations de Yang et Mills, modeles σ a deux dimensions et generalisation*, Progr. Math. **37** (1983), 43–64.

16. A. E. Fischer, *A grand superspace for unified field theories*, General Relativity Gravitation **18** (1986), no. 6, 597–608.

17. J. Frohlich, *A new look at generalized, nonlinear sigma-models and Yang-Mills theory*, Symposium Quantum fields—algebras, processes. Univ. Bielefeld 1978, Springer-Verlag, Berlin, 1980, pp. 361-378.

18. P. L. Garcia, *Connections and 1-jet fibre bundles*, Rend. Sem. Mat. Univ. Padova **47** (1972), 227–242.

19. M. J. Gotay and J. A. Isenberg, *Geometric quantization and gravitational collapse*, Phys. Rev. **D 22** (1980), 235–260.

20. F. Gursey and H. Tze, *Complex and quaternionic analyticity in chiral and gauge theories: I*, Ann. Physics **128** (1980), no. 1, 29–130.

21. H. T. Laquer, *Stability properties of the Yang-Mills functional near the canonical connection*, Michigan Math. J. **31** (1984), no. 2, 139–159.

22. S. L. Lauritzen, *Conjugate connections in statistical theory*, Workshop on Geometrization of Statistical Theory Lancaster 28-31 October 1987 (C. T. J. Dodson, eds.), ULDM Publications, Univ. Lancaster, 1987, pp. 33–52.

23. S. Kobayashi, *Principal fibre bundles with the 1-dimensional toroidal group*, Tôhoku Math. J. **2** (1956), no. 8, 29–45.

24. S. Kobayashi and K. Nomizu, *Foundations of Differential Geometry I*, Interscience, New York, 1963.

25. S. Kobayashi and K. Nomizu, *Foundations of Differential Geometry II*, Interscience, New York, 1969.

26. S. Kumar, *A remark on universal connections*, Math. Ann. **260** (1982), no. 4, 453–462.

27. L. Mangiarotti and M. Modugno, *Fibred spaces, jet spaces and connections for field theories*, International Meeting on Geometry and Physics, Florence 12-15 October 1982 (1983), Pitagora Editrice, Bologna, 135-165.

28. M. Modugno, *Systems of vector valued forms on a fibred manifold and applications to gauge theories*, Conf. Diff. Geom. Meth. in Math. Phys., Salamanca 1985, Lect. Notes in Math., vol. 1251, Springer-Verlag, 1987, pp. 238–264.

29. M. Modugno, *An introduction to systems of connections*, Preprint, Inst. Mat. App. G. Sansone, Florence, 1986.

30. M. Modugno, *Linear overconnections*, Preprint, Inst. Mat. App. G. Sansone, Florence 1987.

31. P. O. Mazur and L. M. Sokolowski, *Teleparallelism as a universal connection on null hypersurfaces in general relativity*, General Relativity Gravitation **18** (1986), no. 8, 781–804.

32. M. S. Narasimhan and S.Ramanan, *Existence of universal connections I*, Amer. J. Math. **83** (1961), 563–572.

33. M. S. Narasimhan and S. Ramanan, *Existence of universal connections II*, Amer. J. Math. **85** (1963), 223–231.

34. K. Nomizu, *Lie Groups and Differential Geometry*, Math. Soc. Japan, Tokyo, 1956.

35. H. Porta and L. Recht, *Classification of linear connections*, J. Math. Anal. and Applic. **118** (1986), no. 2, 547–560.

36. R. F. Picken, *Universal connections, gauge anomolies and Lie group cohomology*, J. Phys. **A 19** (1986), no. 5, L219–L224.

37. T. R. Ramadas, *On the space of maps inducing isomorphic connections*, Ann. Inst. Fourier (Grenoble) **32** (1982), no. 1, viii, 263–276.
38. R. Schlafly, *Universal connections*, Invent. Math. **59** (1980), no. 1, 59–65.
39. R. Schlafly, *Universal connections: the local problem*, Pacific J. Math. **98** (1982), no. 1, 157–171.
40. D. Tischler, *Closed 2-forms and an embedding theorem for symplectic manifolds*, J. Diff. Geom. **12** (1977), no. 2, 229–235.
41. E. Witten, *Physics and geometry*, Int. Congress of Mathematicians, Berkeley, 1986 (A. M. Gleason, eds.), Amer. Math. Soc., Providence, RI, 1987, pp. 267–303.
42. E. Witten, Contribution at the International Congress of Mathematicians, Kyoto, 1990. (cf. also: Topological quantum field theory, Topological σ-models, Topological gravity: Preprints IAS-PUB-IASSNS-HEP-87/72 Princeton, 1988.

DEPARTMENT OF MATHEMATICS, UNIVERSITY OF TORONTO, TORONTO, ONTARIO, CANADA M5S 1A4

E-mail address: dodson@ecf.toronto.edu

DIMENSION THEORY AND THE SULLIVAN CONJECTURE

JERZY DYDAK AND JOHN J. WALSH

University of Tennessee and University of California Riverside

ABSTRACT. The covering dimension of a compact metric space X, denoted $\dim X$, is the least integer n such that every open cover $\mathcal{U}$ of X has an open refinement $\mathcal{V}$ such that, for any $n+2$ distinct elements $V_1, V_2, \cdots, V_{n+2} \in \mathcal{V}$, the intersection $V_1 \cap V_2 \cap \cdots \cap V_{n+2} = \emptyset$. The (integral) cohomological dimension of X, denoted $\dim_{\mathbb{Z}} X$, is the least integer n such that $H^q(A, B; \mathbb{Z}) = 0$ for all closed pairs $B \subset A$ in X and all $q \geq n+1$. In each theory, the concept of infinite dimensionality, denoted $\dim X = \infty$ and $\dim_{\mathbb{Z}} X = \infty$, respectively, is introduced to specify situations when no such integer n exist. During the 1930's, early in the development of these dimension theories, P. S. Alexandroff established the equality $\dim X = \dim_{\mathbb{Z}} X$ *for those spaces X for which $\dim X < \infty$.* The possibility that there might be compact metric spaces X for which $\dim X = \infty$ but $\dim_{\mathbb{Z}} X < \infty$ remained. Finally, in 1987 A. N. Dranishnikov produced compact metric spaces X with $\dim X = \infty$ but $\dim_{\mathbb{Z}} X = 3$. Subsequently, the authors produced compacta X with $\dim X = \infty$ but $\dim_{\mathbb{Z}} X = 2$. In this paper the role that the Sullivan Conjecture (which was shown to be valid by H. Miller) plays in distinguishing the covering theory from the cohomological theory of dimension will be explored.

0. INTRODUCTION

There are several equivalent formulations of covering dimension; see [HW]. The one that is best suited to our purposes is the following characterization which we take to be the definition. For a metric space X, the covering dimension $\dim X \leq n$ provided, for every map $f : A \to S^n$ of a closed subset $A \subset X$ to the n–sphere, there is an extension to a map $F : X \to S^n$. Since Čech cohomology can be defined as homotopy classes of maps to Eilenberg-MacLane complexes $K(\mathbb{Z}, n)$, cohomological dimension has the following formulation. For a metric space X, the integral cohomological dimension $\dim_{\mathbb{Z}} X \leq n$ provided, for every map $f : A \to K(\mathbb{Z}, n)$ of a closed subset $A \subset X$, there is an extension to a map $F : X \to K(\mathbb{Z}, n)$. The statement $\dim X = n$ means that $\dim X \leq n$ but $\dim X \not\leq n-1$ while $\dim X = \infty$ means that $\dim X \not\leq n$ for all n. Similar conventions apply for cohomological dimension.

Since the 1–sphere S^1 is a $K(\mathbb{Z}, 1)$, it is transparent that $\dim X \leq 1$ if and only if $\dim_{\mathbb{Z}} X \leq 1$. The fact that $\dim X = \dim_{\mathbb{Z}} X$ for spaces with $\dim X < \infty$ can be extracted

1991 *Mathematics Subject Classification.* 54F45, 55M10, 55N99, 55Q40, 55P20.

Key words and phrases. Cohomological dimension, dimension, Eilenberg-MacLane complexes, loops spaces, Sullivan Conjecture, cohomology theories.

Second author supported in part by a grant from the National Science Foundation

from the knowledge that, for each n, $K(\mathbb{Z}, n)$ has a representation as a CW-complex containing no $(n + 1)$–cells (see [Wa$_1$]).

Specify Eilenberg-MacLane complexes so that there is an inclusion $S^n \hookrightarrow K(\mathbb{Z}, n)$ inducing an isomorphism of nth–homotopy groups. Suppose that X is a compact metric space for which $\dim X \neq \dim_{\mathbb{Z}} X$. Necessarily, X has infinite covering dimension and finite cohomological dimension, say $\dim_{\mathbb{Z}} X = k_0$. Since $\dim X = \infty$, for each $n \geq k_0 + 1$, there are (many) closed subsets $A \subset X$ and maps $f : A \to S^n$ that are essential, as they do not extend to a maps defined on all of X. Yet, every such map $f : A \to S^n$ composed with the inclusion yields a map $f : A \to K(\mathbb{Z}, n)$ that is inessential, as every map $f : A \to K(\mathbb{Z}, n)$ defined on a closed subset $A \subset X$ is inessential, as it extends to map of the cone of A since $\dim_{\mathbb{Z}} A \leq k_0$ implies that the cone of A has cohomological dimension at most $k_0 + 1$.

Detecting the coexistence of the phenomenon of there being a large class of essential maps to spheres and of there being no essential maps to $K(\mathbb{Z}, n)$'s, suggested the need for an invariant that has some sensitivity for maps to spheres but is totally insensitive for maps to $K(\mathbb{Z}, n)$'s at least for $n \geq k_0 + 1$. The form that such an invariant should take was identified by A. N. Dranishnikov based on his analyses of an explicit construction that evolved from (unpublished) work of R. D. Edwards (see [Ed]) that was set forth in [Wa$_1$]. The construction is described below in §4 and leads to a characterization of compact spaces X with $\dim_{\mathbb{Z}} X \leq k_0$ as inverse limits of sequences of polyhedra that are obtained using "Pontrjagin-like" modifications.

In general terms, Dranishnikov determined that a suitable invariant would be a generalized cohomology theory h^* satisfying:

   (i) $h^*(K(\mathbb{Z}^r, k_0)) = 0$, where $\mathbb{Z}^r$ denotes the sum of r copies of $\mathbb{Z}$;
   (ii) $h^*(S^n) \neq 0$ for some $n > k_0$; and
   (iii) for each finite complex C, $h^*(C)$ is a finite group.

Specifically, such a cohomology theory can be used to produce a compact metric space X with $\dim X = \infty$ and $\dim_{\mathbb{Z}} X = k_0$. The complex K-theory of the Eilenberg-MacLane spaces $K(\mathbb{Z}, n)$ had been worked out by Anderson-Hodgkin [AH] and Buchstaber-Mischenko [BM]. Complex K-theory itself satisfies neither (i) nor (iii) but the theory with finite coefficients $\mathbb{Z}/p$ (p a prime) vanishes on $K(\mathbb{Z}, n)$ for $n \geq 3$ (and, hence, by the Künneth formula on $K(\mathbb{Z}^r, n)$ as well) and the third requirement holds as the coefficients are finite.

In striking fashion, A. N. Dranishnikov exhibited in [Dr$_1$] compact metric spaces X with $\dim X = \infty$ and $\dim_{\mathbb{Z}} X = 3$, thereby answering a question posed by P. S. Alexandroff in 1932 [Al]. These examples combined with work of R. D. Edwards (see [Wa$_1$]) can be used to produce cell-like dimension raising maps and, in turn, homology n–manifolds that are infinite dimensional for $n \geq 7$.

Now, a familiar representation of $K(\mathbb{Z}, 2)$ is infinite complex projective space $\mathbb{C}P^{\infty}$. Complex K-theory and, apparently, other standard generalized cohomology theories do not vanish on $\mathbb{C}P^{\infty}$. Consequently, the invariant that sufficed for constructing the examples in [Dr$_1$] was not available for constructing compact metric spaces X with $\dim X = \infty$ and $\dim_{\mathbb{Z}} X = 2$. (At the time, one had to entertain the possibility that the distinct nature of $K(\mathbb{Z}, 2)$ might preclude the existence of such examples.)

In [DW$_1$], the authors, guided by the work of A. Zabrodsky in [Za], identified a connection between cohomological dimension and the Sullivan conjecture, the latter a theorem due to H. Miller [Mi]. In particular, an invariant was identified that can be used to produce compact metric spaces having infinite covering dimension but finite cohomological dimension. The approach includes the construction of such spaces X with $\dim_{\mathbb{Z}} X = 2$.

The purpose of this paper is to expand on the admittedly sketchy presentation that was given in [DW₁], particularly, regarding the nature of the role of the Sullivan conjecture. In addition, details of variations of the examples constructed in [DW₁] are included.

1. GROUP STRUCTURES ON $[X, Y]$ AND "TRUNCATED COHOMOLOGY THEORIES"

A pointed space is a space together with a designated base point (generally, the latter is not an explicit part of the notation). Maps and homotopies between pointed spaces are assumed to preserve the base point. For pointed spaces X and Y, $[X, Y]$ is the set of equivalence classes where two base point preserving maps are equivalent provided there is a base point preserving homotopy between them. Alternatively, $[X, Y]$ is the set of path components of the function space Y^X of base point preserving maps.

It is necessary to consider certain function spaces, specifically loop spaces, and, hence, the setting must be more general than that consisting of CW-complexes. The category of compactly generated spaces is thoroughly described in [WH; Ch.1] and provides a setting that suffices for our purposes. The reader is referred to the latter for the numerous foundational details that will be omitted.

Generalized cohomology theories can be viewed as arising from a family of pointed spaces $\{Y_n : n \in \mathbb{Z}\}$ for which there are group structures on $[X, Y_n]$ for arbitrary X. Of course, an essential feature of a cohomology theory is that the Y_n's and the associated groups are related.

A group structure on $[X, Y]$ is induced by a "homotopy" group structure on Y, namely, Y being an H-space. While spaces themselves may not carry such a structure, their loop spaces $\Omega Y = [(I, \{0, 1\}), (Y, *)]$ are H-spaces and their second loop spaces $\Omega^2 Y = \Omega(\Omega Y)$ are "homotopy" commutative. Thus, $[X, \Omega Y]$ is a group and $[X, \Omega^2 Y]$ is an abelian group. (*Everything is pointed with the base point of a loop space being the constant loop.*)

The relationship between the various Y_n's that provides a cohomology theory, as well as the abelian group structure on each $[X, Y_n]$, is that $Y_n \simeq \Omega Y_{n+1}$ where "$\simeq$" denotes pointed homotopy equivalent. Of course, iteration leads to $Y_n \simeq \Omega^r Y_{n+r}$, where $\Omega^r \equiv \Omega(\Omega^{r-1})$.

Čech/Alexander-Spanier cohomology can be defined in terms Eilenberg-MacLane spaces: namely, $Y_n \simeq K(\mathbb{Z}, n)$ for $n \geq 0$ and $Y_n = \{*\}$ for $n < 0$. Thus, there is the natural identification of reduced cohomology $\tilde{H}^n(X; \mathbb{Z}) = [X, K(\mathbb{Z}, n)]$.

Complex K-theory has a similar representation that arises from Bott periodicity $\Omega^2 U \simeq U$ where $U = \cup_{n=1}^{\infty} U(n)$ is the unitary group. Setting $Y_{2n+1} = U$ and $Y_{2n} = \Omega U$ and using the further identification $\Omega U \simeq BU \times \mathbb{Z}$ where BU is the classifying space of U, the complex K-theory of a connected space X is $[X, BU]$ in even dimensions and $[X, U]$ in odd dimensions.

Analyzing the interplay between the general method developed by R. D. Edwards for constructing spaces having finite cohomological dimension described in [Wa₁] and the use of complex K-theory to detect their covering dimension in [Dr₁] revealed that most of the structure of a cohomology theory was not needed. The feature that was used in an essential manner was the Mayer-Vietoris sequence. The next lemma places the latter sequence in a context that displays the broad setting in which it is valid. The proof involves completely standard arguments and, since a detailed outline can be found in [DW$_1$; §2], is omitted.

A fundamental concept in homotopy theory is the homotopy extension property. Namely, in many "naturally" occurring situations of $A \subset X$, given a map $h : X \times \{0\} \cup A \times I \to Y$, where $I = [0,1]$ is the unit interval, there is an extension $H : X \times I \to Y$. An inclusion $A \subset X$ for which such an extension exists for all maps h and all spaces Y is called a cofibration. For example, the inclusion of a subcomplex in a CW-complex is a cofibration. Also, as will be needed later, if $p : E \to B$ is a fibration, B_0 is a closed subset, and $B_0 \subset B$ is a cofibration, then $p^{-1}(B_0) \subset E$ is a cofibration.

Lemma 1. *Let L be a pointed CW-complex, X a compactly generated space, $A, B \subset X$ closed subsets with the inclusions $A \cap B \hookrightarrow A$ and $A \cap B \hookrightarrow B$ cofibrations and specify that the base point be in $A \cap B$ (in particular, $A \cap B \neq \emptyset$). Then the sequence*

$$\cdots \to [A, \Omega^{n+1}L] \oplus [B, \Omega^{n+1}L] \to [A \cap B, \Omega^{n+1}L] \to [A \cup B, \Omega^n L] \to$$
$$[A, \Omega^n L] \oplus [B, \Omega^n L] \to [A \cap B, \Omega^n L] \to \cdots$$

is an exact sequence of abelian groups provided $n \geq 2$.

It is possible to associate to a pointed CW-complex L and its iterated loop spaces a structure that might be called a "truncated" generalized cohomology theory. The collection $\{Y_n : -\infty < n \leq -2\}$ where $Y_n = \Omega^{-n}L$, while only indexed by the integers truncated at -2, satisfies $Y_n = \Omega Y_{n+1}$. Setting $h_L^n(X) = [X, Y_n]$ for $n \leq 2$, provides a family of contravariant functors taking values in the category of abelian groups. The Mayer-Vietoris sequence in the preceding lemma takes the familiar form:

$$\cdots \to h_L^{n+1}(A) \oplus h_L^{n+1}(B) \to h_L^{n+1}(A \cap B) \to h_L^n(A \cup B) \to$$
$$h_L^n(A) \oplus h_L^n(B) \to h_L^n(A \cap B) \to \cdots$$

provided $-\infty < n \leq -2$.

While the notation "h_L^n" reflects that often used for designating cohomology theories, it will not be used subsequently as there are no clear advantages of economy for adopting it. At the same time, just as the argument that establishes the preceding lemma is a "standard one" from the development of genuine generalized cohomology theories, so are some that will be needed in the next section.

2. A COMBINATORIAL VIETORIS-BEGLE THEOREM

For Čech cohomology it is a classical result that a proper map $f : X \to Y$ for which the reduced groups $\widetilde{H}^*(f^{-1}(y)) \simeq \widetilde{H}^*(\text{point})$ for each $y \in Y$ induces an isomorphism $f^* : H^*(Y) \simeq H^*(X)$. The comparable result for generalized cohomology theories fails. A "dramatic" illustration can be found in the example due to J. Taylor [Ta] of a cell-like map $f : X \to I^\infty$ from a compactum X to the Hilbert cube where the reduced complex K-theory of X is nontrivial.

However, the following combinatorial version is valid generally and can be extracted using the Mayer-Vietoris sequence of the previous section, induction, and an analysis of the Lim^1-term that arise from computing the cohomology of an infinite complex in terms of the cohomology of its finite subcomplexes (see [Mil]).

Lemma 2. *Let* $\pi : \widetilde{K} \to K$ *be a map from a compactly generated space to a countable simplicial complex such that the inclusion* $\pi^{-1}(A) \hookrightarrow \widetilde{K}$ *is a cofibration for every subcomplex* $A \subset K$. *If* $n_0 \geq 2$ *and* L *is any pointed CW-complex with*

$$[\pi^{-1}(\sigma), \Omega^n L] = 0 \text{ for all simplices } \sigma \in K \text{ and all } n \geq n_0,$$

then the induced map

$$\pi^* : [K, \Omega^n L] \to [\widetilde{K}, \Omega^n L] \text{ is an isomorphism for all } n \geq n_0.$$

Outline of proof. (Additional details can be found in [DW$_1$; §3].) First, consider the case that K is a finite simplicial complex and argue using induction on the number of simplices in K. Write $K = K' \cup \sigma$ where σ is a maximal dimensional simplex; thus, $\sigma \cap K' = \partial\sigma$. The conclusion of the lemma is valid for σ (by assumption) and for both $\partial\sigma$ and K' (by induction).

If $\partial\sigma \neq \emptyset$ (i.e., $\dim \sigma > 0$), then Lemma 1 applies to the pair $A = K'$ and $B = \sigma$ and to the pair $A = \pi^{-1}(K')$ and $B = \pi^{-1}(\sigma)$ and the conclusion of the lemma follows using the Five Lemma (see [Sp; p.185].)

For σ a 0–dimensional simplex, Lemma 1 does not apply (as the explicit requirement in Lemma 1 that $A \cap B \neq \emptyset$ is needed to avoid "base point problems"). Denote by $[\ ,\ ]_f$ "free" homotopy classes; i.e., neither the maps nor the homotopies are required to preserve base points. Choosing the base points in $\pi^{-1}(K')$ and K', respectively, $[K, \Omega^n L] \simeq [K', \Omega^n L] \oplus [\sigma, \Omega^n L]_f$ and $[\pi^{-1}(K), \Omega^n L] \simeq [\pi^{-1}(K'), \Omega^n L] \oplus [\pi^{-1}(\sigma), \Omega^n L]_f$. Furthermore, the induced map π^* is the direct sum of the induced map $[K', \Omega^n L] \to [\pi^{-1}(K), \Omega^n L]$, which is an isomorphism by induction, and the induced map $[\sigma, \Omega^n L]_f \to [\pi^{-1}(\sigma), \Omega^n L]_f$. Clearly, $[\sigma, \Omega^n L]_f$ can be identified as the path components of $\Omega^n L$. It remains to identify $[\pi^{-1}(\sigma), \Omega^n L]_f$ as the path components of $\Omega^n L$ (and observe that the induced map is a bijection). The hypothesis $[\pi^{-1}(\sigma), \Omega^n L] = 0$ assures that either $\Omega^n L$ is path connected or $\pi^{-1}(\sigma)$ is connected. If $\Omega^n L$ is path connected, then $[\pi^{-1}(\sigma), \Omega^n L]_f = 0$ as every map $\pi^{-1}(\sigma) \to \Omega^n L$ is homotopic to a base point preserving map and, hence, is itself null homotopic. Otherwise, $\pi^{-1}(\sigma)$ is connected and $[\pi^{-1}(\sigma), \Omega^n L]_f \simeq \oplus_{P \in \mathcal{P}} [\pi^{-1}(\sigma), P]_f$ where $\mathcal{P}$ is the set of path components of $\Omega^n L$. It remains to establish that each $[\pi^{-1}(\sigma), P]_f = \{\text{point}\}$. Given $f : \pi^{-1}(\sigma) \to P$, choose a loop $\omega \subset P$ and define $\omega^{-1}f$ by setting $\omega f(x) = \omega^{-1} * f(x)$ where $*$ is multiplication of loops. Observe that the image of $\omega^{-1}f$ is contained in the path component of $\Omega^n L$ containing the base point ($=$constant loop) and, thus, $\omega^{-1}f$ is null homotopic as it is homotopic to a base point preserving map. Denoting the null homotopy by $\{h_t\}$, the homotopy $\{\omega h_t\}$ provides a null homotopy of $\omega\omega^{-1}f$. Since the latter is homotopic to f, the latter is null homotopic as well.

The general case that K is a countable CW-complex makes use of the general principal established by Milnor in [Mil] that the cohomology of such a complex can be computed in terms of the cohomology of its finite subcomplexes. While the latter work focused on standard cohomology theories it applies to generalized theories as well as the "truncated" theories needed herein. Following [Mi] and writing $K = \cup_{i=1}^{\infty} K_i$ where each K_i is a subcomplex and $K_i \subset K_{i+1}$, there are short exact sequences

$$0 \to \mathrm{Lim}^1\{[K_i, \Omega^{n+1}L]\} \to [K, \Omega^n L] \to \mathrm{InvLim}\{[K_i, \Omega^n L]\} \to 0$$

and

$$0 \to \mathrm{Lim}^1\{[\pi^{-1}(K_i), \Omega^{n+1}L]\} \to [\widetilde{K}, \Omega^n L] \to \mathrm{InvLim}\{[\pi^{-1}(K_i), \Omega^n L]\} \to 0.$$

Requiring in addition that the K_i's be finite complexes, for $n \geq n_0$ the restrictions of π induces isomorphisms between $[K_i, \Omega^{n+1}L]$ and $[\pi^{-1}(K_i), \Omega^{n+1}L]$ as well as $[K_i, \Omega^n L]$ and $[\pi^{-1}(K_i), \Omega^n L]$ for each i. In turn, these induce isomorphisms between the Lim^1 and InvLim terms and the Five Lemma detects that π itself induces an isomorphism. $\square$

3. SULLIVAN CONJECTURE, $S^n \hookrightarrow K(\mathbb{Z}, n)$, AND $\Omega^k S^m$

In general terms, the validity of the Mayer-Vietoris sequence and, in turn, of the combinatorial version of a Vietoris-Begle theorem for the family of iterated loop spaces of an arbitrary CW-complex L identifies the latter as a variant on an actual cohomology theory that might suffice for distinguishing between covering dimension and cohomological dimension. Of course, the "heart of the matter" is to identify complexes L whose family of loop spaces provides a theory with the "right" properties. In general terms, these consist of a nontriviality property, a vanishing property, and a finiteness property. Namely:

   (i) (Nontriviality) neither L nor any of its loop spaces $\Omega^k L$ should have the weak homotopy type of a point;
  (ii) (Vanishing) for $n \geq 2$ and for $r \geq 1$, $[K(\mathbb{Z}^r, n), \Omega^k L] = 0$ for all but finitely many values of k;
 (iii) (Finiteness) for any finite CW-complex C, the groups $[C, \Omega^k L]$ are finite for all but finitely many values of k.

The "adjointness" of suspending and looping reflected in the natural equivalence $[\Sigma X, Y] \simeq [X, \Omega Y]$ together with the identification $S^{k+j} = \Sigma^k S^j$ detect the isomorphisms $\pi_j(\Omega^k L) \simeq \pi_{k+j}(L)$. Consequently, the nontriviality condition is nothing more that the requirement that infinitely many of the homotopy groups $\pi_j(L)$ are non-zero.

In light of the discussion of the preceding paragraph, the finiteness condition requires that all but finitely many of the homotopy groups $\pi_j(L)$ are finite. In fact, the latter restriction turns out to imply the finiteness condition; see the Lemma 5 below.

Of course, the vanishing condition is the formidable one. Its validity requires one restriction on L beyond the requirement, needed for the finiteness condition, that all but finitely many of the homotopy groups $\pi_j(L)$ are finite. Namely, L should be finite dimensional. However, its proof fundamentally depends on the Sullivan Conjecture whose validity was established by H. Miller in [Mi]. A statement of the Sullivan Conjecture appears below and a concise statement of the vanishing condition is the content of Lemma 4.

Evidently, spheres satisfy the stated restrictions. For $n \geq 2$, S^n has infinitely many nonzero homotopy groups and, for n odd, only $\pi_n(S^n)$ is infinite while, for n even, only $\pi_n(S^n)$ and $\pi_{2n-1}(S^n)$ are infinite; see [Sp, pp.515–516]. Since the goal is to produce examples that distinguish between dimension theories, one based on maps to spheres and the other on maps to $K(\mathbb{Z}, n)$'s, the emergence of the higher homotopy groups of spheres as central "gadgets" is, perhaps, expected.

Following [Mi], a group G is said to be locally finite provided every finitely generated subgroup is finite. For an abelian group, local finiteness translates to each element of

G having finite order, i.e., G is a torsion group. While the Sullivan Conjecture focuses on locally finite groups, the specific group in that class of interest herein is the infinitely generated torsion abelian group $\mathbb{Q}^t/\mathbb{Z}^t$.

Sullivan Conjecture. (Proved by H. Miller in [Mi]) *If G is a locally finite group and L is a finite dimensional CW-complex, then the space of pointed maps from $K(G,1) \to L$, denoted $map_*(K(G,1), L)$, is weak homotopy equivalent to a point.*

The conclusion that the space of pointed maps has the weak homotopy type of a point translates to the homotopy groups of the mapping space being trivial. In turn, there is a natural equivalence $\pi_k(map_*(X, L)) \simeq [X, \Omega^k L]$; see [Wh; p.105]. The specific form of the Sullivan Conjecture that is used below is that $[K(G,1), \Omega^k L] = 0$ for all k, where G is locally finite and L is finite dimensional.

Detailed proofs of the following results can be found in [DW$_1$; §3 and §4]. Hence, only outlines are included. The strategy it to exploit the fibration $K(\mathbb{Q}^t/\mathbb{Z}^t, 1) \hookrightarrow K(\mathbb{Z}^t, 2) \to K(\mathbb{Q}^t, 2)$ that arises from the short exact sequence $\mathbb{Z}^t \hookrightarrow \mathbb{Q}^t \to \mathbb{Q}^t/\mathbb{Z}^t$. The Sullivan Conjecture provides the necessary information about $K(\mathbb{Q}^t/\mathbb{Z}^t, 1)$. Information about $K(\mathbb{Q}^t, 2)$ is gathered from the path fibration $K(\mathbb{Q}^t, 1) \hookrightarrow \mathcal{P} \to K(\mathbb{Q}^t, 2)$, once the fiber $K(\mathbb{Q}^t, 1)$ is analyzed directly.

Once an adequate vanishing result is obtained for $K(\mathbb{Z}^t, 2)$, induction and the Combinatorial Vietoris-Begle Theorem applied to the path fibration $K(\mathbb{Z}^t, s) \hookrightarrow \mathcal{P} \to K(\mathbb{Z}^t, s+1)$ extracts the vanishing result for $K(\mathbb{Z}^t, s)$ for $s \geq 3$ as well.

Lemma 3. *If L is a pointed CW-complex and $r_0 \geq 3$ is an integer such that $\pi_k(L)$ is finite for $k \geq r_0$, then $[K(\mathbb{Q}^t, s), \Omega^k L] = 0$ for $k \geq r_0 - 1$ and $t, s \geq 1$.*

Proof. The proof is by induction on both t and s. Fix $k \geq r_0 - 1$.

First, consider $[K(\mathbb{Q}, 1), \Omega^k L]$. A model for $K(\mathbb{Q}, 1)$ is the 2–dimensional complex that is the infinite telescope associated with the sequence of maps from $S^1 \to S^1$ where the nth map has degree $n!$. Using this model, express $K(\mathbb{Q}, 1) = \cup_{i=1}^{\infty} T_i$ as an increasing union of finite subcomplexes each homotopy equivalent to S^1. Since $\pi_1(\Omega^k L) = \pi_{k+1}(L)$ and the latter is a finite group, any map $f : K(\mathbb{Q}, 1) \to \Omega^k L$ is trivial on π_1 and, hence, its restriction to each T_i is null homotopic. According to Lemma 6 below the map f itself must be null homotopic as well.

Next, apply induction (on t) and the Combinatorial Vietoris-Begle Theorem to the projection

$$K(\mathbb{Q}^t, 1) = K(\mathbb{Q}^{t-1}, 1) \times K(\mathbb{Q}, 1) \to K(\mathbb{Q}^{t-1}, 1)$$

in order to deduce that $[K(\mathbb{Q}^t, 1), \Omega^k L] = 0$ for all $t \geq 1$.

Finally, apply induction (on s) and the Combinatorial Vietoris-Begle Theorem to the path fibration $K(\mathbb{Q}^t, s-1) \hookrightarrow \mathcal{P} \to K(\mathbb{Q}^t, s)$ in order to deduce that $[K(\mathbb{Q}^t, s), \Omega^k L] = 0$ for all s and t. $\square$

Lemma 4. *If L is a finite dimensional pointed CW-complex and $r_0 \geq 3$ is an integer such that $\pi_k(L)$ is finite for $k \geq r_0$, then $[K(\mathbb{Z}^t, s), \Omega^k L] = 0$ for $k \geq r_0 - 1$, $s \geq 2$, and $t \geq 1$.*

Proof. Consider the fibration $K(\mathbb{Q}^t/\mathbb{Z}^t, 1) \hookrightarrow K(\mathbb{Z}^t, 2) \to K(\mathbb{Q}^t, 2)$ that arises from $\mathbb{Z}^t \hookrightarrow \mathbb{Q}^t \to \mathbb{Q}^t/\mathbb{Z}^t$. The Sullivan Conjecture applies to establish that $[K(\mathbb{Q}^t/\mathbb{Z}^t, 1), \Omega^k L] = 0$

and Lemma 3 states that $[K(\mathbb{Q}^t, 2), \Omega^k L] = 0$. The Combinatorial Vietoris-Begle Theorem applies to detect that $[K(\mathbb{Z}^t, 2), \Omega^k L] = 0$.

Induction (on s) and the Combinatorial Vietoris-Begle Theorem applied to the path fibration $K(\mathbb{Z}^t, s-1) \hookrightarrow \mathcal{P} \to K(\mathbb{Z}^t, s)$ leads to the conclusion that $[K(\mathbb{Z}^t, s), \Omega^k L] = 0$. $\square$

Recorded above is that the spheres themselves satisfy the conditions imposed on L in the preceding lemma, the minimum value for r_0 depending on the parity of the dimension of the sphere.

Corollary 1. *For n odd, $[K(\mathbb{Z}^t, s), \Omega^k S^n] = 0$ for $k \geq n$, $s \geq 2$, and $t \geq 1$ and for n even, $[K(\mathbb{Z}^t, s), \Omega^k S^n] = 0$ for $k \geq 2n - 1$, $s \geq 2$, and $t \geq 1$.*

Lemma 5. *If L is a pointed CW-complex and $r_0 \geq 2$ is an integer such that the homotopy groups $\pi_k(L)$ are finite for $k \geq r_0$, then $[C, \Omega^k L]$ is finite for every finite complex C and every $k \geq r_0$.*

Proof. The proof is by induction on the dimension and number of cells in C. For C 0-dimensional, the $[C, \Omega^k L]$ is finite since $\pi_0(\Omega^k L) = \pi_k(L)$ is finite. If $C = C' \cup \sigma$ where $\dim \sigma = m \geq 1$ and $C' \cap \sigma \simeq S^{m-1}$, then the Mayer-Vietoris sequence becomes

$$\cdots \to \pi_{m-1}(\Omega^{k+1} L) \to [C, \Omega^k L] \to [C', \Omega^k L] \to \cdots .$$

The left most group is finite by assumption and the right most is finite by induction. Exactness forces the middle group to be finite as well. $\square$

A map $f : K \to X$ defined on a CW-complex K is said to be *phantom* provided the restriction of f to each finite subcomplex of K is null homotopic. A critical step in the construction of infinite dimensional compacta having finite cohomological dimension involves replacing an essential map defined on an infinite complex with its restriction to a finite subcomplex, the latter chosen so that the restriction remains essential. Evidently, successfully completing this step requires knowing that the original map is not phantom. In [Dr₁] the use of complex K-theory with *finite* coefficients yields the necessary information. The next result records that the finiteness of the higher homotopy groups of spheres provides that information for our purposes.

Lemma 6. *Let L be a pointed CW-complex and $r_0 \geq 3$ an integer such that the homotopy groups $\pi_k(L)$ are finite for $k \geq r_0$ and let $K = \cup_{i=1}^{\infty} K_i$ be a countable CW-complex expressed as the increasing union of finite complexes with base point $* \in K_1$. If $k \geq r_0 - 1$ and $f : K \to \Omega^k L$ is a map whose restriction to each K_i represents the trivial element in $[K_i, \Omega^k L]$, then f represents the trivial element in $[K, \Omega^k L]$.*

Proof. The analysis from [Mil] provides the exact sequence

$$0 \to \mathrm{Lim}^1\{[K_i, \Omega^{k+1} L]\} \to [K, \Omega^k L] \to \mathrm{InvLim}\{[K_i, \Omega^k L]\} \to 0.$$

Lemma 5 states that each of the groups $[K_i, \Omega^{k+1} L]$ is finite and Lim^1 of an inverse sequence of finite groups is trivial. $\square$

4. "Pontrjagin" Construction and Edwards-Walsh Complexes

Consider the following construction due to Pontrjagin. Specify a prime q and produce an inverse sequence as follows. Let $P_0 = S^2$ and specify a triangulation τ_0 of P_0. Obtain P_1 by removing each 2-simplex of τ_0 and sewing it back in using the degree q map on its boundary. Let $p_1 : P_1 \to P_0$ be defined so that $p_1^{-1}(P_1^{(1)}) = P_1^{(1)}$ and the restriction of p_1 to this 1-skeleton is the identity and so that the p_1 maps the interior of each resewn 2-simplex to itself. Specify a triangulation τ_1 of P_1. Obtain P_2 by removing each 2-simplex of τ_1 and sewing it back in using a degree q map on its boundary and specify a map $p_2 : P_2 \to P_1$ similarly. Recursively, produce an inverse sequence in this manner choosing the triangulations so that for each k the size of the images $p_{k+1} \circ p_{k+2} \circ \cdots \circ p_j(\sigma)$, for σ a simplex in τ_j, converges to 0 as j goes to infinity. Let X_q denote the inverse limit of such a system.

Pontrjagin introduced the spaces X_q to illustrate that the product of a pair of 2-dimensional compacta may only be three dimensional. While $\dim X_q = 2$ for each q, for distinct primes q and r, $\dim(X_q \times X_r) = 3$. This classical construction serves to illustrate two things. First, while the construction outlined below and appearing in detail in [Wa$_1$] is significantly more complicated, it is one of many variations on the Pontrjagin construction that has proven useful. Second, while the focus of this paper is to distinguish between the covering and cohomological theories of dimension, one of the most important uses of cohomological dimension theory is to compute the covering dimension of finite dimensional spaces. In particular, the computation $\dim(X_q \times X_r) = 3$ is achieved by analyzing its cohomological dimension with respect to various coefficients thereby deducing that its integral cohomological dimension is 3. In turn, the agreement of the theories for finite dimensional spaces reveals that its covering dimension is 3 as well.

For each $n \geq 1$, associated to a pair (P, τ) consisting of a compact polyhedron P and a triangulation τ are a CW-complex $P_{\mathbb{Z},n}$ and a map $\pi : P_{\mathbb{Z},n} \to P$ satisfying:

(a) $\pi^{-1}(C)$ is a subcomplex of $P_{\mathbb{Z},n}$ for each subcomplex $C \subset (P, \tau)$;
(b) $\pi^{-1}(P^{(n)}) = P^{(n)}$ and the restriction of π to $\pi^{-1}(P^{(n)})$ is the identity;
(c) for each $\sigma \in \tau$ with $\dim \sigma \geq n+1$, $\pi^{-1}(\sigma) \simeq K(\pi_n(\sigma^{(n)}), n)$ and the inclusion $\sigma^{(n)} \hookrightarrow \pi^{-1}(\sigma)$ induces an isomorphism of nth homotopy groups; observe that the group $\pi_n(\sigma^{(n)}) \simeq \mathbb{Z}^t$ where the value of t depends on the integer n and the dimension of σ.

For $n = 1$, $P_{\mathbb{Z},1} = P^{(1)}$ and π is the inclusion. If $\dim P \leq n$, then $P_{\mathbb{Z},n} = P$ and π is the identity. If $\dim P = n+1$, then $P_{\mathbb{Z},n}$ is obtained by removing the interior of each n-simplex and replacing it with a $K(\mathbb{Z}, n)$, say by starting with the boundary (n-1)-sphere of the n-simplex and attaching cells to produce a $K(\mathbb{Z}, n)$. The map π can be obtained by mapping the interior of each attached cell to the interior of the n-simplex. Generally, $P_{\mathbb{Z},n}$ is constructed by working inductively through the skeleta of P. For example, if $\dim P = n+2$, first construct $(P^{(n+1)})_{\mathbb{Z},n}$ and the map $\pi : (P^{(n+1)})_{\mathbb{Z},n} \to P^{(n+1)}$ as just described. For each (n+2)-simplex σ in P, check that the inclusion $\sigma^{(n)} \hookrightarrow \pi^{-1}(\sigma^{(n+1)})$ induces isomorphisms of homotopy groups through dimension n. Consequently, cells can be added to $\pi^{-1}(\sigma^{(n+1)})$ to produce a $K(\pi_n(\sigma^{(n)}), n)$ and the map π extended by mapping the interior of each attached cell to the interior of σ. A detailed construction of these $P_{\mathbb{Z},n}$'s can be found in [Wa₁]. While some authors label these complexes as "Walsh complexes", the reference "Edwards-Walsh complexes" more accurately reflects the source of their introduction.

Consider the following recursive construction. Fix an integer n. Let P_0 be a compact polyhedron and let τ_0 be a triangulation. Let $\pi : P_{Z,n} \to P$ be an Edwards-Walsh complex for (P_0, τ_0). Let P_1 be a finite subcomplex of $P_{Z,n}$ and let $\pi_1 : P_1 \to P_0$ be the restriction of π. Specify a triangulation τ_1 of P_1 (at this point it is necessary to state that the CW-complexes encountered are to have the property that finite subcomplexes are polyhedra). Let $\pi : (P_1)_{Z,n} \to P_1$ be an Edwards-Walsh complex for (P_1, τ_1). Let P_2 be a finite subcomplex of $(P_1)_{Z,n}$ and let $\pi_2 : P_2 \to P_1$ be the restriction of π. Let τ_2 be a triangulation of P_2. Generally, P_i is a finite subcomplex of an Edwards-Walsh complex $\pi : (P_{i-1})_{Z,n} \to P_{i-1}$, π_i is the restriction of π, and τ_i is a triangulation of P_i.

Suppose that the triangulations τ_i are specified so that, for each $k \geq 1$ and $\epsilon > 0$, there is an $i \geq k$ so that for each $j \geq i$ and each $\sigma \in \tau_j$ the diameter of $\pi_k \circ \pi_{k-1} \circ \cdots \circ \pi_j(\sigma) < \epsilon$. Then, setting $X = \text{InvLim}\{(P_i, \pi_i)\}$, $\dim_Z X \leq n$. Details of this computation can be found in [Dr₁], [DW₁], or [Wa₁]. The latter reference includes an analyses leading to a characterization of compacta X with $\dim_Z X \leq n$ as limits of inverse sequences of polyhedra produced essentially as described in the preceding paragraph.

In the intervening years between the appearance of [Wa₁] and [Dr₁], the problem was to determine that such a construction could produce a limit X whose covering dimension was not also $\leq n$.

5. EXAMPLES WITH $\dim X = \infty$ AND $\dim_Z X < \infty$

First, let us construct an inverse sequence $\{\pi_i : P_i \to P_{i-1}\}$ of compact polyhedra whose limit X satisfies $\dim_Z X = 2$ but $\dim X > 2$. (Recall that necessarily $\dim X = \infty$.) Since $\pi_k(\Omega^3 S^2) \simeq \pi_{k+3}(S^2)$, there is an essential map $f : S^m \to \Omega^3 S^2$ for some $m \geq 3$. Set $P_0 = S^m$ and let τ_0 be a triangulation of P_0. Consider an Edwards-Walsh complex $\pi : (P_0)_{Z,2} \to P_0$. The structure of Edwards-Walsh complexes recorded as properties (a), (b), (c) in §4 combined with Corollary 1 in §3 permit the use of the Combinatorial Vietoris-Begle Theorem (Lemma 2 in §2) to conclude that $\pi^* : [P_0, \Omega^3 S^2] \to [(P_0)_{Z,2}, \Omega^3 S^2]$ is an isomorphism. In particular, the composition $f \circ \pi$ remains essential. Since $\pi_2(S^2)$ and $\pi_3(S^2)$ are the only infinite homotopy group of S^2, Lemma 6 in §3 insures the existence of a finite subcomplex $P_1 \subset (P_0)_{Z,2}$ such that, setting π_1 equal to the restriction of π, the composition $f \circ \pi_1 : P_1 \to \Omega^3 S^2$ is essential. Let τ_1 be a triangulation of P_1 and repeat the construction. Namely, specify an Edwards-Walsh complex $\pi : (P_1)_{Z,2} \to P_1$, observe that the composition $f \circ \pi_1 \circ \pi$ is essential, and specify a finite subcomplex $P_2 \subset (P_1)_{Z,2}$ such that, setting π_2 equal to the restriction of π, the composition $f \circ \pi_1 \circ \pi_2 : P_2 \to \Omega^3 S^2$ is essential. Evidently, this process can be repeated indefinitely. Specifically, suppose that compact polyhedra and maps

$$P_{i-1} \xrightarrow{\pi_{i-1}} P_{i-2} \xrightarrow{\pi_{i-2}} \cdots \xrightarrow{\pi_1} P_0 \xrightarrow{f} \Omega^3 S^2$$

have been specified so that the composition $f \circ \pi_1 \circ \cdots \circ \pi_{i-1}$ is essential. Choose a triangulation τ_{i-1} of P_{i-1} so that the image of each simplex of τ_{i-1} in each P_j for $j < i-1$

has diameter less than $1/i$. Then $\pi_i : P_i \to P_{i-1}$ can be chosen by restricting to a finite subcomplex the map of an Edwards-Walsh complex $\pi : (P_{i-1})_{\mathbf{Z},2} \to P_{i-1}$ for (P_{i-1}, τ_{i-1}), the finite complex chosen so that the composition $f \circ \pi_1 \circ \cdots \circ \pi_{i-1} \circ \pi_i$ remains essential. Setting $X = \mathrm{InvLim}\{(P_i, \pi_i)\}$, $\dim_{\mathbf{Z}} X \leq 2$ but the induced map $\pi_{\infty,0} : X \to P_O = S^m$ is essential, since its composition $f \circ \pi_{\infty,0}$ is essential, and, hence, $\dim X \geq m \geq 3$. While the equality $\dim_{\mathbf{Z}} X = 2$ is detectable by direct inspection of the construction, it is an immediate consequence of the equivalence $\dim_{\mathbf{Z}} X \leq 1$ iff $\dim X \leq 1$. The equality $\dim X = \infty$ is forced by the equivalence of the theories whenever the covering dimension is finite. This construction produces the example whose existence is asserted in the next result.

Theorem 0. *There is a compact metric space X with $\dim_{\mathbf{Z}} X = 2$ and $\dim X = \infty$.*

The construction just described contains "parameters" that can be varied. Any one of the loop spaces $\Omega^k S^{2n+1}$, where $k \geq 2n + 1$ and $n \geq 1$, or $\Omega^k S^{2n}$, where $k \geq 4n - 1$ and $n \geq 1$, could serve in place of $\Omega^3 S^2$. Recall that the distinction between even dimensional spheres is needed as, for n odd, only $\pi_n(S^n)$ is infinite while, for n even, both $\pi_n(S^n)$ and $\pi_{2n-1}(S^n)$ are infinite. Of course, working with iterated loop spaces whose higher homotopy groups are finite precludes the existence of phantom maps; see Lemma 5 in §3. Specifically, the knowledge that there are no phantom maps is needed to restrict the essential map defined on each Edwards-Walsh complex introduced to a finite subcomplex on which it remains essential. Finally, examples with $\dim_{\mathbf{Z}} X = q$ and $\dim X = \infty$ can be constructed for any $q \geq 2$ by choosing an essential map of S^m where $m > q$ and using Edwards-Walsh complexes $(P_i)_{\mathbf{Z},q}$ at each stage in the construction. The next result encapsulates these variations.

Theorem 1. *Let $n \geq 2$ and let $k \geq n$ if n is odd and $k \geq 2n - 1$ if n is even. For $m > q \geq 2$ and essential map $f : S^m \to \Omega^k S^n$, there is a compact metric space X with $\dim X = \infty$ and $\dim_{\mathbf{Z}} X = q$ for which there is an essential map $\pi : X \to S^m$ such that the composition $\pi \circ f : X \to \Omega^k S^n$ is essential.*

The next result records that the fact that these examples with $\dim_{\mathbf{Z}} X = q$ but $\dim X = \infty$ have their covering dimension detected by essential maps $X \to S^m$ for some $m > q$ which have (possibly, several) suspensions that are also essential. This property has some relation to the notion of the "stable cohomotopy dimension" of a compactum discussed in §7. This contrasts with the construction in $[\mathrm{Dr}_1]$ that produces such maps that evidently have all suspensions essential (see $[\mathrm{Dr}_3$; proof of Theorem 4]).

Theorem 2. *For each $k \geq 1$, $m \geq 3$, and $m > q \geq 2$, there is a compact metric space X with $\dim_{\mathbf{Z}} X = q$ and $\dim X = \infty$ and an essential map $\pi : X \to S^m$ whose k^{th} suspension $\Sigma^k \pi : \Sigma^k X \to S^{m+k}$ is essential.*

Proof. Theorem 1 provides a compactum X with $\dim_{\mathbf{Z}} X = q$ and $\dim X = \infty$ together with maps $\pi : X \to S^m$ and $f : S^m \to \Omega^k S^n$ such that the composition $f \circ \pi$ is essential. Under the natural equivalence $[X, \Omega^k S^n] \simeq [\Sigma X, \Omega^{k-1} S^n]$, the class of $f \circ \pi$ in $[X, \Omega^k S^n]$ corresponds to the class of a map, denoted $(f \circ \pi)^*$, in $[\Sigma X, \Omega^{k-1} S^n]$. Properly interpreted, $(f \circ \pi)^*([(x,t)]) = (f \circ \pi(x))(t)$. Similarly, the class of f in $[S^m, \Omega^k S^n]$ corresponds to the class of a map, denoted f^*, in $[S^{m+1}, \Omega^{k-1} S^n]$, where $S^{m+1} \equiv \Sigma S^m$ and $f^*([(x,t)]) =$

$(f(x))(t)$. For the suspension $\Sigma\pi : \Sigma X \to S^{m+1}$, $\Sigma\pi([(x,t)]) = [(\pi(x),t)]$. The computation $f^* \circ \Sigma\pi([(x,t)]) = f^*([(\pi(x),t)]) = (f \circ \pi(x))(t) = (f \circ \pi)^*([(x,t)])$ detects that $f^* \circ \Sigma\pi = (f \circ \pi)^*$. The essentiality of $(f \circ \pi)^*$ reveals that $f^* \circ \Sigma\pi$ and, hence, $\Sigma\pi$ are essential.

Repeating the argument starting with $f^* \circ \Sigma\pi$ shows that $(f^*)^* \circ \Sigma^2\pi$ and, hence, $\Sigma^2\pi$ are essential. Further repetitions lead to the desired conclusion that $\Sigma^k\pi : \Sigma^k X \to S^{m+k}$ is essential. □

The final result is a refinement providing an example that has "minimal" cohomological dimension and yet has infinite covering dimension. First, the notion of cohomological dimension with respect to an arbitrary abelian group G is needed. Namely, $\dim_G X \le n$ provided, for every map $f : A \to K(G,n)$ of a closed subset $A \subset X$, there is an extension to a map $F : X \to K(G,n)$. Specific groups that are of fundamental interest are $\mathbb{Z}/p$ and $\mathbb{Q}$. First, notice that Corollary 1 is valid with $[K(\mathbb{Z}^t,s),\Omega^k S^n]$ replaced by either $[K(\mathbb{Q}^t,1),\Omega^k S^n]$ (see Lemma 3) or $[K((\mathbb{Z}/p)^t,1),\Omega^k S^n]$ (use the Sullivan Conjecture). Second, there is a notion of Edwards-Walsh complex for these groups and there is a development of Edwards-Walsh complexes for these groups comparable to that given in §4 for $\mathbb{Z}$. A detailed account will be contained in [DW$_2$].

Theorem 3. *There is a compactum X with $\dim_{\mathbb{Z}} X = 2$, $\dim_{\mathbb{Q}} X = 1$, $\dim_{\mathbb{Z}/p} X = 1$ for each prime p, and $\dim X = \infty$. In particular, $\dim_{\mathbb{Z}} X \times X = 3$.*

Outline of Proof. Observe that the conclusion $\dim_{\mathbb{Z}} X \times X = 3$ is a consequence of the Bockstein inequalities (see [Ku]) since $\dim_{\mathbb{Q}} X \times X = 2$ and $\dim_{\mathbb{Z}/p} X \times X = 2$ for each prime, as both $\mathbb{Q}$ and the $\mathbb{Z}/p$'s are fields.

The construction of X follows that described above incorporating the following adjustments. Partition the natural numbers into infinite sets, one for each of the groups $\mathbb{Z}$, $\mathbb{Q}$, and $\mathbb{Z}/p$ for p a prime. Denote by N_G the set of natural numbers associated to each group G. Start with an essential map $f : S^m \to \Omega^3 S^3$ for some $m \ge 3$ and set $P_0 = S^3$. Suppose that $\pi_j : P_j \to P_{j-1}$ has been specified for $j \le i - 1$ according to the following recipe. Denote by G the group amongst $\mathbb{Z}$, $\mathbb{Q}$, and the $\mathbb{Z}/p$'s for which $i \in N_G$. If $G = \mathbb{Z}$, then choose P_i to be a finite subcomplex of an Edwards-Walsh complex $\pi : (P_{i-1})_{\mathbb{Z},2} \to P_{i-1}$ chosen so the, setting π_i equal to the restriction of π, the composition $f \circ \pi_1 \circ \cdots \circ \pi_i$ is essential. If $G \ne \mathbb{Z}$, then choose P_i to be a finite subcomplex of an Edwards-Walsh complex $\pi : (P_{i-1})_{G,1} \to P_{i-1}$ chosen so that, setting π_i equal to the restriction of π, the composition $f \circ \pi_1 \circ \cdots \circ \pi_i$ is essential.

Setting X equal to the limit of such an inverse sequence, the fact that there are Edwards-Walsh complexes of each of the types $(\mathbb{Z},2)$, $(\mathbb{Q},1)$, and $(\mathbb{Z}/p,1)$ for each prime p occurring infinitely often in the construction leads to the conclusions $\dim_{\mathbb{Z}} X = 2$, $\dim_{\mathbb{Q}} X = 1$, $\dim_{\mathbb{Z}/p} X = 1$ for each prime p.

6. DIMENSION OF HOMOLOGY 4–MANIFOLDS AND RELATED PROBLEMS

Recall that map $f : A \to Y$ between compact metric spaces is said to be *cell-like* provided each point-inverse $f^{-1}(y)$ has trivial shape. The latter condition is the requirement that every map $f^{-1}(y) \to P$ is null homotopic for every polyhedron P and, for $\dim f^{-1}(y) < \infty$, is equivalent to there being an embedding $f^{-1}(y) \hookrightarrow \mathbb{R}^n$ for some n such that the image has arbitrarily small neighborhoods homeomorphic to an n–cell. A central question, often referred to as the Cell-Like Mapping Problem, was whether there was a cell-like map f : $X \to Y$ where $\dim X < \infty$ and $\dim Y = \infty$. The classical Vietoris-Begle Mapping Theorem detects $\dim_Z Y \leq \dim_Z X$ and the latter is always $\leq \dim X$. Consequently, the existence of a cell-like dimension raising map implies the existence of an infinite dimensional compactum having finite cohomological dimension. In [Ed], Edwards established the converse: if X is a compact metric space with $\dim_Z X \leq n$, then there is a compact metric space $\widetilde{X}$ with $\dim \widetilde{X} \leq n$ and a cell-like map $f : \widetilde{X} \to X$. (Details of the Edwards construction can be found in [Wa$_1$].)

In turn, the Cell-Like Mapping Problem had direct bearing on the question of whether homology n–manifolds must have finite covering dimension. If $f : X \to Y$ is a cell-like map with $\dim X \leq n$ and $\dim Y = \infty$, then an infinite dimensional homology $(2n + 1)$–manifold can be constructed as follows. Embed $X \hookrightarrow S^{2n+1}$ and form the adjunction space $S^{2n+1} \cup_f Y$. The induced map $S^{2n+1} \to S^{2n+1} \cup_f Y$ is cell-like and, hence, the latter is an homology $(2n + 1)$–manifold but, as it contains Y, is infinite dimensional.

It is a classical result that homology 1–manifolds and 2–manifolds are "honest" manifolds. While there are various constructions of homology 3–manifolds that produce non manifolds, it was shown in [Wa$_2$] that homology 3–manifolds have finite covering dimension. In the presence of the discussion above, the examples of compacta X with $\dim_Z X = 3$ and $\dim X = \infty$ in [Dr$_1$] can be used to produce homology 7–manifolds that are infinite dimensional. Similarly, the examples in [DW$_1$] described above in §5 of compacta X with $\dim_Z X = 2$ and $\dim X = \infty$ provide examples of homology 5–manifolds that are infinite dimensional. The situation in dimension four remains unsettled.

Question 1. Is there a cell-like dimension raising map defined on a 4–manifold?

Question 2. Is there an acyclic dimension raising map defined on a 4–manifold?

Question 3. Is there an infinite dimensional homology 4–manifold?

A few words are in order concerning the notion of homology n–manifold. There is the general class defined in terms of Borel-Moore homology, denoted $H(X; \mathbb{Z})$, and a restricted class defined in terms of singular homology, denoted $_sH(X, \mathbb{Z})$. Generally, a locally compact metric space X is a homology n–manifold provided: (i) X is hlc$^\infty$, i.e., locally homologically connected with respect to Borel-Moore homology; (ii) $\dim_Z X = n$; and (iii) the local homology with compact supports $H_*^c(X, X \backslash \{x\}) \simeq H_*^c(\mathbb{R}^n, \mathbb{R}^n \backslash \{0\})$. If the same definition is stated using singular homology (which is compactly supported by definition), one gets a restricted notion of homology n–manifold. For example, the image of a cell-like map defined on a manifold is an homology manifold of the latter type. While the image of an acyclic map (i.e., the reduced Čech cohomology of each point-inverse vanishes in each dimension) defined on a manifold may only be a homology manifold of the former type.

7. DIMENSION AND GENERALIZED COHOMOLOGY THEORIES

In [Dr$_3$], Dranishnikov analyses dimension defined with respect to an arbitrary generalized cohomology theory. Not all such theories give rise to "reasonable" dimension theories. Periodic theories such as both real and complex K-theory cause difficulties. Nevertheless, there is a class of suitable theories identified in [Dr$_3$].

Question 4. Is there a generalized cohomology theory whose associated dimension theory agrees with the covering theory of dimension?

S. Nowak proposed the "stable cohomotopy dimension" as a possible candidate. The following is a brief description of the generalized cohomology theory that determines this dimension theory. For additional details and references, the reader can consult [Dr₃].

Nominally, a spectrum is a family of pointed spaces $\{Y_n : n \in \mathbb{Z}\}$ and base point preserving maps $e_i : \Sigma Y_i \to Y_{i+1}$. The spheres form a "natural" spectrum: for $n < 0$, $Y^n = \{point\}$ and e_n is the inclusion and for $n \geq 0$, $Y_n = S^n$ and $e_i : \Sigma S^n \equiv S^{n+1}$. There are two ways to describe the generalized cohomology theory associated to the sphere spectrum. The first is to define $S^n(X) = \mathrm{DirLim}_k\{[\Sigma^k X, S^{n+k}]\}$ where the map $[\Sigma^k X, S^{n+k}] \to [\Sigma^{k+1} X, S^{n+k+1}]$ is induced by suspension. The second is to replace the sphere spectrum by an Ω-spectrum (one with $Y_n \simeq \Omega Y_{n+1}$) and proceed as discussed in §2. Namely, set $\mathcal{Y}_n = \mathrm{DirLim}_k\{\Omega^k S^{n+k}\}$ where the map $\Omega^k S^{n+k} \to \Omega^{k+1} S^{n+k+1}$ is $\Omega^k(\tilde{e}_{n+k})$ where $\tilde{e}_{n+k} : S^{n+k} \to \Omega S^{n+k+1}$ is corresponds to e_{n+k} under the equivalence $[\Sigma S^{n+k}, S^{n+k+1}] \simeq [S^{n+k}, \Omega S^{n+k+1}]$. The family $\{\mathcal{Y}_n : n \in \mathbb{Z}\}$ forms an Ω-spectrum and $S^n(X) = [X, \mathcal{Y}_n]$ as well.

The *stable cohomotopy dimension* of a space X, denoted $\dim_S X$, is that determined by the theory associated to the sphere spectrum. Disentangling these involved developments leads to: for a compactum X, $\dim_S X \leq n$ provided, for every map $f : A \to S^n$ of a closed subset $A \subset X$, there is a $k \geq 0$ so that the suspension $\Sigma^k f : \Sigma^k A \to S^{n+k}$ extends to a map $F : \Sigma^k X \to S^{n+k}$. The reader might be interested in showing that $\dim_S X \leq n$ implies that $\dim_S X \leq n + 1$.

In any case, the sphere spectrum is one which is covered by results in [Dr₃] that yield the estimates $\dim_{\mathbb{Z}} X \leq \dim_S X \leq \dim X$. In particular, the theories coincide for the class of spaces having finite covering dimension. The construction in [Dr₁] is used in [Dr₃; Theorem 4] to produce a compactum X with $\dim_{\mathbb{Z}} X = 3$ but $\dim X = \dim_S X = \infty$.

Question 5. (S. Nowak) Does $\dim_S X = \dim X$ for all compact metric spaces X?

In contrast with the examples in [Dr₁] that have maps to spheres for which it is easy to detect that all the suspensions are essential, those from [DW₁] described above in §5 only come equipped with maps to spheres with a certain number of suspensions known to be essential.

Question 6. Is there a compact metric space X with $\dim_{\mathbb{Z}} X = 2$ but $\dim_S X = \infty$?

REFERENCES

[Al]   P. S. Alexandroff, *Dimensionstheorie, ein Beitrag zur Geometrie der abgeschlossenen Mengen*, Math. Ann. **106** (1932), 161–238.

[AH]   D. W. Anderson and L. Hodgkin, *The K-theory of Eilenberg-MacLane complexes*, Topology **7** (1968), 317–329.

[Bo]   P. L. Bowers, *General position properties satisfied by finite products of dendrites*, Trans. Amer. Math. Soc. **288** (1985), 739–753.

[BM]   V. M. Buchstaber and A. S. Mischenko, *K-Theory on the category of infinite cell complexes*, Mathematics of the USSR Izvestija **32 No.2** (1968), 560–604.

[Dr₁]   A. N. Dranishnikov, *On a problem of P. S. Alexandroff*, Matem. Sbornik **135** (1988), 551–556.

[Dr₂]   ――――, *K-theory of Eilenberg-MacLane spaces and cell-like problem*, preprint.

[Dr₃]   ――――, *Generalized cohomological dimension of compact metric spaces*, preprint.

[DK]   J. Dydak and G. Kozlowski, *Vietoris-Begle theorem and spectra*, to appear, Proc. Amer. Math. Soc..

[DW₁]J. Dydak and J. J. Walsh, *Infinite dimensional compacta having cohomological dimension two: an application of the Sullivan conjecture*, preprint.

[DW2]_____, *Aspects of cohomological dimension theory for principal ideal domains*, in preparation (title tentative).

[Ed] R. D. Edwards, *A theorem and question related to cohomological dimension and cell-like mappings*, Notices of the Amer. Math. Soc. **25**, A-259–A-260.

[HW] W. Hurewicz and H. Wallman, *Dimension Theory*, Princeton University Press, Princeton, NJ, 1948.

[Ko] Y. Kodama, *Note on absolute neighborhood extensor for metric spaces*, Math. Soc. of Japan **8** (1956), 206–215.

[Ku] W. I. Kuzminov, *Homological dimension theory*, Russian Math. Surveys **23** (1968), 1–45.

[Mi] H. Miller, *The Sullivan conjecture on maps from classifying spaces*, Annals of Mathematics **120** (1984), 39–87.

[Mil] J. W. Milnor, *An axiomatic homology theory*, Pacific J. of Math. **12** (1962), 337–341.

[RS] L. R. Rubin and P. J. Schapiro, *Cell-like maps onto noncompact spaces of finite cohomological dimension*, Topology and Its Appl. **27** (1987), 221–244.

[Sp] E. Spanier, *Algebraic Topology*, McGraw-Hill, New York, NY, 1966.

[Ta] J. L. Taylor, *A counterexample in shape theory*, Bull. Amer. Math. Soc. **81** (1975), 629–632.

[Wa1] J. J. Walsh, *Dimension, cohomological dimension, and cell-like mappings*, Lecture Notes in Mathematics Vol. 870, Springer-Verlag, New York, NY, 1981, pp. 105–118.

[Wa2] _____, *The finite dimensionality of integral homology 3-manifolds*, Proc. Amer. Math. Soc. **88** (1983), 154–156.

[Wh] George W. Whitehead, *Elements of Homotopy Theory*, Graduate Texts in Mathematics, Springer-Verlag, New York, NY, 1978.

[Za] A. Zabrodsky, *On phantom maps and the theorem of H. Miller*, Israel Journal of Math. **870** (1987), 129–143.

UNIVERSITY OF TENNESSEE, KNOXVILLE, TN 37996 AND UNIVERSITY OF CALIFORNIA, RIVERSIDE, CA 92521

E-mail address: walsh@ucrmath.ucr.edu

GENERIC PROPERTIES OF FIRST ORDER
PARTIAL DIFFERENTIAL EQUATIONS

S. Izumiya

Hokkaido University

Dedicated to Professor Noboru Tanaka on his 60th birthday

ABSTRACT. We consider some generic properties of first order partial differential equations for real-valued functions. The main aim is the study of behaviours of solutions at a point of the discriminant set.

0. INTRODUCTION

In this paper we will describe some generic properties of first order partial differential equations. Our first result asserts that almost all first order partial differential equations have no singular solutions and the discriminant set consists of singularities of solutions. (cf. Theorem 2.2). In the case of ordinary differential equations, this result has been observed by many people. (Darboux [3], Dyck [4], Thom [11], Dara [2] and Fukuda [5]). But their methods depend on the classification theory of mappings from the plane to the plane. In our case we can not apply the classification theory of stable mappings without the nice range in the sense of Mather. Moreover, we will treat first order partial differential equations with singular solution under a certain generic condition. We can prove that such an equation has a complete solution and the singular solution is an envelope of such a family of solutions. (cf. Corollary 3.3).

All map germs and diffeomorphisms considered here are differentiable of class C^∞, unless stated otherwise.

1. FORMULATIONS AND PRELIMINARIES

The aim of this section is to describe the structures connected with non-linear first order equations.

Let $J^1(\mathbb{R}^n, \mathbb{R})$ be the 1-jet bundle of functions $f : \mathbb{R}^n \to \mathbb{R}$. Since we only consider the local situation, the jet bundle $J^1(\mathbb{R}^n, \mathbb{R})$ may be considered as $\mathbb{R}^{2n+1}$ with natural coordinates given by

$$(x_1, \ldots, x_n, z, p_1, \ldots, p_n).$$

We have the canonical projection $\pi : J^1(\mathbb{R}^n, \mathbb{R}) \to \mathbb{R}^n \times \mathbb{R}$, $\pi(x, z, p) = (x, z)$.

Let θ be the canonical contact form on $J^1(\mathbb{R}^n, \mathbb{R})$ which is given by $\theta = dz - \sum_{i=1}^n p_i dx_i$. Throughout the remainder of this paper, we shall consider $J^1(\mathbb{R}^n, \mathbb{R})$ as a contact manifold whose contact structure is given by the contact form θ. Using this approach, a first order partial differential equation is most naturally interpreted as being a closed subset of $J^1(\mathbb{R}^n, \mathbb{R})$. Unless the contrary is specifically stated, we use the following definition. *A first order partial differential equation germ (or, briefly an equation)* is a smooth submersion germ $F : (J^1(\mathbb{R}^n, \mathbb{R}), (x_0, z_0, p_0)) \to (\mathbb{R}, 0)$.

The notion of solutions of an equation is given by the philosophy of Lie. A submanifold $i : L \subset J^1(\mathbb{R}^n, \mathbb{R})$ is said to be *a Legendrian submanifold* if $\dim L = n$ and $i^*\theta = 0$. A *solution* of $F = 0$ is a Legendrian submanifold germ $i : (L, q_0) \subset J^1(\mathbb{R}^n, \mathbb{R})$ such that $i(L) \subset F^{-1}(0)$. We can show that a solution $i : (L, q_0) \subset J^1(\mathbb{R}^n, \mathbb{R})$ is given by a jet extension $j^1 f$ of a function germ f if and only if $\pi \circ i$ is a non-singular map.

We now define the notion of singularities of solutions and equations. Let $i : (L, q_0) \subset J^1(\mathbb{R}^n, \mathbb{R})$ be a Legendrian submanifold germ. We say that q_0 is *a Legendrian singular point* if q_0 is a singular point of $\pi \circ i$. Let $F : (J^1(\mathbb{R}^n, \mathbb{R}), q_0) \to (\mathbb{R}, 0)$ be an equation. Then q_0 is *a contact singular point* if

$$F = F_{p_1} = \cdots = F_{p_n} = F_{x_1} + p_1 F_z = \cdots = F_{x_n} + p_n F_z = 0$$

at q_0, where $F_{p_i} = \frac{\partial F}{\partial p_i}$, $F_{x_i} = \frac{\partial F}{\partial x_i}$, $F_z = \frac{\partial F}{\partial z}$ for $i = 1, \ldots, n$. We also say that q_0 is *a π-singular point* if

$$F = F_{p_1} = \cdots = F_{p_n} = 0.$$

It is easy to show that (x, z, p) is a contact singular point if and only if

$$\theta(T_{(x,z,p)} F^{-1}(0)) = 0.$$

Let $\Sigma(F)$ be the set of π-singular points. We say that $D_F = \pi(\Sigma(F))$ is *the discriminant set* of the equation $F = 0$. The notion of a singular solution is as follows. If the π-singular set $\Sigma(F)$ is a Legendrian submanifold, then we call it *a singular solution* of $F = 0$. In this case, the discriminant set D_F is the graph of the singular solution.

Since the solution is a Legendrian submanifold of $J^1(\mathbb{R}^n, \mathbb{R})$, we also need an explicit characterization of the isotropic points of an n-dimensional submanifold in $J^1(\mathbb{R}^n, \mathbb{R})$.

Lemma 1.1. *Suppose that $(f_1, \ldots, f_{n+1})$ is a submersion on $J^1(\mathbb{R}^n, \mathbb{R})$ and N is an n-dimensional submanifold which is given by equations $f_1 = \cdots = f_{n+1} = 0$. Then N is isotropic at $(x_0, z_0, p_0) \in N$ (i.e. $\theta(T_{(x_0, z_0, p_0)} N) = 0$) if and only if*

$$\mathrm{rank} \begin{pmatrix} f_{jx_i} + p_i f_{jz} \\ f_{jp_i} \end{pmatrix} < n+1$$

at (x_0, z_0, p_0).

The proof is given by the fact that N is isotropic at (x_0, z_0, p_0) if and only if there exists $(\lambda_1, \ldots, \lambda_{n+1}) \neq (0, \ldots, 0)$ in $\mathbb{R}^{n+1}$ such that

$$\theta \wedge (\lambda_1 df_1 + \cdots + \lambda_{n+1} df_{n+1}) = 0$$

at (x_0, z_0, p_0).

By this lemma, we have a characterization of the singular solution. Let $F : (J^1(\mathbb{R}^n, \mathbb{R}), (x_0, z_0, p_0)) \to (\mathbb{R}, 0)$ be an equation. Since $\Sigma(F)$ is given by equations $F = F_{p_1} = \cdots = F_{p_n} = 0$, we have the following corollary.

Corollary 1.2. *Suppose that* $(F, F_{p_1}, \ldots, F_{p_n})$ *is a submersion at* (x_0, z_0, p_0). *Then* $\Sigma(F)$ *is isotropic at* (x_0, z_0, p_0) *if and only if*

$$\operatorname{rank} \begin{pmatrix} F_{x_i} + p_i F_z & F_{p_i x_j} + p_j F_{p_i z} \\ 0 & F_{p_i p_j} \end{pmatrix} < n + 1$$

at (x_0, z_0, p_0), *where* $\ell < k$.

The Legendre transformation is the most famous contact diffeomorphism. Let

$$(x_1, \ldots, x_n, z, p_1, \ldots, p_n)$$

be a coordinate system of $\mathbb{R}^{2n+1}$ with the contact structure given by $\theta = dz - \sum_{i=1}^n p_i dx_i$. We adopt another coordinate system $(X_1, \ldots, X_n, Z, P_1, \ldots, P_n)$ of $\mathbb{R}^{2n+1}$ whose contact structure is given by $\Theta = dZ - \sum_{i=1}^n P_i dX_i$. We now define a diffeomorphism $*L : \mathbb{R}^{2n+1} \to \mathbb{R}^{2n+1}$ by

$$X_1 = p_1, \ldots, X_n = p_n, Z = \sum_{i=1}^n x_i p_i - z, P_1 = x_1, \ldots, P_n = x_n.$$

It is easy to see that $*L$ is a contact diffeomorphism

$$*L : (\mathbb{R}^{2n+1}, \theta) \to (\mathbb{R}^{2n+1}, \Theta).$$

We call $*L$ *the Legendre transformation.* Let $F : (\mathbb{R}^{2n+1}, (x_0, z_0, p_0)) \to (\mathbb{R}, 0)$ be an equation. If we apply the Legendre transformation $*L$ to our hypersurface $F = 0$ we obtain a new hypersurface

$$G(X_1, \ldots, X_n, Z, P_1, \ldots, P_n) = F(P_1, \ldots, P_n, \sum_{i=1}^n P_i X_i - Z, X_1, \ldots, X_n).$$

If we calculate partial differentials at the point $(X_0, Z_0, P_0,)$ corresponding to (x_0, z_0, p_0), we can obtain the following:

$$G_{P_i}(X_0, Z_0, P_0) = (F_{x_i} + p_i F_z)(x_0, z_0, p_0),$$
$$G_Z(X_0, Z_0, P_0) = -F_z(x_0, z_0, p_0),$$
$$G_{X_i}(X_0, Z_0, P_0) = (F_{p_i} + x_i F_z)(x_0, z_0, p_0)$$

for $i = 1, \ldots, n$.

It follows from these formulas that we obtain the following proposition.

Proposition 1.3. *Let* $F : (\mathbb{R}^{2n+1}, (x_0, z_0, p_0)) \to (\mathbb{R}, 0)$ *be an equation. Then we have*

(1) $F_z \neq 0$ *at* (x_0, z_0, p_0) *if and only if* $G_Z \neq 0$ *at* (X_0, Z_0, P_0).

(2) *Suppose that* $F^{-1}(0)$ *is* π-singular *at* (x_0, z_0, p_0). *Then it is contact regular if and only if* $G^{-1}(0)$ *is* π-regular *at* (X_0, Z_0, P_0).

(3) *If* $F^{-1}(0)$ *is contact singular at* (x_0, z_0, p_0) *then* $G^{-1}(0)$ *is* π-singular *at* (X_0, Z_0, P_0).

2. First order Partial Differential Equations of general type

In this section we will determine the framework of the class of first order partial differential equations which will be studied in the remainder of this paper. Let

$$F : (J^1(\mathbb{R}^n, \mathbb{R}), (x_0, z_0, p_0)) \to (\mathbb{R}, 0)$$

be an equation. Then we say that $F = 0$ is *an equation of general type* if $\Sigma(F)$ is a smooth submanifold and fold points of $\pi_{|F^{-1}(0)} : F^{-1}(0) \to \mathbb{R}^n \times \mathbb{R}$ are dense in $\Sigma(F)$ in some neighbourhood of the point (x_0, z_0, p_0). Here, $(x, z, p) \in \Sigma(F)$ is a *fold point of* $\pi_{|F^{-1}(0)}$ if $\det(F_{p_i p_j}) \neq 0$ at (x, z, p).

We shall now consider the genericity of the above properties. We adopt the notion of genericity as follows. Let U be an open set in $J^1(\mathbb{R}^n, \mathbb{R})$ and $C^\infty(U, \mathbb{R})$ be the space of smooth functions on U with the Whitney C^∞ topology. We denote by $\mathcal{R}(U, \mathbb{R})$ the subset of $C^\infty(U, \mathbb{R})$ which consists of smooth functions on U such that $0 \in \mathbb{R}$ is regular value of these. We may consider that $\mathcal{R}(U, \mathbb{R})$ is the space of equations on U. A subset of $\mathcal{R}(U, \mathbb{R})$ is called *generic* if it is an open and dense subset in $\mathcal{R}(U, \mathbb{R})$. We say that *almost all equations* $F : (J^1(\mathbb{R}^n, \mathbb{R}), (x_0, z_0, p_0)) \to (\mathbb{R}, 0)$ *have a property* P if for some neighbourhood U the set

$$\mathcal{P}(U) = \{G \in \mathcal{R}(U, \mathbb{R}) \mid \text{The germ } G : (U, (x, z, p)) \to (\mathbb{R}, 0)$$
$$\text{has the property } P \text{ for any point } (x, z, p) \in U\}$$

is generic in $\mathcal{R}(U, \mathbb{R})$. Then we have the following proposition.

Proposition 2.1. *Almost all equations*

$$F : (J^1(\mathbb{R}^n, \mathbb{R}), (x_0, z_0, p_0)) \to (\mathbb{R}, 0)$$

are general types.

Proof. We now define a subset Σ of $J^2(J^1(\mathbb{R}^n, \mathbb{R}), \mathbb{R})$ by

$$\{j^2 f(x, z, p) \mid f(x, z, p) = f_{p_1}(x, z, p) = \cdots = f_{p_n}(x, z, p) = 0\}.$$

Then Σ is a linear subspace of $J^2(J^1(\mathbb{R}^n, \mathbb{R}), \mathbb{R})$. We also define a subset Σ_1 of $J^2(J^1(\mathbb{R}^n, \mathbb{R}), \mathbb{R})$ by

$$\{j^2 f(x, z, p) \in \Sigma \mid \det(f_{p_i p_j}) = 0\}.$$

Then Σ_1 is a proper algebraic subset of Σ. Hence $\Sigma - \Sigma_1$ is a open, dense subset of Σ.

Let $F : (J^1(\mathbb{R}^n, \mathbb{R}), (x_0, z_0, p_0)) \to (\mathbb{R}, 0)$ be an equation. The set germ $j^2 F^{-1}(\Sigma - \Sigma_1)$ consists of fold points of F. If $j^2 F$ is transverse to $\Sigma - \Sigma_1$ and "the stratified set" Σ_1, then F is a general type. This completes the proof.

Our first theorem is the following.

Theorem 2.2. *Almost all equations*

$$F : (J^1(\mathbb{R}^n, \mathbb{R}), (x_0, z_0, p_0)) \to (\mathbb{R}, 0)$$

have no singular solutions and the π-singular set germ Σ consists of Legendrian singular points except at most the point (x_0, z_0, p_0).

For the proof of the above theorem, we need some properties of equations of general type with singular solution.

Proposition 2.3. *Let $F : (J^1(\mathbb{R}^n, \mathbb{R}), (x_0, z_0, p_0)) \to (\mathbb{R}, 0)$ be an equation of general type with singular solution, then the equation is contact singular at any $(x, z, p) \in \Sigma(F)$.*

Proof. By the definition, on a dense subset of $\Sigma(F)$, $\det(F_{p_i p_j}) \neq 0$. It follows that

$$\mathrm{rank} \begin{pmatrix} F_{x_i} + p_i F_z & * \\ 0 & F_{p_i p_j} \end{pmatrix} \geq n$$

on such a subset of $\Sigma(F)$.

By Corollary 1.2, we have

$$F_{x_i} + p_i F_z = 0 \qquad (i = 1, \ldots, n, 1 \leq k < \ell \leq n)$$

on such a point. By the density of the fold points in $\Sigma(F)$, these equalities are satisfied on all the point of $\Sigma(F)$. Then the equation is contact singular at such a point.

The following corollary is not used for the proof of Theorem 2.2, but it will be useful in the later section.

Corollary 2.4. *If $F : (J^1(\mathbb{R}^n, \mathbb{R}), (x_0, z_0, p_0)) \to (\mathbb{R}, 0)$ is an equation of general type with singular solution, then $F_z \neq 0$ at (x_0, z_0, p_0).*

Proof. By the above proposition, we have

$$F = F_{p_1} = \cdots = F_{p_n} = F_{x_1} + p_1 F_z = \cdots = F_{x_n} + p_n F_z = 0$$

at (x_0, z_0, p_0). If $F_z = 0$ at (x_0, z_0, p_0), then

$$F = F_{p_1} = \cdots = F_{p_n} = F_{x_1} = \cdots = F_{x_n} = 0$$

at (x_0, z_0, p_0). This contradicts the fact that $\mathrm{grad}\, F \neq 0$ at (x_0, z_0, p_0).

We can prove Theorem 2.2.

Proof of Theorem 2.2. We now define a subset of $J^1(J^1(\mathbb{R}^n, \mathbb{R}), \mathbb{R})$ by

$$A = \{j^1 f(x, z, p) \mid f(x, z, p) = f_{p_1}(x, z, p) = \cdots = f_{p_n}(x, z, p)$$
$$= (f_{x_1} + p_1 f_z)(x, z, p) = \cdots = (f_{x_n} + p_n f_z)(x, z, p) = 0\}.$$

Then A is a submanifold of codim $A = 2n + 1$.

Let $F : (J^1(\mathbb{R}^n, \mathbb{R}), (x_0, z_0, p_0)) \to (\mathbb{R}, 0)$ be an equation of general type such that $j^1 F$ is transverse to A at (x_0, z_0, p_0). Since

$$\dim J^1(\mathbb{R}^n, \mathbb{R}) = 2n + 1$$

the image of $J^1 F$ and A intersect at most at discrete set of points. By Proposition 2.3, such a point is only a point on which $\Sigma(F)$ is isotropic. Then the π-singular set $\Sigma(F)$ cannot be a Legendrian submanifold.

For the latter half of the statement, we may assume that $\Sigma(F)$ is non-isotropic at (x, z, p). It follows that there exists a vector $v \in T_{(x,z,p)}\Sigma(F)$ such that $\theta(v) \neq 0$. By Proposition 1.3 (2) and the classical existence theorem of solutions, there is a Legendrian submanifold $L \subset F^{-1}(0)$ through (x, z, p). By the definition, v is not contained in $T_{(x,z,p)}L + Ker \, d\pi$. If $\pi_{|L}$ is non-singular at (x,z,p), then $\dim d\pi(T_{(x,z,p)}L) = n$. Since $\Sigma(F)$ and L are submanifolds of $F^{-1}(0)$, then $\langle v \rangle_{\mathbb{R}} + T_{(x,z,p)}L$ is a linear subspace of $T_{(x,z,p)}F^{-1}(0)$. Applying $d\pi$ to both vector spaces, then we have

$$d\pi(T_{(x,z,p)}F^{-1}(0)) = T_{(x,z,p)}\mathbb{R}^n \times \mathbb{R}.$$

This contradicts the fact that (x, z, p) is an element of $\Sigma(F)$. This completes the proof.

Theorem 2.2 is a first step in the study of singularities of equations and solutions. We remark that we can prove a more detailed assertion as follows:

If we define a subset B of $J^2(J^1(\mathbb{R}^n, \mathbb{R}), \mathbb{R}))$ by

$$\{j^2 f(x, z, p) \mid f(x, z, p) = f_{p_1}(x, z, p) = \cdots = f_{p_n}(x, z, p)$$
$$= (f_{x_1} + p_1 f_z)(x, z, p) = \cdots = (f_{x_n} + p_n f_z)(x, z, p) = \det(f_{p_i p_j}) = 0\}.$$

Then B is an algebraic subset of $\dim B = 2n + 2$. Hence, we can generically avoid this set. It follows that the following three cases are generic:

(1) (x_0, z_0, p_0) is a π-regular point,
(2) (x_0, z_0, p_0) is a fold point of $\pi \mid F^{-1}(0)$ and an isolated contact singular point,
(3) (x_0, z_0, p_0) is a π-singular point and a contact regular point.

We have some examples which suggest the above result.

Example 2.5. Consider the case where $n = 2$. The following are defined near the origin $0 \in \mathbb{R}^5 = J^1(\mathbb{R}^2, \mathbb{R})$.

(1) The origin is a fold point of $\pi_{|F^{-1}(0)}$ and an isolated contact singular point:

$$F(x_1, x_2, z, p_1, p_2) = p_1^2 + p_2^2 + x_1^2 + x_2^2 - z.$$

(2) The origin is a fold point of $\pi_{|F^{-1}(0)}$ and a contact regular point:

$$F(x_1, x_2, z, p_1, p_2) = p_1^2 + p_2^2 + x_1 + x_2 p_2 + z p_1.$$

(3) The origin is a cusp point of $\pi_{|F^{-1}(0)}$ and a contact regular point:

$$F(x_1, x_2, z, p_1, p_2) = p_1^3 + p_2^2 + zp_1 + x_1.$$

(4) The origin is a swallowtail point of $\pi_{|F^{-1}(0)}$ and a contact regular point:

$$F(x_1, x_2, z, p_1, p_2) = p_1^4 + p_2^2 + zp_1^2 + x_1p_1 + x_2.$$

Since equations (2), (3) and (4) are contact regular at the origin, they have complete integrals. By Theorem 2.2, these complete integrals have Legendrian singularities at a point of $\Sigma(F)$. We will develop classification theories of these Legendrian singularities in forthcoming papers ([7], [9]).

3. First Order Partial Differential Equations with Singular Solutions

In this section we study properties of equations of general type with singular solution. By Corollary 2.4 and the implicit function theorem, if

$$F : (J^1(\mathbb{R}^n, \mathbb{R}), (x_0, z_0, p_0)) \to (\mathbb{R}, 0)$$

is an equation of general type with singular solution, then there exists a function germ $h : (T^*\mathbb{R}^n, (x_0, p_0)) \to \mathbb{R}$ such that $F(x, z, p) = z - h(x, p)$. In the terminology of Kossowski [10], the equation of the above form is called *a graphlike equation*. We now define a map germ

$$\mathrm{graph}(h) : (T^*\mathbb{R}^n, (x_0, p_0)) \to (J^1(\mathbb{R}^n, \mathbb{R}), (x_0, z_0, p_0))$$

by

$$\mathrm{graph}(h)(x, p) = (x, h(x, p), p).$$

We set a 1-form on $T^*\mathbb{R}^n$ by

$$\theta_h = \mathrm{graph}(h)^*\theta = dh - \sum_{i=1}^{n} p_i dx_i.$$

Then we have the following one to one correspondence.

$$\{L \mid L \text{ is a solution of } z - h(x, p) = 0\}$$
$$\mathrm{graph}(h) \uparrow \quad \downarrow \Pi$$
$$\{L \mid i : L \subset T^*\mathbb{R}^n \text{ is a maximal integral submanifold of } \theta_h = 0\}$$

By this reason, a solution of graphlike equation $z - h(x, p) = 0$ regarded as a maximal isotropic submanifold of $(T^*\mathbb{R}^n, \theta_h)$. Since $-d\theta_h = \sum_{i=1}^n dp_i \wedge dx_i$ is the canonical symplectic two form, then a solution of $z - h(x, p) = 0$ is a Lagrangian submanifold of $(T^*\mathbb{R}^n, \omega)$, where $\omega = -d\theta_h$. For the definition and properties of Lagrangian submanifolds, see [1]. Hence, in our case,

$$\Sigma(h) = \{(x, p) \mid h_{p_1} = \cdots = h_{p_n} = 0\}$$

is a Lagrangian submanifold of $(T^*\mathbb{R}^n, \omega)$. We now refer the following very important result.

Theorem 3.1. *(Kostant-Sternberg [6]) Let (P, ω) be a symplectic manifold, L a Lagrangian submanifold and α a smooth 1-form on P with $\alpha_{|L} = 0$ and $d\alpha = \omega$. Then there exists a tubular neighbourhood V of L in P, and a unique vector bundle isomorphism $K : V \to (T^*L, \theta_L)$ such that K is the identity on L and $K^*\theta_L = \alpha$. Here, θ_L is the canonical 1-form on T^*L.*

Applying this result to our situation, we have the following existence theorem of solutions.

Theorem 3.2. *Let $F : (J^1(\mathbb{R}^n, \mathbb{R}), (x_0, z_0, p_0)) \to (\mathbb{R}, 0)$ be an equation of general type with singular solution. Then there exists a foliation by solutions on $F^{-1}(0)$ such that*

(1) *Leaves are transverse to $\Sigma(F)$.*
(2) *Leaves are Legendrian non-singular at points of $\Sigma(F)$.*

Proof. By the above argument, there exists a function germ $h : (T^*\mathbb{R}^n, 0) \to \mathbb{R}$ such that $F(x, z, p) = z - h(x, p)$. By the assumption, $\Sigma(h)$ is a Lagrangian submanifold of $(T^*\mathbb{R}^n, \omega)$. We may apply the Kostant-Sternberg theorem to conclude that there exists a tubular neighbourhood V of $\Sigma(h)$ in $T^*\mathbb{R}^n$ and a unique vector bundle isomorphism $K : V \to (T^*\Sigma(h), \theta_{\Sigma(h)})$ such that K is the identity on $\Sigma(h)$ and $K^*\theta_{\Sigma(h)} = -\theta_h$. Since the fibre of the cotangent bundle $T^*\Sigma(h) \to \Sigma(h)$ is a maximal integral submanifold of $\theta_h = 0$, these fibres make a foliation whose leaves are solutions of $F^{-1}(0)$ and transverse to $\Sigma(F)$. This completes the proof of (1).

Without loss of generality we may assume that $\Sigma(F)$ is parametrized by the local coordinate $(p_1, \ldots, p_n)$ near p_0. (See the proof of Theorem 3.1.3 in [8]). We now apply the Legendre transformation $*L$ to our hypersurface $F = 0$. By Propositions 1.2 and 2.3, we have $*L(\Sigma(F)) = \Sigma(G)$, where

$$G(X_1, \ldots, X_n, Z, P_1, \ldots, P_n) = F(P_1, \ldots, P_n, \sum_{i=1}^n P_i X_i - Z, X_1, \ldots, X_n).$$

The assumption that $\Sigma(F)$ is parametrized by $(p_1, \ldots, p_n)$ means that $\Sigma(G)$ is Legendrian non-singular near the point (X_1, Z_1, P_1) corresponding to (x_1, z_1, p_1). It is equivalent to the fact that (X_1, Z_1, P_1) is a fold type singular point of $\Pi_{|G^{-1}(0)}$, where $\Pi(X, Z, P) = (X, Z)$. It follows from ([10], Corollary 3.2) that the solution of $G^{-1}(0)$ has a Morse type generating function germ. The original solution of $F^{-1}(0)$ is a pullback of this solution by the Legendre transformation. Then this is Legendrian non-singular. This completes the proof.

We say that an n-parameter family of function germs

$$f : (\mathbb{R}^n \times \mathbb{R}^n, (x_0, t_0)) \to (\mathbb{R}, 0)$$

is *a complete solution* of the equation

$$F : (J^1(\mathbb{R}^n, \mathbb{R}), (x_0, z_0, p_0)) \to (\mathbb{R}, 0)$$

if

$$F(x_1, \ldots, x_n, f(x,t), f_{x_1}(x,t), \ldots, f_{x_n}(x,t)) = 0$$

and

$$\text{rank}(f_{t_i}, f_{t_i x_j}) = n.$$

Let $f : (\mathbb{R}^n \times \mathbb{R}^n, (x_0, t_0)) \to (\mathbb{R}, 0)$ be a complete solution of $F = 0$. We now define an immersion germ

$$j_*^1 f : (\mathbb{R}^n \times \mathbb{R}^n, (x_0, t_0)) \to J^1(\mathbb{R}^n, \mathbb{R})$$

by

$$j_*^1 f(x,t) = j^1 f_t(x), \text{where } f_t(x) = f(x,t).$$

Since $\dim F^{-1}(0) = 2n$, then the above map $j_*^1 f$ gives a local parametrization of $F^{-1}(0)$ and $j_*^1 f_{|\mathbb{R}^n \times t}$ is a Legendrian immersion germ for any $t \in (\mathbb{R}^n, t_0)$. By this fact, we have the following assertion.

Corollary 3.3. *Let $F : (J^1(\mathbb{R}^n, \mathbb{R}), (x_0, z_0, p_0)) \to (\mathbb{R}, 0)$ be an equation of general type with singular solution. Then it has a complete solution and $\Sigma(F)$ is an envelope of such a family of solutions.*

We now present two examples which suggest the above result.

1) *The Clairaut's equation.* The following is the classical example of a partial differential equation with singular solution.

Consider the following partial differential equation on $J^1(\mathbb{R}^n, \mathbb{R})$,

$$z = x_1 p_1 + \cdots + x_n p_n + f(p_1, \ldots p_n).$$

For generic function germs $f(p_1, \ldots, p_n)$, these equations are the general type. A leaf of the solution foliation in Theorem 2.2 is given by the equation

$$z = x_1 t_1 + \cdots + x_n t_n + f(t_1, \ldots, t_n),$$

where $t_1, \ldots, t_n$ are parameters.

The envelope of this family is the singular solution of the equation.

2) *"Free particle" in the plane.* Consider the following partial differential equation on $J^1(\mathbb{R}^2, \mathbb{R})$,

$$z^2 + p_1^2 + p_2^2 - 1 = 0.$$

By the definition of the π-singular set,

$$\Sigma(F) = \left\{ (x_1, x_2, \pm 1, 0, 0) \in J^1(\mathbb{R}^2, \mathbb{R}) \mid (x_1, x_2) \in \mathbb{R}^2 \right\}.$$

It is easy to show that $\Sigma(F)$ is the singular solution of the equation and it consists of only fold points.

Remark. By Theorem 2.2, the set of first order partial differential equations with singular solutions is not generic in the space of all first order partial differential equations. On the other hand, a first order partial differential equation of general type with singular solution has a "classical" complete solution. We can define the notion of "abstract" complete solutions, that is defined as an n-parameter family of Legendrian submanifolds. In [9] we have studied equations with "abstract" complete solutions which are called completely integrable equations. And we have shown that the set of completely integrable equations with singular solution is generic in the space of all completely integrable equations. These results clarify the situation of the set of equations with singular solution (such as Clairaut's equation).

REFERENCES

1. V.I. Arnol'd, S.M. Gusein-Zade and A.N. Varchenko, *Singularities of differentiable maps, Vol. 1*, Monographs in Math. 82, Birkhäuser, 1985.

2. L. Dara, *Singularités générique des équations differentielles multiformes*, Bol. Soc. Brasil Mat. **6** (1975), 95–128.

3. M. G. Darboux, *Sur les solutions singulières des équations aux derivees ordinaires du premier ordere*, Bull. Sciences Math. and Astron. **IV** (1873), 158–176.

4. W. Dyck, *Über die Gestaltlich Verhältnisse der durch eine Differentialgleichung erster Ordung zwischen zwei Variabeln definierten Curvensysteme*, Bay. Akad. der Wissenschaften München Sizungberichte der Math. Phys. Classe. **Band XXI** (1891), 23–57.

5. M. Fukuda and T. Fukuda, *Singular solutions of ordinary differential equations*, The Yokohama Math. Jour. **15** (1977), 41–58.

6. M. Golubitsky and V. Guillemin, *Contact equivalence for Lagrange manifolds*, Adv. In Math. **15** (1975), 375–387.

7. A. Hayakawa, G. Ishikawa, S. Izumiya and K. Yamaguchi, *Classification of generic integral diagram and first order ordinary differential equations*, preprint.

8. L. Hörmander, *Fourier integral operators I*, Acta Math. **127** (1971), 79–183.

9. S. Izumiya, *The theory of Legendrian unfoldings and first order differential equations*, In preparation.

10. M. Kossowski, *First order partial differential equations with singular solution*, Indiana Univ. Math. Jour. **35** (1986), 209–223.

11. R. Thom, *Sur les equations différentielles multiformes et leurs integrals singulieres*, Bol. da Soc. Brasil Mat. **3** (1972), 1–11.

DEPARTMENT OF MATHEMATICS, FACULTY OF SCIENCE, HOKKAIDO UNIVERSITY, SAPPORO 060, JAPAN

PERIODIC ORBITS ON SURFACES
VIA NIELSEN FIXED POINT THEORY

BOJU JIANG

Peking University

ABSTRACT. Suppose V is a vector field on a surface M, V depends periodically on time. Suppose a set C of periodic solutions of V is already known. We estimate the number and homological behavior of other periodic orbits of V in terms of the braiding of C in $M \times \mathbb{R}$. The approach is based upon the Nielsen fixed point theory. A zeta function containing the desired dynamical information is computed from some matrix representation of the braid group of M. When M is the infinite cylinder (the tangent bundle of the circle) and V is 1-periodic in time, it is shown that one 2-periodic orbit or two 1-periodic orbits "almost always" imply the existence of an infinite number of periodic orbits, with a list of simplest braids as the only possible exceptions.

1. INTRODUCTION

Let M be an orientable surface of finite type (i.e. M can be smoothly imbedded into a closed surface $\widehat{M}$ so that $\widehat{M} \setminus M$ is finite). Consider a differential equation on M

$$(1) \qquad \frac{dz}{dt} = V(t, z), \qquad t \in \mathbb{R}, \ z \in M,$$

where $V_t := V(t, \cdot)$ are vector fields on M, and V is C^1 on $\mathbb{R} \times M$. We assume that:

- (i) $V(t+1, z) = V(t, z)$ for all $t \in \mathbb{R}$, $z \in M$. In other words, the vector field V_t is 1-periodic in the time variable t.
- (ii) The general solution $z = \phi(t; z_0)$ to Eq.(1) with initial condition $z|_{t=0} = z_0$ exists on $0 \le t < \infty$ for any initial position $z_0 \in M$.
- (iii) $C = \{c_1(t), c_2(t), \cdots, c_n(t)\}$ is a given set of n periodic solutions of the equation, such that $\{c_1(0), c_2(0), \cdots, c_n(0)\} = \{c_1(1), c_2(1), \cdots, c_n(1)\}$.

1991 *Mathematics Subject Classification*. Primary 55M20, 58F22, 34C25.
Partially supported by an NSFC grant

This set C of known solutions gives rise to a geometric braid

$$S = \big\{ \big(c_i(t), t\big) \mid 0 \leq t \leq 1,\ 1 \leq i \leq n \big\}$$

in the space $M \times I$. This information often enables us to predict the existence and behavior of other periodic solutions of Eq.(1).

This line of research was started by Matsuoka who in [M1], [M2], [M3] studied Eq.(1) on $M = \mathbb{R}^2$ as a sophisticated application of the Lefschetz Fixed Point Theorem. Huang and Jiang [HJ] recast Matsuoka's results via Nielsen fixed point theory applied to study the periodic points of the Poincaré map $\phi_1 : M \to M$. This paves the way for generalization to other surfaces, and a theory for the compact surface $M = T^2$ is worked out in [HJ]. The present paper is an exposition of this approach with more emphasis on noncompact M.

After an account of the relevant fixed point theory in §2, we shall explain in §3 our general approach to the homological estimation of periodic orbits and the key role played by braid representations. As an example of the principles, a theory for the infinite cylinder $M = S^1 \times \mathbb{R}$ is presented in §4.

The estimates of §4 is further specialized to the case $n = 2$ in §5 in order to show the quantitative and algorithmic features of this approach. We confirm the generic existence of an infinite number of periodic orbits, as expected by Thurston's theory. (The generic Poincaré map should be pseudo-Anosov hence with positive entropy and an infinite number of periodic points.) But the present approach is computable and provides more information on the number and homological behavior of periodic orbits. The conjugacy classes of the exceptional braids that allow finitely many periodic orbits are explicitly listed.

2. An account of Nielsen fixed point theory

General references for the following notions and facts of Nielsen fixed point theory are [J1], [F1]. See also [HJ, §1].

§2.1. Nielsen number, Lefschetz number, and zeta function.

Let X be a compact connected polyhedron, $f : X \to X$ be a map. The fixed point set is denoted by $\operatorname{Fix} f := \{ x \in X \mid x = f(x) \}$. Let x_0 be the base point in X, and let $\pi = \pi_1(X, x_0)$. For the sake of simplicity, we assume $f(x_0) = x_0$.

Suppose H is a *commutative* (multiplicative) group. Suppose $\mu : \pi \to H$ is a homomorphism such that the diagram

$$(*) \qquad
\begin{array}{ccc}
\pi & \xrightarrow{\ f_* \ } & \pi \\
{\scriptstyle \mu} \big\downarrow & & \big\downarrow {\scriptstyle \mu} \\
H & =\!\!=\!\!= & H
\end{array}$$

commutes. (Equivalently, H is a quotient of $\operatorname{Coker}(1 - f_* : H_1(X) \to H_1(X))$ but written multiplicatively, and $\mu : \pi \to H$ is the natural projection.) It extends to a homomorphism of the group-rings $\mu : \mathbb{Z}\pi \to \mathbb{Z}H$. The theory described below corresponds to the "mod K Nielsen theory" of [J, Chap.III] with $K = \text{Ker}\mu$.

For every $x \in \text{Fix}\, f$, its H-*coordinate* $\text{cd}_H(x, f) \in H$ is defined as follows: Pick a path c from x_0 to x and form the loop $(f \circ c)c^{-1}$ at x_0. Then $\text{cd}_H(x, f) := \mu\langle(f \circ c)c^{-1}\rangle$ where $\langle \cdot \rangle$ denotes the loop class in the fundamental group. It is independent of the choice of the path c because of the assumption ($*$). (Note that this definition differs from that of [J, p.26] by an inversion.)

The fixed points split into H-*classes* according to their H-coordinates. Each H-class F is an isolated subset of $\text{Fix}\, f$ hence its index $\text{ind}(F, f) \in \mathbb{Z}$ is defined. The H-coordinate $\text{cd}_H(F, f)$ of an H-class F is defined to be the common H-coordinate of its members. Define the H-*Lefschetz number* (cf. [FH], [F1]) as

$$L_H(f) := \sum_{H\text{-classes } F} \text{ind}(F, f) \cdot \text{cd}_H(F, f) \in \mathbb{Z}H.$$

Define the H-*Nielsen number* $N_H(f)$ as the number of non-zero terms in $L_H(f)$, i.e. the number of H-classes with non-zero index. Obviously $N_H(f)$ is a lower bound for the number of fixed points of f.

Periodic points of f are fixed points of iterates of f. Their H-coordinates have the following properties:

(1) If x is a fixed point of f^p, then $\text{cd}_H(x, f^p) = \text{cd}_H(f(x), f^p)$, so that the whole f-orbit $\{x, f(x), \cdots, f^{p-1}(x)\}$ lies in one H-class of f^p.

(2) If x is a fixed point of f^q and q divides p, then

$$\text{cd}_H(x, f^p) = \left(\text{cd}_H(x, f^q)\right)^{p/q},$$

so that the number of periodic orbits of least period p is bounded below by the number of elements $\gamma \in H$ appearing in $L_H(f^p)$ such that γ is not a q-th power in H for any prime factor q of p.

For the H-Lefschetz number of all the iterates of f, Fried [F1] introduced a generating function called the *Lefschetz zeta function*.

$$\zeta_H(f) := \exp\left(\sum_{p=1}^{\infty} \frac{L_H(f^p)}{p} t^p\right)$$

in the multiplicative subgroup $1 + t\mathbb{Z}H[[t]]$ of the formal power series ring $\mathbb{Z}H[[t]]$.

Homotopy invariance. If $f \simeq f' : X \to X$, then $N_H(f') = N_H(f)$ and

$$L_H(f') = \omega L_H(f) \quad \in \mathbb{Z}H,$$
$$\zeta_H(f')(t) = \zeta_H(f)(\omega t) \in 1 + t\mathbb{Z}H[[t]],$$

for some $\omega \in H$.

If the homotopy fixes the base point x_0 then $L_H(f') = L_H(f)$ and $\zeta_H(f')(t) = \zeta_H(f)(t)$.

Homotopy type invariance. *Suppose* $h: X, x_0 \to X', x_0'$ *is a homotopy equivalence. Suppose* $f': X', x_0' \to X', x_0'$ *is a map such that the diagram*

$$
\begin{array}{ccc}
X, x_0 & \xrightarrow{\ f\ } & X, x_0 \\
h \downarrow & & \downarrow h \\
X', x_0' & \xrightarrow{\ f'\ } & X', x_0'
\end{array}
$$

commutes up to homotopy. If we identify $\pi_1(X', x_0')$ *with* $\pi_1(X, x_0)$ *via* h_*, *then* $N_H(f') = N_H(f)$, $L_H(f') = L_H(f)$ *and* $\zeta_H(f')(t) = \zeta_H(f)(t)$.

§2.2. The trace formula.

The invariant $L_H(f)$ is sometimes called the *Reidemeister trace* because it can be computed as an alternating sum of traces on the chain level, similar to that for the ordinary Lefschetz number (cf. [R], [W]).

Take any cellular decomposition $\{e_j^d\}$ of X. It lifts to a π-invariant cellular structure on the universal covering $\widetilde{X}$ of X, where π is identified with the group of covering translations of $\widetilde{X}$. Choose an arbitrary lift $\tilde{e}_j^d$ for each e_j^d. They constitute a free $\mathbb{Z}\pi$-basis for the cellular chain complex of $\widetilde{X}$. Without loss we assume f to be a cellular map, and consider its lift $\tilde{f}: \widetilde{X} \to \widetilde{X}$ with $\tilde{f}(\tilde{x}_0) = \tilde{x}_0$. In every dimension d, the cellular chain map $\tilde{f}_d$ gives rise to a $\mathbb{Z}\pi$-matrix $\widetilde{F}_d$ with respect to the above basis. Taking the μ-image of every element, we get a $\mathbb{Z}H$-matrix $\widetilde{F}_d^\mu$. Then we have

$$
L_H(f) = \sum_d (-1)^d \mathrm{tr}\widetilde{F}_d^\mu \in \mathbb{Z}H.
$$

It follows that

$$
L_H(f^p) = \sum_d (-1)^d \mathrm{tr}\big(\widetilde{F}_d^\mu\big)^p \in \mathbb{Z}H.
$$

Therefore we have ([F1, p.269 Theorem 2])

$$
\zeta_H(f) = \prod_d \det(I - t\widetilde{F}_d^\mu)^{(-1)^{d+1}},
$$

where I is the identity matrix.

§2.3. Homology of periodic orbits.

Another interpretation of the Lefschetz zeta function is suggested by [F2] and [J2].

Let $\mathbb{R}_+$ stand for the real interval $[0, \infty)$. The *mapping torus* T_f of $f: X \to X$ is the space obtained from $X \times \mathbb{R}_+$ by identifying $(x, s+1)$ with $(f(x), s)$ for all $x \in X$, $s \in \mathbb{R}_+$. On T_f there is a natural semi-flow ("sliding along the rays")

$$
\varphi: T_f \times \mathbb{R}_+ \to T_f, \qquad \varphi_\tau(x, s) = (x, s+\tau) \text{ for all } \tau \geq 0.
$$

We may identify X with the cross-section $X \times 0 \subset T_f$, then the map $f : X \to X$ is just the return map of the semi-flow φ.

Take the base point x_0 of X (a fixed point of f) as the base point of T_f. Let $\bar{\pi} = \pi_1(T_f, x_0)$. By the van Kampen Theorem, $\bar{\pi}$ is obtained from π by adding a new generator t represented by the closed orbit $\{\varphi_\tau(x_0)\}_{0 \le \tau \le 1}$, and adding the relations $t^{-1}\alpha t = f_*(\alpha)$ for all $\alpha \in \pi$. Let $\bar{H}$ be the direct product of H and the infinite cyclic group $\mathbb{Z}$ generated by t. In view of the assumption $(*)$, the homomorphism $\mu : \pi \to H$ extends to a homomorphism $\bar{\mu} : \bar{\pi} \to \bar{H}$. Since $\bar{H}$ is commutative, $\bar{\mu}$ factors through the homology group $H_1(T_f)$.

Each point $x \in X$ determines an orbit $\gamma_x := \{\varphi_\tau(x)\}_{\tau \ge 0} \subset T_f$. A point x is a periodic point of f if and only if γ_x is closed. Suppose x is a fixed point of f^p. We shall write $\gamma_{x,p}$ for the closed curve $\{\varphi_\tau(x)\}_{0 \le \tau \le p}$ which represents a homology class $[\gamma_{x,p}] \in H_1(T_f)$. Pick any path c in X from x_0 to x. The loop class $\langle c\gamma_{x,p}c^{-1}\rangle = \langle \gamma_{x_0,p}(f^p \circ c)c^{-1}\rangle = t^p\langle(f^p \circ c)c^{-1}\rangle$. Hence $\bar{\mu}[\gamma_{x,p}] \in \bar{H}$ equals $\mathrm{cd}_H(x, f^p)t^p$. It follows that $x, x' \in \mathrm{Fix}\, f^p$ are in the same H-class of f^p if and only if $\bar{\mu}[\gamma_{x,p}] = \bar{\mu}[\gamma_{x',p}]$ in $\bar{H}$.

The Lefschetz zeta function can thus be regarded as the formal sum counting the closed orbits of the natural flow φ on T_f, classified according to the image in $\bar{H}$ of their homology class. Fried [F2] has shown that the Lefschetz zeta function of f is indeed the Reidemeister torsion of the mapping torus T_f.

§2.4. Surface maps.

Let X be a surface with boundary. Then X has a deformation retract X_0 which is a bouqué of circles at the base point x_0, so π has a free basis $\{\alpha_1, \cdots, \alpha_r\}$. Since X is aspherical, the homotopy class of a map $f : X, x_0 \to X, x_0$ is determined by the homomorphism $f_* : \pi \to \pi$, hence by the system of elements $f_i := f_*(\alpha_i) \in \pi$, $i = 1, \cdots, r$.

With respect to the obvious cellular decomposition of X_0, the chain map $\tilde{f} : \tilde{X}_0 \to \tilde{X}_0$ is described by the Jacobian matrix in Fox calculus (cf. [FH])

$$\tilde{F}_0 = 1,$$

$$\tilde{F}_1 = \left(\frac{\partial f_i}{\partial \alpha_j}\right)_{r \times r}$$

Denote

$$D_H(f) = \left(\frac{\partial f_i}{\partial \alpha_j}\right)^\mu$$

where the superscript μ means taking the μ-image of the elements of the matrix. Then

$$L_H(f) = 1 - \mathrm{tr}D_H(f),$$

$$\zeta_H(f) = \frac{\det(I - tD_H(f))}{1 - t}.$$

3. Homological estimation of periodic orbits

In the setting of §1, define $\phi_t : M \to M$ by $\phi_t(z) = \phi(t; z)$. The map $\phi_1 : M \to M$ is usually called the Poincaré transformation of the differential equation (1).

Clearly, for a natural number p, $z(t)$ is a p-periodic solution of the differential equation $\Longleftrightarrow z(0)$ is a p-periodic point of ϕ_1. By "p-periodic" we always mean "of minimal integer period p". So, finding periodic solutions other than the given $\{c_i(t)\}$ is equivalent to finding periodic points of the map $\phi'_1 := \phi_1 | M' : M' \to M'$ where $M' := M \setminus \{c_1(0), c_2(0), \cdots, c_n(0)\}$ is an open surface. It should be compactified in order that fixed point theory be applicable.

§3.1. Compactifying M'.

Compactify M into a closed surface $\widehat{M}$ by adding a "point at infinity" ∞_j to each end of M, $1 \le j \le e$, where e is the number of ends of M. And compactify M' into X, a surface with boundary, by adding the "circle of directions" Δ_i to each puncture $c_i(0)$ and adding a "circle at infinity" Γ_j (i.e. the "circle of directions" at the corresponding $\infty_j \in \widehat{M}$) to each end of M. We would like to extend $\phi'_1 : M' \to M'$ to X.

The general solution $z = \phi(t; z_0)$ and the related map ϕ'_1 on M' can be extended to Δ_i according to their tangent maps at $c_i(0)$ because they are C^1 there. On the other hand, if we extend ϕ to $\widehat{M}$ by defining $\phi(t; \infty_j) = \infty_j$ for all $t \ge 0$ and all $1 \le j \le e$, this extension is not guaranteed continuous there because ϕ_t are not necessarily homeomorphisms of M (the general solution is not assumed to exist for negative t). Even when it is continuous, it would not be differentiable, so there is no natural way to extend ϕ onto Γ_j.

In this respect, compact M ($e = 0$) is easier to deal with, and [HJ, §3] has already given a good example for the torus. In this paper we shall concentrate on noncompact M ($e > 0$).

§3.2. Modifying the Poincaré map.

Equip $\widehat{M}$ with a Riemannian metric. For $1 \le j \le e$ and $\epsilon > 0$, denote $D_j^{(\epsilon)} := \{ z \in \widehat{M} \mid d(z, \infty_j) < \epsilon \}$. For some sufficiently small $\epsilon_0 > 0$, $\{D_j^{(\epsilon_0)}\}_{1 \le j \le e}$ are disjoint open disks away from the curves $\{c_i(t)\}_{t \ge 0}$. Let $\beta : \mathbb{R} \to [0,1]$ be a smooth non-decreasing function which is identically 0 near 0 and identically 1 near 1.

For $0 < \epsilon < \epsilon_0$, define

$$V^{(\epsilon)}(t, z) = \begin{cases} V(t, z) & \text{if } z \notin \bigcup_{j=1}^e D_j^{(\epsilon)}, \\ \beta\left(\dfrac{d(z, \infty_j)}{\epsilon}\right) V(t, z) & \text{if } z \in D_j^{(\epsilon)} \setminus \infty_j, \\ 0 & \text{if } z = \infty_j, \end{cases}$$

and consider on $\widehat{M}$ the differential equation

$$(1') \qquad\qquad\qquad \frac{dz}{dt} = V^{(\epsilon)}(t, z).$$

Let $\phi^{(\epsilon)}(t;x)$ be the general solution of Eq.(1'), and let $\phi_1^{(\epsilon)} : \widehat{M} \to \widehat{M}$ be the Poincaré map. Then, restricted to M, Eq.(1') also satisfies the conditions (i)–(iii), and

$$\phi(t;z) = \lim_{\epsilon \to 0} \phi^{(\epsilon)}(t;z) \qquad \text{and} \qquad \phi_1(z) = \lim_{\epsilon \to 0} \phi_1^{(\epsilon)}(z).$$

For all $0 < \epsilon < \epsilon_0$, the map $\phi_1^{(\epsilon)'} : M' \to M'$ naturally extends to a homeomorphism $f_\epsilon : X \to X$. In fact, f_ϵ is always the identity near Γ_j. Now fixed point theory can be applied to study the periodic points of f_ϵ. Since all these maps are homotopic to each other rel ∂X, we shall use the generic notation $f : X \to X$ for any one of them.

We shall identify the periodic orbits of $f : X \to X$ which are not periodic orbits of $\phi_1' : M' \to M'$. We say a fixed point $x \in X$ of f is f-related to Γ_j if there is a path a in X from x to Γ_j such that the loop $(f \circ a)^{-1}a$ can be deformed into Γ_j.

Choose the base point x_0 in a sufficiently small neighborhood of Γ_e (where f is the identity), and we identify $\pi = \pi_1(X, x_0) = \pi_1(M', x_0)$.

Proposition. *Suppose $x \in M' = X \setminus \partial X$ is a fixed point of $f^p : X \to X$.*

(1) *If x is not f^p-related to any Γ_j, then x is a fixed point of $\phi_p' : M' \to M'$.*

(2) *If x is f^p-related to Γ_j then $\mathrm{cd}_H(x, f^p)$ is of the form $\ell_j^p m_j^k$, where $m_j := \mu[\Gamma_j]$ (defined up to an inversion), $\ell_j := \mathrm{cd}_H(\Gamma_j, f)$, and $k \in \mathbb{Z}$.*

Proof. (1) Let us denote $Y^{(\epsilon)} := M \setminus \bigcup_{j=1}^{e} D_j^{(\epsilon)}$, a compact subset of M, and $A_j^{(\epsilon)} := D_j^{(\epsilon)} \setminus \infty_j$, an open annulus in M'.

Fix a natural number p. Choose $0 < \delta < \epsilon_0$ such that $\phi_p\left(Y^{(\epsilon_0)}\right) \subset Y^{(\delta)}$, and choose $\epsilon > 0$ such that $\bigcup_{0 \le t \le p} \phi_t\left(Y^{(\delta)}\right) \subset Y^{(\epsilon)}$. Obviously $\phi_t^{(\epsilon)}(x) = \phi_t(x)$ for all $x \in Y^{(\delta)}$, $0 \le t \le p$, so that $\phi_p^{(\epsilon)}$ and ϕ_p have the same fixed points on $Y^{(\delta)}$.

Now suppose x is a fixed point of $f_\epsilon^p = \phi_p^{(\epsilon)}$ on $A_j^{(\delta)}$. We have $A_j^{(\delta)} \subset \phi_p^{(\epsilon)} A_j^{(\epsilon_0)}$ because $A_j^{(\delta)} \cap \phi_p^{(\epsilon)} Y_j^{(\epsilon_0)} = A_j^{(\delta)} \cap \phi_p Y_j^{(\epsilon_0)} = \emptyset$. Thus $A_j^{(\delta)} \cup f_\epsilon^p A_j^{(\delta)} \subset \phi_p^{(\epsilon)} A_j^{(\epsilon_0)}$. It follows that x is f_ϵ^p-related to Γ_j. This proves (1).

(2) By definition, if x is f^p-related to Γ_j then $\mathrm{cd}_H(x, f^p) = \mathrm{cd}_H(\Gamma_j, f^p) \cdot m_j^k$ for some $k \in \mathbb{Z}$. Then use the identity $\mathrm{cd}_H(\Gamma_j, f^p) = \ell_j^p$. $\square$

§3.3. Computing $f_* : \pi \to \pi$ via braids.

Before we can apply the theory of §2 to the map $f : X \to X$, we need the following set-up.

(1) Choose the abelian group H and the homomorphism $\mu : \pi_1(X, x_0) \to H$ satisfying the condition (∗). This choice depends on the behavior of $f_* : \pi \to \pi$, but is otherwise rather flexible. A smaller H means less information out of less work.

(2) Calculate $f_* : \pi_1(X, x_0) \to \pi_1(X, x_0)$ and the associated matrix $D_H(f)$ from the given solutions C. Braid theory (as presented in [B, Chaps.1 and 3]) comes in here.

Since M is not a closed surface, $\pi_1(M)$ is a free group and $\pi_1(X) = \pi_1(M')$ is also free with n more generators. Suppose $\alpha_1, \cdots, \alpha_r$ is a basis of π.

Let S be the geometric braid $\{(c_i(t), t) \mid 0 \le t \le 1, \ 1 \le i \le n\} \subset M \times I$. It is not hard to see that the following diagram commutes:

$$
\begin{array}{ccccc}
\pi = \pi_1(X, x_0) & = & \pi_1(M', x_0) & \xrightarrow{\ i_{0*}\ } & \pi_1\big(M \times I \setminus S, (x_0, 0)\big) \\[2mm]
\Big\downarrow{\scriptstyle f_*} & & \Big\downarrow{\scriptstyle \phi'_{1*}} & & \Big\downarrow{\scriptstyle u_*} \\[2mm]
\pi = \pi_1(X, x_0) & = & \pi_1(M', x_0) & \xrightarrow{\ i_{1*}\ } & \pi_1\big(M \times I \setminus S, (x_0, 1)\big)
\end{array}
$$

where $i_t : x \mapsto (x, t)$ is the obvious inclusion map, u_* is the isomorphism induced by the vertical path $x_0 \times I$ joining the base points.

It is then clear that $f_* : \pi \to \pi$ is invariant under any isotopy of S, so it depends only on the braid $\sigma \in \pi_1\big(B_{0,n}M, \{c_1(0), \cdots, c_n(0)\}\big)$ represented by the geometric braid S (cf. [B, p.6]).

In this way, every braid $\sigma \in \pi_1 B_{0,n} M$ gives rise to an automorphism $\sigma : \pi \to \pi$, $\alpha \mapsto \alpha\sigma$. What we get is actually a homomorphism from $\pi_1 B_{0,n} M$ to $\operatorname{Aut} \pi$, i.e. $(\alpha\sigma)\sigma' = \alpha(\sigma\sigma')$. The effect on π of the standard generators of the braid group can be read off from pictures. So $f_* = \sigma : \pi \to \pi$ can be routinely computed once we write σ as a product of standard generators.

For a braid $\sigma \in \pi_1 B_{0,n} M$ let

$$
D(\sigma) = \left(\frac{\partial(\alpha_i \sigma)}{\partial \alpha_j} \right) \in GL(r, \mathbb{Z}\pi).
$$

The chain rule in Fox calculus takes the form (cf. [B, p.116])

$$
(2) \qquad\qquad D(\sigma\sigma') = D(\sigma)^{\sigma'} D(\sigma'),
$$

where as before the superscript σ' means taking the σ'-image of all elements of the matrix. This rule is very useful in computations. The matrix $D_H(f) \in GL(r, \mathbb{Z}H)$ of §2.4 is the μ-image of $D(\sigma)$.

§3.4. Homological interpretation of $\zeta_H(f)$.

After the preparation of §§3.1–3, the theory of §2 is readily applicable to calculate $\zeta_H(f)$ which provides information on the periodic orbits of f. It remains to reinterpret it in terms of periodic orbits of the differential system (1).

Consider the product manifold $\mathfrak{M} := M \times S^1$, regarded as the quotient space of $M \times \mathbb{R}_+$ by identifying $(z, s+1)$ to (z, s) for all $z \in M$, $s \in \mathbb{R}_+$. Eq.(1) gives rise to a differential system on $\mathfrak{M}$:

$$
\begin{cases}
\dfrac{dz}{dt} = V(s, z), \\[3mm]
\dfrac{ds}{dt} = 1,
\end{cases}
\qquad z \in M, \ s \in \mathbb{R}_+,
$$

which determines a semi-flow $\overline{\phi}$ on $\mathfrak{M}$. The given set C of periodic solutions corresponds to a set $\mathfrak{C} = \{(c_i(s), s) \mid s \in \mathbb{R}_+, \ 1 \le i \le n\} \subset \mathfrak{M}$ of closed orbits of $\overline{\phi}$.

We observe that there is a homeomorphism from the mapping torus $T_{f'}$ of the map $f' := f_\epsilon | M' : M' \to M'$ onto $\mathfrak{M} \setminus \mathfrak{C}$, defined by $(z, s) \mapsto \left(\phi_s^{(\epsilon)}(z), s \right)$. It transforms the natural flow on $T_{f'}$ to the flow $\overline{\phi}^{(\epsilon)}$ on $\mathfrak{M} \setminus \mathfrak{C}$ which differs from $\overline{\phi}$ only near $\mathfrak{C}$. But $T_{f'}$ is the interior of the mapping torus T_f. So T_f can be regarded as the compactification of $\mathfrak{M} \setminus \mathfrak{C}$ by adding some tori for $\mathfrak{C}$ and adding a torus $\Gamma_j \times S^1$ for every end $\infty_j \times S^1$. Thus the terms in $\zeta_H(f)$ correspond to the $\overline{\mu}$-images of the homology classes represented by the closed orbits of $\overline{\phi}^{(\epsilon)}$ in $\mathfrak{M} \setminus \mathfrak{C}$.

In this language, the Proposition of §3.2 says that: A closed orbit of $\overline{\phi}^{(\epsilon)}$ in $\mathfrak{M} \setminus \mathfrak{C}$ is actually an orbit of $\overline{\phi}$ if in T_f it cannot be deformed into any $\Gamma_j \times S^1$. The meridian and longitude of the torus $\Gamma_j \times S^1$ are mapped by $\overline{\mu}$ into m_j and $\ell_j t$ respectively.

4. The theory for the cylinder

We now concentrate on the theory for the infinite cylinder $M = S^1 \times \mathbb{R}$ which, as the tangent bundle of the circle, is often encountered in applications. For example, a second order differential equation $\ddot{\theta} = a(t, \theta, \dot{\theta})$ of an angular variable θ can be converted into a differential equation

$$\begin{cases} \dfrac{d\theta}{dt} = \omega, \\[2mm] \dfrac{d\omega}{dt} = a(t, \theta, \omega), \end{cases}$$

on the (θ, ω) plane, where $a(t, \theta, \omega)$ is a C^1 function 2π-periodic in the variable θ. If we change to a complex variable $z = e^{\omega + i\theta} \in \mathbb{C}^* := \mathbb{C} \setminus \{0\}$, the above equation takes the form

$$(3) \qquad\qquad \frac{dz}{dt} = V(t, z), \qquad t \in \mathbb{R}, \ z \in \mathbb{C}^*$$

where $V : \mathbb{R} \times \mathbb{C}^* \to \mathbb{C}$ is a C^1 function. This is nothing but Eq.(1) for $M = \mathbb{C}^*$. We always assume it satisfies the conditions (i)–(iii) of §1. In particular, a set $C = \{c_1(t), \cdots, c_n(t)\}$ of periodic solutions to Eq.(3) is already given. For convenience we define $c_0(t) \equiv 0$.

According to the theory of §3, when $M = \mathbb{C}^*$, $\widehat{M} = \mathbb{C}^* \cup 0 \cup \infty = \mathbb{C} \cup \infty = S^2$ and we denote $\infty_1 := 0$, $\infty_2 := \infty$. The space X is obtained from $M' = \mathbb{C}^* \setminus \{c_1(0), \cdots, c_n(0)\}$ by adding Γ_1 (the circle of directions at 0), Γ_2 (the circle at infinity), and the circles Δ_i of directions at the punctures $c_i(0)$. The base point x_0 is chosen near ∞. The group π has a basis $\alpha_0, \alpha_1, \cdots, \alpha_n$ depicted in [B, Fig.5] or [HJ, Fig.1]. Let H be the free abelian group with two basis elements u, v and define the homomorphism $\mu : \pi \to H$ by $\mu(\alpha_0) = u$, $\mu(\alpha_i) = v$ for $1 \le i \le n$. This μ always satisfies the condition (*).

§4.1. Braids and their matrix representations.

The braid theory for $\mathbb{C}^*$ is classical. Let S be the geometric braid $\{(c_i(t), t) \mid 0 \le t \le 1, 1 \le i \le n\} \subset \mathbb{C}^* \times I$, and let $\sigma_c \in \pi_1\big(B_{0,n}\mathbb{C}^*, \{c_1(0), \cdots, c_n(0)\}\big)$ be the braid represented by S. The geometric n-braid S determines a geometric $(n+1)$-braid $S' = \{(c_i(t), t) \mid 0 \le t \le 1, 0 \le i \le n\} \subset \mathbb{C} \times I$. Thus, we naturally regard $\pi_1\big(B_{0,n}\mathbb{C}^*, \{c_1(0), \cdots, c_n(0)\}\big)$ as a subgroup of the $(n+1)$-braid group $\pi_1\big(B_{0,n+1}\mathbb{C}, \{c_0(0), c_1(0), \cdots, c_n(0)\}\big)$ of $\mathbb{C}$. It is well known that (cf. [B, pp.5–11]) $\pi_1 B_{0,n+1}\mathbb{C}$ has a standard presentation with generators $\sigma_0, \sigma_1, \cdots, \sigma_{n-1}$, and relations $\sigma_i \sigma_{i+1} \sigma_i = \sigma_{i+1} \sigma_i \sigma_{i+1}$ and $\sigma_i \sigma_j = \sigma_j \sigma_i$ when $|i-j| \ge 2$. It is not hard to see that the subgroup $\pi_1 B_{0,n}\mathbb{C}^*$ of $\pi_1 B_{0,n+1}\mathbb{C}$ is generated by the generators $\sigma_1, \cdots, \sigma_{n-1}$ and $\rho := \sigma_0^2$.

The action of $\pi_1 B_{0,n+1}\mathbb{C}$ on $\mathbb{Z}\pi$ is given on [B, p.25] or [HJ, p.112]. The group-ring $\mathbb{Z}H$ is nothing but $\mathbb{Z}\big[u^{\pm 1}, v^{\pm 1}\big]$, the integral Laurent polynomial ring in two variables u, v. As indicated on [HJ, p.112], by a change of basis in π, the representation $D_H : \pi_1 B_{0,n}\mathbb{C}^* \to GL\big(n+1, \mathbb{Z}\big[u^{\pm 1}, v^{\pm 1}\big]\big)$, $\sigma \mapsto D_H(\sigma)$ can be reduced to the representation $B : \pi_1 B_{0,n}\mathbb{C}^* \to GL\big(n, \mathbb{Z}\big[u^{\pm 1}, v^{\pm 1}\big]\big)$ defined by

$$B(\rho) = \begin{pmatrix} uv & 1-u & \\ 0 & 1 & \\ & & I_{n-2} \end{pmatrix},$$

$$B(\sigma_j) = \begin{pmatrix} I & & & & \\ & 1 & 0 & 0 & \\ & v & -v & 1 & \\ & 0 & 0 & 1 & \\ & & & & I \end{pmatrix} \leftarrow (j+1)\text{-th row}, \qquad 1 \le j \le n-1.$$

Every n-braid $\sigma \in \pi_1 B_{0,n}\mathbb{C}^*$ is a product of the generators $\rho, \sigma_1, \cdots, \sigma_{n-1}$ and their inverses, hence $B(\sigma)$ can be computed as a product of $B(\rho), B(\sigma_i)$ and their inverses. The reduction from D_H to B tells us

$$(4) \qquad \operatorname{tr} D_H(\sigma) = 1 + \operatorname{tr} B(\sigma), \qquad \det\big(I - tD_H(\sigma)\big) = (1-t)\det\big(I - tB(\sigma)\big)$$

for all $\sigma \in \pi_1 B_{0,n}\mathbb{C}^*$.

Define $\lambda_p(\sigma) := -\operatorname{tr} B(\sigma)^p \in \mathbb{Z}\big[u^{\pm 1}, v^{\pm 1}\big]$ and $\zeta_\sigma(t) := \det(I - tB(\sigma))$. From matrix theory we know

$$(5) \qquad \sum_{p=1}^{\infty} \frac{\lambda_p(\sigma)}{p} t^p = \log \zeta_\sigma(t).$$

Another invariant of the braid $\sigma \in \pi_1 B_{0,n}\mathbb{C}^*$ is the sum $s(\sigma)$ of the exponents of ρ when σ is written as a product of $\rho, \sigma_1, \cdots, \sigma_{n-1}$ and their inverses. The geometric interpretation of this integer: Suppose $C = \{c_1(t), \cdots, c_n(t)\}$ contains m periodic orbits, and the k-th orbit in $\mathbb{C}^*$ winds around the origin s_k times in the positive sense. Then $s(\sigma_c) = s_1 + \cdots + s_m$.

§4.2. The main estimate.

Theorem. *Let p be a natural number. Let $NO(C,p)$ be the number of p-periodic orbits of Eq.(3) not contained in C. Let k_p be the the number of periodic orbits in C whose least period divides p. In the finite sum*

$$\lambda_p(\sigma_C) = \sum_{j,k} r_{j,k}^{(p)} u^j v^k$$

let $K(\sigma_C, p)$ be the number of terms satisfying the following three conditions:

    (a) *the greatest common divisor $(p, j, k) = 1$,*
    (b) *$k \neq nj$,*
    (c) *$k \neq ps(\sigma_C)$.*

Then $NO(C,p) \geq K(\sigma_C, p) - k_p$.

Proof. By the theory of §§2–3 and Eq.(4), $D_H(f) = D_H(\sigma_C)$ and

$$L_H(f^p) = 1 - \operatorname{tr} D_H(\sigma_C) = -\operatorname{tr} B(\sigma_C) = \lambda_p(\sigma_C),$$

$$\zeta_H(f) = \frac{\det(I - t D_H(\sigma_C))}{1 - t} = \det(I - tB(\sigma_C)) = \zeta_{\sigma_C}.$$

Hence $\lambda_p(\sigma_C)$ counts the H-classes of fixed points of f^p.

The condition (a) guarantees p to be the least integer period of the counted orbit, as explained in §2.1.

We apply the Proposition of §3.2 to exclude periodic orbits of f that are not real ones of ϕ'. Observe that we now have

$$m_1 = u, \qquad \ell_1 = v^{s(\sigma_C)},$$
$$m_2 = uv^n, \qquad \ell_2 = 1.$$

Therefore the conditions (b) and (c) serve to rule out the fake periodic points near ∞ and 0.

Finally, on Δ_i there can be fixed points of f^p only when the least period of $c_i(t)$ divides p. In that case such fixed points must be in the same class because f^p is an orientation preserving homeomorphism there. Moreover, if $c_{i'}(t) = c_i(t + s)$ for some integer s, then the fixed points of f^p on Δ_i and $\Delta_{i'}$ must be in the same f-orbits hence also in the same H-class. Thus $\bigcup_{i=1}^n \Delta_i$ can intersect at most k_p H-classes of f^p. □

Example. Suppose c is a solution of period 2 whose orbit is shown in Fig.1. Let $c_1(t) = c(t)$, $c_2(t) = c(t+1)$, $C = \{c_1(t), c_2(t)\}$. The corresponding geometric braid is shown in Fig.2. Thus the braid $\sigma_C \in \pi_1 B_{0,2}\mathbb{C}^*$ is $\sigma_C = \sigma_1 \sigma_0^2 \sigma_1^2 = \sigma_1 \rho \sigma_1^2$.

Simple calculation shows

$$B(\sigma_C) = \begin{pmatrix} 1 & 0 \\ v & -v \end{pmatrix} \begin{pmatrix} uv & 1-u \\ 0 & 1 \end{pmatrix} \begin{pmatrix} 1 & 0 \\ v & -v \end{pmatrix}^2$$

$$= \begin{pmatrix} v - v^2 + uv^2 & v^2 - uv^2 \\ uv^3 & -uv^3 \end{pmatrix},$$

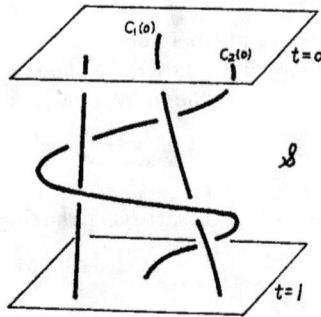

FIGURE 1

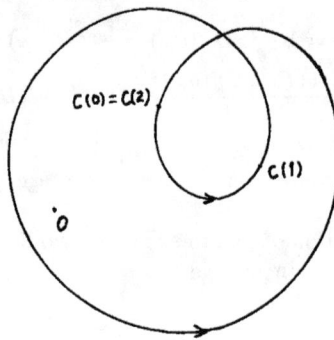

FIGURE 2

hence

$$\zeta_{\sigma_C}(t) = \det\left(I - tB(\sigma_C)\right) = 1 - (v - v^2 + uv^2 - uv^3)t - uv^4 t^2.$$

Using the formal power series expansion

$$-\log(1-z) = z + \frac{z^2}{2} + \frac{z^3}{3} + \cdots$$

and (5) we obtain

$$\lambda_1 = -v + v^2 - uv^2 + uv^3,$$
$$\lambda_2 = -v^2 - v^4 - u^2v^4 - u^2v^6 + 2(v^3 - uv^3 + uv^4 - uv^5 + u^2v^5),$$
$$\lambda_3 = -v^3 + v^6 - u^3v^6 + u^3v^9$$
$$\qquad + 3(v^4 - uv^4 - v^5 + 2uv^5 - u^2v^5 - 2uv^6$$
$$\qquad\qquad + 2u^2v^6 + uv^7 - 2u^2v^7 + u^3v^7 + u^2v^8 - u^3v^8),$$

etc.

Since $s(\sigma_c) = 1$, we have

$$K(\sigma_c, 1) = 2, \quad K(\sigma_c, 2) = 5, \quad K(\sigma_c, 3) = 12, \quad \text{etc.}$$

As $k_{\text{odd}} = 0$ and $k_{\text{even}} = 1$, we get

$$NO(C, 1) \geq 2, \quad NO(C, 2) \geq 4, \quad NO(C, 3) \geq 12, \quad \text{etc.}$$

5. THE SPECIAL CASE $n = 2$

We now apply the theory of §4 to show that on the cylinder, one 2-periodic orbit or two 1-periodic orbits "almost always" imply the existence of an infinite number of periodic orbits, with only a few simplest cases as exceptions. This analysis is in the same spirit as [M1, Theorem 1] but more elegant.

§5.1. Matrices.

Let

$$A = \begin{pmatrix} x & a \\ 0 & 1 \end{pmatrix}, \qquad B = \begin{pmatrix} 1 & 0 \\ b & y \end{pmatrix}.$$

Let $N = (n_1, n_2, \cdots, n_{2k-1}, n_{2k})$ be a sequence of integers of even length. For later use we need a formula for the product matrix $A^{n_1} B^{n_2} \cdots A^{n_{2k-1}} B^{n_{2k}}$ and its trace.

Notations. For any integer n, define the Laurent polynomial $D_n(z)$ to be the Fox derivative $D_n(z) = \dfrac{\partial z^n}{\partial z}$, i.e.

$$D_n(z) = \begin{cases} 1 + z + \cdots + z^{n-1} & \text{if } n > 0, \\ 0 & \text{if } n = 0, \\ -z^{-1} - \cdots - z^n & \text{if } n < 0. \end{cases}$$

Useful properties: $D_0(z) = 0$, $D_1(z) = 1$, and $D_{n+m}(z) = D_n(z) + z^n D_m(z)$. It is easy to verify that for any n,

$$A^n = \begin{pmatrix} x^n & aD_n(x) \\ 0 & 1 \end{pmatrix}, \qquad B^n = \begin{pmatrix} 1 & 0 \\ bD_n(y) & y^n \end{pmatrix}.$$

Now fix a sequence $N = (n_1, \cdots, n_{2k})$. For $1 \leq i \leq 2k$ denote

$$P_i^N := \begin{cases} aD_{n_i}(x) & \text{for odd } i, \\ bD_{n_i}(y) & \text{for even } i, \end{cases}$$

$$Q_i^N := \begin{cases} x^{n_i} & \text{for odd } i, \\ y^{n_i} & \text{for even } i, \end{cases}$$

A set of integers is *sparse* if no pair of elements is adjacent. For $1 \leq s \leq t \leq 2k$, consider the sparse subsets $E = \{i_1, i_2, \cdots, i_r\}$ of the integers $\{s, s+1, \cdots, t-1\}$. The family of all such subsets will be denoted $\mathcal{E}_{s,t}$. Define

$$\Psi_{s,t}^{N} := \sum_{E \in \mathcal{E}_{s,t}} \left(\prod_{\substack{s \leq i \leq t \\ i-1, i \notin E}} P_i^{N} \right) \left(\prod_{i \in E} Q_i^{N} \right).$$

Note that by definition, when $s = t$ we have $\Psi_{s,s}^{N} = P_s^{N}$; if the sequence N' is an extension of N (hence $k < k'$), then $\Psi_{s,t}^{N} = \Psi_{s,t}^{N'}$ when $s \leq t \leq 2k$. From definition follows the recurrence ($\Psi_{s,s-1}^{N} := 1$)

$$\Psi_{s,t+1}^{N} = \Psi_{s,t}^{N} P_{t+1}^{N} + \Psi_{s,t-1}^{N} Q_t^{N} \qquad \text{if } s < t+1 \leq 2k.$$

Define $\Phi^{N} := \Psi_{1,2k}^{N} + \Psi_{2,2k-1}^{N} Q_{2k}^{N}$. An alternative definition follows. Let $\mathbb{Z}_{2k}$ be the set $\{1, 2, \cdots, 2k\}$ of integers mod $2k$. A subset E of $\mathbb{Z}_{2k}$ is said to be sparse if no pair of elements of E is adjacent with respect to the obvious cyclic order of $\mathbb{Z}_{2k}$. Let $\mathcal{E}$ be the family of all sparse subsets of $\mathbb{Z}_{2k}$. Then the above definition is equivalent to

$$\Phi^{N} := \sum_{E \in \mathcal{E}} \left(\prod_{i-1, i \notin E} P_i^{N} \right) \left(\prod_{i \in E} Q_i^{N} \right).$$

Lemma. *For any sequence of integers* $N = (n_1, n_2, \cdots, n_{2k-1}, n_{2k})$,

$$A^{n_1} B^{n_2} \cdots A^{n_{2k-1}} B^{n_{2k}} = \begin{pmatrix} \Psi_{1,2k}^{N} & \Psi_{1,2k-1}^{N} Q_{2k}^{N} \\ \Psi_{2,2k}^{N} & \Psi_{2,2k-1}^{N} Q_{2k}^{N} \end{pmatrix}.$$

Hence

$$\operatorname{tr}\left(A^{n_1} B^{n_2} \cdots A^{n_{2k-1}} B^{n_{2k}}\right) = \Phi^{N}.$$

Proof. For $k = 1$,

$$A^{n_1} B^{n_2} = \begin{pmatrix} Q_1 & P_1 \\ 0 & 1 \end{pmatrix} \begin{pmatrix} 1 & 0 \\ P_2 & Q_2 \end{pmatrix} = \begin{pmatrix} P_1 P_2 + Q_1 & P_1 Q_2 \\ P_2 & Q_2 \end{pmatrix} = \begin{pmatrix} \Psi_{1,2} & \Psi_{1,1} Q_2 \\ \Psi_{2,2} & \Psi_{2,1} Q_2 \end{pmatrix},$$

The formula is verified.

Let $N = (n_1, \cdots, n_{2k})$, $N' = (n_1, \cdots, n_{2k}, n_{2k+1}, n_{2k+2})$. Using the inductive hypothesis, $A^{n_1} B^{n_2} \cdots A^{n_{2k-1}} B^{n_{2k}} A^{n_{2k+1}} B^{n_{2k+2}}$ equals

$$\begin{pmatrix} \Psi_{1,2k} & \Psi_{1,2k-1} Q_{2k} \\ \Psi_{2,2k} & \Psi_{2,2k-1} Q_{2k} \end{pmatrix} \begin{pmatrix} P_{2k+1} P_{2k+2} + Q_{2k+1} & P_{2k+1} Q_{2k+2} \\ P_{2k+2} & Q_{2k+2} \end{pmatrix}$$

$$= \begin{pmatrix} \Psi_{1,2k+2} & \Psi_{1,2k+1} Q_{2k+2} \\ \Psi_{2,2k+2} & \Psi_{2,2k+1} Q_{2k+2} \end{pmatrix}$$

by the above recurrence formula for $\Psi_{s,t}^{N}$.

This completes the induction. $\square$

§5.2. Braids.

Braid groups. (I) *The braid group* $\pi_1 B_{0,3}\mathbb{C} = \langle \sigma_0, \sigma_1 \mid \sigma_0\sigma_1\sigma_0 = \sigma_1\sigma_0\sigma_1 \rangle$. *Its center is the infinite cyclic group generated by* ω^2, *where* $\omega := \sigma_0^2\sigma_1$.

(II) *The braid group* $\pi_1 B_{0,2}\mathbb{C}^* = \langle \rho, \sigma_1 \mid \rho\sigma_1\rho\sigma_1 = \sigma_1\rho\sigma_1\rho \rangle = \langle \omega, \sigma_1 \mid \omega^2\sigma_1 = \sigma_1\omega^2 \rangle$, *where* $\rho = \sigma_0^2$, $\omega = \rho\sigma_1$. *Its center is* $\langle \omega^2 \rangle$.

(III) *The factor group* $\pi_1 B_{0,2}\mathbb{C}^*/\langle \omega^2 \rangle = \langle \omega, \sigma_1 \mid \omega^2 = 1 \rangle = \mathbb{Z}_2 * \mathbb{Z}$. *Every conjugacy class has a representative of the form* $1, \omega, \sigma_1^m$, *or* $\omega\sigma_1^{m_1} \cdots \omega\sigma_1^{m_k}$, *and the cyclic order of the nonzero integers* $m_1, \cdots, m_k$ *is uniquely determined.*

Proof. (I) follows from [B, pp.18,28]. (II) $\pi_1 B_{0,2}\mathbb{C}^*$ is the subgroup of $\pi_1 B_{0,3}\mathbb{C}$ generated by ρ, σ_1, the presentation is obtained by means of the standard coset enumeration algorithm (cf. [Jo, §13]). (III) is a corollary of (II). The solution of the conjugacy problem in a free product is well known. □

Proposition. *Suppose* $\sigma, \sigma' \in \pi_1 B_{0,2}\mathbb{C}^*$ *have conjugate projections in the factor group* $\pi_1 B_{0,2}\mathbb{C}^*/\langle \omega^2 \rangle$. *Then* $K(\sigma,p) = K(\sigma',p)$ *for every* p, *where* $K(\sigma,p)$ *is defined in the Theorem of §4.2.*

Proof. By assumption $\sigma' = \omega^{2m}\tau^{-1}\sigma\tau = \tau^{-1}\omega^{2m}\sigma\tau$, where m is an integer, $\tau \in \pi_1 B_{0,2}\mathbb{C}^*$. Since $B(\omega^2) = uv^2 I$, so $B(\sigma')$ is similar to $B(\omega^{2m}\sigma) = (uv^2)^m B(\sigma)$, hence $\lambda_p(\sigma') = u^{pm}v^{2pm}\lambda_p(\sigma)$. On the other hand, $s(\sigma') = s(\sigma) + 2m$, $n = 2$. Now the desired equality follows from the definition of $K(\sigma,p)$. □

§5.3. Estimates.

Suppose a 2-periodic solution or a pair of 1-periodic solutions to Eq.(3) is known. It forms a set $C = \{c_1(t), c_2(t)\}$ satisfying the condition (iii) with $n = 2$, and determines a braid $\sigma \in \pi_1 B_{0,2}\mathbb{C}^*$.

In view of the Proposition and the fact (III) above, we suppose

$$\sigma = \omega\sigma_1^{m_1} \cdots \omega\sigma_1^{m_k} = \rho\sigma_1^{m_1+1} \cdots \rho\sigma_1^{m_k+1}, \quad k \geq 1.$$

Then

$$B(\sigma) = B(\rho)B(\sigma_1)^{m_1+1} \cdots B(\rho)B(\sigma_1)^{m_k+1},$$

where

$$B(\rho) = \begin{pmatrix} uv & 1-u \\ 0 & 1 \end{pmatrix}, \qquad B(\sigma_1) = \begin{pmatrix} 1 & 0 \\ v & -v \end{pmatrix}.$$

We shall use the formula of §5.1 to calculate the trace of $B(\sigma)$. Let N be the sequence $(1, m_1+1, 1, m_2+1, \cdots, 1, m_k+1)$. Since $n_i = 1$ for all odd i, according to §5.1 we set

$$P_i := \begin{cases} 1-u & \text{for odd } i, \\ vD_{m_j+1}(-v) & \text{if } i = 2j, \end{cases}$$

$$Q_i := \begin{cases} uv & \text{for odd } i, \\ (-v)^{m_j+1} & \text{if } i = 2j. \end{cases}$$

The Lemma in §5.1 then gives

$$
(6) \qquad \operatorname{tr}B(\sigma) = \Phi = \sum_{E \in \mathcal{E}} \left(\prod_{i-1, i \notin E} P_i \right) \left(\prod_{i \in E} Q_i \right).
$$

Hence

$$
\zeta_\sigma(t) = \det(I - tB(\sigma)) = 1 - (\operatorname{tr}B(\sigma))t + (\det B(\sigma))t^2
$$

$$
(7) \qquad = 1 - \Phi t + (-1)^{k+m} u^k v^{2k+m} t^2, \qquad \text{where } m = \sum_{j=1}^{k} m_j.
$$

From formula (6) we see that the u-exponent in $\operatorname{tr}B(\sigma)$ is at most k and at least 0. Moreover, the terms with u-exponent k come from those E consisting solely of odd numbers, namely from $\prod_{j=1}^{k}(P_{2j-1}P_{2j} + Q_{2j-1})$; the terms with u-exponent 0 come from those E consisting solely of even numbers, namely from $\prod_{j=1}^{k}(P_{2j}P_{2j+1} + Q_{2j})$. In view of the identity $D_{n+1}(z) = 1 + zD_n(z) = D_n(z) + z^n$, we see the part with u-exponent k is

$$
\prod_{j=1}^{k} \left[-uvD_{m_j+1}(-v) + uv \right] = (uv)^k \prod_{j=1}^{k} \left[1 - D_{m_j+1}(-v) \right]
$$

$$
= u^k v^{2k} \prod_{j=1}^{k} D_{m_j}(-v) = u^k v^{2k} R(v),
$$

and the part with u-exponent 0 is

$$
\prod_{j=1}^{k} \left[vD_{m_j+1}(-v) + (-v)_{m_j+1} \right] = (-v)^k \prod_{j=1}^{k} \left[(-v)^{m_j} - D_{m_j+1}(-v) \right]
$$

$$
= v^k \prod_{j=1}^{k} D_{m_j}(-v) = v^k R(v),
$$

where $R(v) := \prod_{j=1}^{k} D_{m_j}(-v)$. Thus (7) becomes

$$
\zeta_\sigma(t) = 1 - \left[u^k v^{2k} R(v) + \cdots + v^k R(v) \right] t + (-1)^{k+m} u^k v^{2k+m} t^2,
$$

in which the dots stand for terms with u-exponent from $k-1$ down to 1. Now from the formal series

$$
-\sum_{p=1}^{\infty} \frac{\lambda_p(\sigma)}{p} t^p = -\log \zeta_\sigma(t)
$$

$$
= \sum_{q=1}^{\infty} \frac{t^q}{q} \left\{ \left[u^k v^{2k} R(v) + \cdots + v^k R(v) \right] - (-1)^{k+m} u^k v^{2k+m} t \right\}^q,
$$

it is easily seen that in $\lambda_p(\sigma) \in \mathbb{Z}\left[u^{\pm 1}, v^{\pm 1}\right]$, the u-exponent runs from pk down to 0, the part with highest u-exponent is $u^{pk} v^{2pk} R(v)^p$, and the part with lowest u-exponent is $v^{pk} R(v)^p$.

A one-variable Laurent polynomial can always be written in decreasing order of the degree of terms. Its *span* is the difference between its highest and lowest degrees. It is *alternating* if no intermediate term is missing and the sign of the terms alternates. Since $m_1, \cdots, m_k$ are nonzero, $R(v)$ is an alternating Laurent polynomial in v, of span $h := \sum_{j=1}^{k} |m_j| - k$. Therefore $R(v)^p$ is alternating of span ph.

Hence $\lambda_p(\sigma)$ has at least $2(ph+1)$ terms and, by the definition of $K(\sigma, p)$,

$$K(\sigma, p) \geq 2h\phi(p),$$

where $\phi(\cdot)$ is the number-theoretic Euler function, i.e. $\phi(p)$ is the number of positive integers $\leq p$ that are coprime to p.

Therefore when $h > 0$ we get $K(\sigma, p) \geq 2\phi(p)$.

When $h = 0$, we shall take a closer look. Every m_j must be ± 1. Let us assume the m_j's are not all the same (leaving out the case of all $m_j = 1$ and the case of all $m_j = -1$, i.e. the case of σ being a power of ρ). Since $\omega \sigma_1^{-1} = \rho$ and $\omega \sigma_1 = \rho\tau$ where $\tau = \sigma_1^2$, the braid σ can be written as $\rho^{n_1} \tau \rho^{n_2} \tau \cdots \rho^{n_d} \tau$, where $d \geq 1$, $n_1, n_2, \cdots, n_d$ are positive integers, $h'(> 0)$ of these numbers being greater than 1. Now

$$B(\rho) = \begin{pmatrix} uv & 1-u \\ 0 & 1 \end{pmatrix}, \qquad B(\tau) = \begin{pmatrix} 1 & 0 \\ v - v^2 & v^2 \end{pmatrix}.$$

We can analyse $\mathrm{tr}B(\sigma)$ as before, but focus on the extreme exponents of v instead of u. We state the result but omit the details: The part of $\mathrm{tr}B(\sigma)$ with lowest v-exponent is $v^d (1-u)^{h'}$, while the highest part is $v^{k+d} u^{k-h'} (u-1)^{h'}$. From this follows the estimate $K(\sigma, p) \geq 2h'\phi(p) \geq 2\phi(p)$.

So far we have left out the braids whose projections in $\pi_1 B_{0,2}\mathbb{C}^* / \langle \omega^2 \rangle = \langle \omega \mid \omega^2 = 1 \rangle * \langle \sigma_1 \rangle$ are conjugate to ω, σ_1^m or ρ^m, $m \in \mathbb{Z}$. Thus from the Theorem of §4.2 we get

Theorem. *Suppose the braid* $\sigma \in \pi_1 B_{0,2}\mathbb{C}^*$ *is not conjugate to the braids* ω^k, $\omega^{2k}\rho^m$ *and* $\omega^{2k}\sigma_1^m$, $k, m \in \mathbb{Z}$. *Then for any natural number* p, *Eq.(3) has at least* $2\phi(p) - 2$ *other* p-*periodic orbits. In particular, it has an infinite number of periodic orbits.* $\square$

Remark 1. Each exceptional braid ω^k, $\omega^{2k}\rho^m$ or $\omega^{2k}\sigma_1^m$ is realizable by an equation with finitely many periodic orbits, hence so are all braids conjugate to them.

Remark 2. The above Theorem cannot be derived from [M1, Theorem 1], because Eq.(3) is not the restriction of a dissipative equation on $\mathbb{C}$.

REFERENCES

[B]   Birman, J.S., *Braids, Links, and Mapping Class Groups*, Ann. Math. Studies vol. 82, Princeton Univ. Press, Princeton, 1974.

[FH]   Fadell, E., Husseini, S., *The Nielsen number on surfaces*, Topological Methods in Nonlinear Functional Analysis (S.P. Singh et al., eds.), Contemp. Math. vol. 21, Amer. Math. Soc., Providence, 1983, pp. 59–98.

[F1]   Fried, D., *Periodic points and twisted coefficients*, Geometric Dynamics (J. Palis Jr., ed.), Lecture Notes in Math. vol. 1007, Springer-Verlag, Berlin, Heidelberg, New York, 1983, pp. 261–293.

[F2]   Fried, D., *Homological identities for closed orbits*, Invent. Math. **71** (1983), 419–442.

[HJ]   Huang, H.-H., Jiang, B.-J., *Braids and periodic solutions*, Topological Fixed Point Theory and Applications (B. Jiang, ed.), Lecture Notes in Math. vol. 1411, Springer-Verlag, Berlin, Heidelberg, New York, 1989, pp. 107–123.

[J1]   Jiang, B., *Lectures on Nielsen Fixed Point Theory*, Contemp. Math. vol. 14, Amer. Math. Soc., Providence, 1983.

[J2]   Jiang, B., *A characterization of fixed point classes*, Fixed Point Theory and its Applications (R.F. Brown, ed.), Contemp. Math. vol. 72, Amer. Math. Soc., Providence, 1988, pp. 157–160.

[Jo]   Johnson, D.L., *Topics in the Theory of Group Presentations*, London Math. Soc. Lecture Note Series vol. 42, Cambridge Univ. Press, Cambridge, 1980.

[M1]   Matsuoka, T., *The number and linking of periodic solutions of periodic systems*, Invent. Math. **70** (1983), 319–340.

[M2]   Matsuoka, T., *Waveform in dynamical systems of ordinary differential equations*, Japan. J. Appl. Math. **1** (1984), 417–434.

[M3]   Matsuoka, T., *The number and linking of periodic solutions of non-dissipative systems*, J. Diff. Eqs. **76** (1988), 190–201.

[R]   Reidemeister, K., *Automorphismen von Homotopiekettenringen*, Math. Ann. **112** (1936), 586–593.

[W]   Wecken, F., *Fixpunktklassen. II*, Math. Ann. **118** (1942), 216–234.

DEPARTMENT OF MATHEMATICS, PEKING UNIVERSITY, BEIJING 100871, CHINA

LEFSCHETZ NUMBERS AND NIELSEN NUMBERS
FOR HOMEOMORPHISMS ON ASPHERICAL MANIFOLDS

BOJU JIANG AND SHICHENG WANG

Peking University

ABSTRACT. Let f be a self-homeomorphism of an aspherical closed 3-manifold M. Let $L(f)$ be its Lefschetz number and $N(f)$ be its Nielsen number. We show that $L(f) \leq N(f)$, and that $|L(f)| \leq N(f)$ when M is orientable and f is orientation preserving. The assumption is that M is "good" in a certain sense. According to Thurston's famous Geometrization Conjecture, all aspherical closed 3-manifolds are "good" in this sense. For higher dimensional aspherical manifolds, there are easy counter-examples to these inequalities.

SECTION 0. INTRODUCTION

Let $L(f)$ and $N(f)$ be the Lefschetz number and Nielsen number of a self-map f of a compact polyhedron. A space is said to be *aspherical* if its homotopy groups $\pi_i = 0$ for all $i > 1$. Our result is

Theorem 0.1. *Suppose a closed aspherical 3-manifold M is finitely covered by an orientable 3-manifold which is either a Seifert manifold, or a hyperbolic 3-manifold, or admits a non-trivial torus decomposition in the sense of Jaco-Shalen-Johannson. Suppose $f : M \to M$ is a homeomorphism. Then*

(1) *the index of each fixed point class of f is not greater than 1, hence $L(f) \leq N(f)$;*
(2) *furthermore, if M is orientable and f is orientation preserving, then the index of each essential fixed point class of f is ± 1, hence $|L(f)| \leq N(f)$.*

For each $n > 3$, there is a homeomorphism f on a closed aspherical n-manifold such that $L(f) > N(f)$.

A direct consequence is

1991 *Mathematics Subject Classification*. Primary 55M20, 57N10.
Both authors partially supported by NSFC.

Corollary 0.2. *For any self-homeomorphism f of an aspherical closed 3-manifold M satisfying the condition of Theorem 0.1, $L(f)$ (respectively, $|L(f)|$) is a lower bound for the number of fixed points in the homotopy class of f.* $\square$

Remark 1. It follows from Thurston's famous Geometrization Conjecture (see [S, §6] for a discussion) that *every* closed aspherical 3-manifold M satisfies the condition of Theorem 0.1.

Remark 2. Part (1) of Theorem 0.1 is true for dimension 1 and 2 (see [J1] and [J2]). The condition "closed" is posed here only for simplicity. However, the condition "aspherical" can not be removed, and "homeomorphism" can not be replaced by "map". Examples for these statements are given in the last section.

Remark 3. Our result and proof are inspired by [JG] (announced in [J2]) on dimension 2. The proof of Theorem 0.1 relies heavily on the achievement in the last fifteen years on the geometry and topology of 2- and 3-manifolds.

Suppose M is a compact aspherical 3-manifold. We say that M is a Seifert manifold, if M is a union of disjoint simple closed curves, called fibers, such that each fiber has a solid torus neighborhood consisting of a union of fibers. We say that M is a hyperbolic 3-manifold, if the interior of M, int$M$, admits a complete Riemannian metric of negative constant curvature and finite volume. We say that M is a 3-manifold with nontrivial (Jaco-Shalen-Johannson) torus decomposition, if M can be cut along a non-empty minimal collection of two-sided incompressible tori so that each piece is either a Seifert manifold or a hyperbolic manifold. (A 2-sided surface $S \subset M$ is incompressible if the inclusion induces a monomorphism on the fundamental group.) The minimal collection of decomposition tori in the definition is known to be unique up to isotopy.

About the terminologies and facts, see [B] and [J1] for fixed point theory; see [T3] for 2-manifolds; see [H] for basic facts on 3-manifolds; see [S] for Seifert manifolds; see [M], [T1] and [T2] for hyperbolic 3-manifolds; see [J] for the Jaco-Shalen-Johannson torus decomposition theorem.

There are five sections after the introduction. Section 1 deals with hyperbolic 3-manifolds. Section 2 deals with aspherical Seifert 3-manifolds. In these two sections, we not only consider closed manifolds, but also give a "nice form" for homeomorphisms on manifolds with boundary, to be used in Section 4. Section 3 gives a "nice form" for homeomorphisms on the neighborhood of the decomposition tori, also useful later. Section 4 combines the "nice forms" of the previous sections to deal with closed 3-manifolds covered by 3-manifolds with non-trivial torus decomposition. Section 5 gives examples for the higher dimensional part of Theorem 0.1 and the examples mentioned in Remark 2.

The following notions will be used frequently.

Let $f : M \to M$ be a map. Then Fix(f) denotes the fixed point set of f. If $C \subset M$ is both open and closed in Fix(f), then index(f, C) denotes the fixed point index of C. If x is a fixed point of f, then $\langle x \rangle$ denotes the Nielsen fixed point class containing x. A subset $A \subset M$ is said to be f-*invariant* if $f(A) \subset A$. Two f-invariant subsets A_0, A_1 are said to be f-*related* if there is a path $c : I \to M$ such that $c \sim f \circ c : I, 0, 1 \to M, A_0, A_1$. The cardinality of a set A will be denoted by $\#A$

The following two lemmas are very useful in simplifying our discussion.

Lemma 0.1. *Suppose $p' : \tilde{M}' \to M$ is a finite covering between compact manifolds. Then there is a covering $q : \tilde{M} \to \tilde{M}'$ such that the composition $p = p' \circ q : \tilde{M} \to M$ is a regular finite covering and every homeomorphism $f : M \to M$ can be lifted to a homeomorphism $\tilde{f} : \tilde{M} \to \tilde{M}$.*

Proof. Let n be the index of $p'_* \pi_1(\tilde{M}')$ in $\pi_1(M)$. As a finitely generated group, $\pi_1(M)$ has only finitely many subgroups of index n. Let G be the intersection of all subgroups of index n in $\pi_1(M)$. Then G is a finite index characteristic subgroup of $\pi_1(M)$. Furthermore, $G \subset p'_* \pi_1(\tilde{M}')$ and $f_*(G) = G$ for every homeomorphism f on M. Now the finite covering $p : \tilde{M} \to M$ corresponding to G satisfies the requirements. $\square$

Lemma 0.2. *Let $p : \tilde{M} \to M$ be a finite regular covering and let $f : M \to M$ be a map that can be lifted to $\tilde{M} \to \tilde{M}$. If for every lift $\tilde{f} : \tilde{M} \to \tilde{M}$, the index of each fixed point class of $\tilde{f}$ is not greater than 1 (not less than -1), then the index of each fixed point class of f is not greater than 1 (not less than -1).*

Proof. Since indices of fixed point classes are homotopy invariants of the map f, via perturbation, we may assume that all fixed points of f are isolated.

Pick $x \in \text{Fix}(f)$, $\tilde{x} \in p^{-1}(x)$ and a lift $\tilde{f}$ of f so that $\tilde{x} \in \text{Fix}(\tilde{f})$. To prove Lemma 0.2, we need only to show the equality $n \times \text{index}(f, \langle x \rangle) = \text{index}(\tilde{f}, \langle \tilde{x} \rangle)$ for some positive integer n. From the definition of fixed point classes, we have $\langle \tilde{x} \rangle = p^{-1}(\langle x \rangle) \cap \text{Fix}(\tilde{f})$, If $y \in \langle x \rangle$, by the uniqueness of path lifting of the covering space, we have $\#(p^{-1}(x) \cap \text{Fix}(\tilde{f})) = \#(p^{-1}(y) \cap \text{Fix}(\tilde{f}))$. Now set $n = \#(p^{-1}(x) \cap \text{Fix}(\tilde{f}))$. Then the equality follows from the fact $\text{index}(f, x) = \text{index}(\tilde{f}, \tilde{x})$. $\square$

SECTION 1. HYPERBOLIC 3-MANIFOLDS

Theorem 1.1. *Suppose M is a compact connected 3-manifold (closed or with boundary) such that $\text{int} M$ admits a complete hyperbolic structure with finite volume. Suppose $f : M \to M$ is a homeomorphism. Then the index of each fixed point class of f is not greater than 1, hence $L(f) \leq N(f)$. Furthermore, if M is orientable and f is orientation preserving, then the index of each essential fixed point class of f is 1, hence $L(f) = N(f)$.*

By the Mostow Rigidity Theorem (cf. [M, p.54]), there is on $\text{int} M$ a unique isometry g homotopic to f.

Let M_ϵ be the submanifold of $\text{int} M$ consisting of the points x such that there is an embedded hyperbolic open ball of radius ϵ centered at x. There exists an $\epsilon_0 > 0$ such that for any $\epsilon \leq \epsilon_0$, M_ϵ is homeomorphic to M and each boundary component (if M is not closed) is a horosphere modulo a rank two abelian group of parabolic motions. Since g is an isometry on $\text{int} M$, $g(M_\epsilon) = M_\epsilon$. (See [T1, 5.11].)

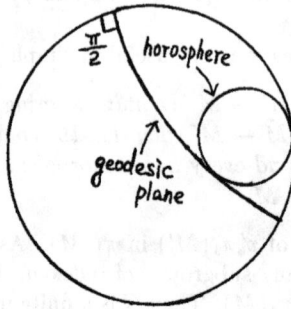

FIGURE 1

Now there is a homeomorphism $j : M \to M_\epsilon$ that is a homotopy inverse to the inclusion and $i : M_\epsilon \to M$. Pullback the hyperbolic metric of M_ϵ to M via j, then $j^{-1} \circ g \circ j$ becomes an isometry. Since $j^{-1} \circ g \circ j$ is homotopic to f, from now on, we may assume that f is an isometry.

Lemma 1.2. *Let one of A_0, A_1 be a point, the other either a point or a boundary component of M. Then every homotopy class of paths $c : I, 0, 1 \to M, A_0, A_1$ contains a unique shortest geodesic.*

Proof. The conclusion is clear when we lift the situation to the Poincaré model of the hyperbolic space, as depicted in Figure 1 where a supporting geodesic plane of the horosphere is also shown. $\square$

Lemma 1.3. *Let one of A_0 and A_1 be a fixed point of f, the other be either a fixed point of f or an f-invariant component of ∂M. If there is a path $c : I, 0, 1 \to M, A_0, A_1$ such that $f \circ c \sim c : I, 0, 1 \to M, A_0, A_1$, then there is a path γ in $\mathrm{Fix}(f)$ such that $\gamma \sim c : I, 0, 1 \to M, A_0, A_1$.*

Proof. Take γ to be the unique shortest geodesic in the homotopy class of c. Since f is an isometry, $f \circ \gamma$ is also a shortest geodesic, then by the uniqueness in Lemma 1.2 we have $f \circ \gamma = \gamma$, hence γ is in $\mathrm{Fix}(f)$. $\square$

Proof of Theorem 1.1. By Lemma 0.2, we need only to prove Theorem 1.1 for orientable M.

By Lemma 1.3, every fixed point class of f is connected, so a fixed point class of f is nothing but a component of $fix(f)$. Since f is an isometry, each component A of $fix(f)$ is a properly embedded totally geodesic submanifold of M and $\mathrm{index}(f, A) = \chi(A)$, the Euler characteristic of A. Now A is either a point, or a geodesic arc, or a geodesic circle, or a properly embedded geodesic surface, or A is M itself. The surfaces with $\chi > 0$ are ruled out because they do not admit hyperbolic metrics (with horocyclic boundary). $\chi(M) = 0$ because the boundary components of M must be tori. So we always have $\chi(A) \leq 0$. If M is orientable and f is orientation preserving, dim$A$ (being the dimension of the linear subspace fixed by the tangent map df) is 1 or 3, so $\chi(A) = 0$ or 1. $\square$

For a complete hyperbolic 3-manifold of finite volume, we know every isometry is of finite order. Therefore our f is a periodic map, hence $f|_{\partial M}$ is also a periodic map.

For the use in Section 4, we shall say such an isometric f is in "nice form", except when f is the identity map of M. In this latter case, we deform further (without insisting f to be isometric) so that f is fixed point free on M and is periodic on ∂M. Such a deformation is easily constructed by means of a suitable non-singular vector field. In any case, Lemma 1.3 holds for the nice forms.

Remark. For f in nice form, any surface component of Fix(f) has index≤ 0.

SECTION 2. SEIFERT MANIFOLDS

Theorem 2.1. *Suppose M is a compact connected aspherical Seifert manifold (closed or with boundary) and $f : M \to M$ is a homeomorphism. Then the index of each fixed point class of f is not greater than 1, hence $L(f) \leq N(f)$. Furthermore, if M is orientable and f is orientation preserving, then the index of each essential fixed point class of f is ± 1, hence $|L(f)| \leq N(f)$.*

By [S, Theorem 3.9], f is isotopic to a fiber preserving homeomorphism except possibly when M is finitely covered by a manifold homotopy equivalent to S^1, $T^2 = S^1 \times S^1$, or $T^3 = S^1 \times S^1 \times S^1$. For these exceptional cases, by the homotopy type invariance of the fixed point invariants and by Lemma 0.1 and 0.2, we only have to verify the conclusion for S^1, T^2 and T^3. But for these spaces it is well known that all fixed point classes have the same index and $N(f) = |L(f)|$ (cf. [J1, p.33]). So it remains to consider fiber-preserving homeomorphisms.

A further reduction is made possible by the following

Lemma 2.2. *Suppose M is an aspherical Seifert manifold. Then there is a finite regular covering $p : \tilde{M} \to M$ such that*

(1) *$\tilde{M}$ is an orientable circle bundle over an orientable surface F, and*
(2) *every homeomorphism $f : M \to M$ can be lifted to $\tilde{f} : \tilde{M} \to \tilde{M}$.*

Furthermore, if $\chi(F) > 0$ or F is the annulus, then $\tilde{M}$ is homeomorphic to $S^1 \times D^2$ or $T^2 \times I$.

Proof. It follows from [S, Lemma 3.1 and p.432 second paragraph] that for an aspherical Seifert manifold M, the base orbifold X is "good" in the sense that it is covered by a surface. Then by [S, Theorem 2.5], X is finitely covered by a surface. Thus M is finitely covered by a circle bundle over a surface. By taking at most two further double covers, we can obtain an orientable circle bundle over an orientable surface. Since every finite cover of the latter is still an orientable circle bundle over an orientable surface, (1) and (2) follow from Lemma 0.1.

124 BOJU JIANG AND SHICHENG WANG

The only orientable surfaces with $\chi(F) > 0$ are the 2-sphere S^2 and the disc D^2. S^2 is ruled out because $\tilde{M}$ is aspherical. When $F = D^2$, obviously $\tilde{M} = S^1 \times D^2$. When F is an annulus, it is clear that $\tilde{M} = T^2 \times I$. $\square$

By Lemma 0.2 and Lemma 2.2, Theorem 2.1 is now reduced to the following

Proposition 2.3. *Suppose M is an orientable circle bundle over an orientable compact surface F with $\chi(F) < 0$ or $F = T^2$, and $f : M \to M$ is a fiber preserving homeomorphism. Then the index of each fixed point class of f is not greater than 1. Furthermore, if f is orientation preserving, then the index of each essential fixed point class of f is ± 1.*

The remaining part of this section is devoted to the proof of Proposition 2.3.

We first study homeomorphisms on the base surface. For a compact surface F with $\chi(F) < 0$, a standard form of self-homeomorphisms is introduced in [JG, Sect.3]. When F is the torus T^2 represented as $\mathbb{R}^2/\mathbb{Z}^2$, a self-homeomorphism is said to be in standard form if it is covered by a linear map $\mathbb{R}^2 \to \mathbb{R}^2$.

Lemma 2.4. *Let F be a compact orientable surface with $\chi(F) < 0$ or $F = T^2$. Then every homeomorphism on F can be isotoped to a standard form $\varphi : F \to F$ such that $\varphi|_{\partial F} : \partial F \to \partial F$ is a periodic map and such that each fixed point class C' of φ is of one of the following types. (Figure 2 gives schematic pictures.)*

(1) *the center of a $2\pi\frac{k}{p}$ rotation, index = 1;*
(2) *a p-prong pseudo-Anosov fixed point, index = $1 - p$;*
(3) *a p-prong pseudo-Anosov fixed point with $2\pi\frac{k}{p}$ rotation, index = 1;*
(4) *a p-prong pseudo-Anosov fixed point with a reflection, index = -1, 0 or 1;*
(5) *a fixed point on a reflected p-prong pseudo-Anosov boundary component, index = 0 or 1;*
(6) *a fixed arc with two sides interchanged, index = -1, 0, or 1;*
(7) *a fixed circle which is a p-prong pseudo-Anosov boundary component, index = $-p$;*
(8) *a fixed circle which is a p-prong pseudo-Anosov boundary component from one side and a fixed boundary of an annular twist from another side, index = $-p$;*
(9) *a fixed circle as the axis of an annular reflection, index = 0;*
(10) *a fixed circle of an annular twist, index = 0;*
(11) *a fixed circle of an annular flip-twist, index = 0;*
(12) *a subsurface with boundary circles in ∂F or in $\text{int} F$, index = $\chi(C') \leq 0$.*

Thus every fixed point class C' of φ is connected, its $\text{index}(\varphi, C') \leq 1$. Furthermore $|\text{index}(\varphi, C')| \leq 1$ when φ is orientation reversing.

Proof. When $\chi(F) < 0$, this list is obtained from [JG, Lemma 3.6]. When φ is orientation reversing, only types (4), (5), (6), (9), (11) can occur, so $|\text{index}| \leq 1$. When F is the torus, the conclusion is clear. $\square$

Lemma 2.5. *Suppose F is an orientable surface with $\chi(F) < 0$, or F is the torus. Suppose $\varphi : F \to F$ is in standard form. Let one of A_0 and A_1 be a fixed point of*

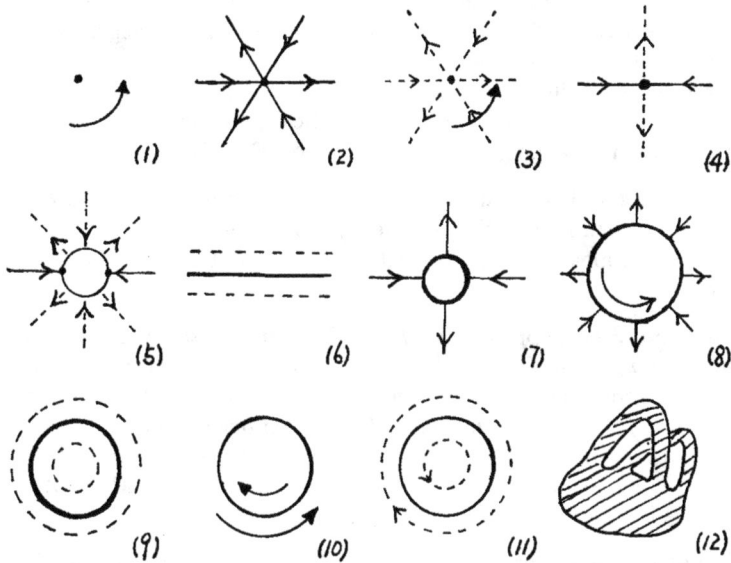

FIGURE 2

φ, the other either a fixed point or a φ-invariant component of ∂F. If there is a path $c: I, 0, 1 \to F, A_0, A_1$ such that $f \circ c \sim c: I, 0, 1 \to F, A_0, A_1$, then there is a path γ in $\mathrm{Fix}(\varphi)$ such that $\gamma \sim c: I, 0, 1 \to F, A_0, A_1$.

Proof. When $\chi(F) < 0$, this is [JG, Lemma 3.4]. When F is the torus, it is obvious. $\square$

Lemma 2.6. *Suppose M is an aspherical orientable circle bundle over an orientable surface. Then all fibers can be coherently oriented so that they represent the same element of infinite order in the center of $\pi_1(M)$.*

Proof. This is clear from the presentation of $\pi_1(M)$ given in [H, Chap. 12]. $\square$

Let a be the element of $\pi_1(M)$ represented by the fiber. We shall say that f *preserves* (or *reverses*) the fiber orientation if $f_*(a) = a$ (or $f_*(a) = -a$).

Lemma 2.7. *Suppose $q : M \to F$ is an orientable circle bundle over a compact orientable surface F where $\chi(F) < 0$ or $F = T^2$, and $f : M \to M$ is a fiber preserving homeomorphism. Then f can be deformed in a fiber preserving way so that the new map induces a map on F in standard form. Furthermore, for an f-invariant boundary torus T of M, if the restriction $f|_T : T \to T$ is isotopic to a periodic map, then we can require that after deformation the new f is periodic on T.*

Proof. Since the induced map on F can be deformed into standard form (Lemma 2.4), the first part is a direct consequence of the classical covering homotopy theorem for fiber bundles (cf. [St, p.50]).

Now that the induced map on the boundary circle $q(T)$ of F is periodic, the second requirement can be achieved by vertically sliding fibers near the boundary torus T (without changing the induced map on F). □

Remark. Observe that when f preserves the fiber orientation, in the second part of Lemma 2.7 we can require that the new f is fixed point free on T.

The next proposition will be used in the proof of Proposition 2.3 and also in Section 4.

Proposition 2.8. *Suppose $q : M \to F$ is an orientable circle bundle over an orientable compact surface F where $\chi(F) < 0$ or F is the torus. Suppose $f : M \to M$ is a fiber preserving homeomorphism that reverses the fiber orientation and the induced map $f' : F \to F$ is in standard form. Let one of A_0 and A_1 be a fixed point of f, the other either a fixed point or an f-invariant component of ∂M. If there is a path $c : I, 0, 1 \to M, A_0, A_1$ such that $f \circ c \sim c : I, 0, 1 \to M, A_0, A_1$, then there is a path γ in Fix(f) such that $\gamma \sim c : I, 0, 1 \to M, A_0, A_1$.*

Proof. Let $c' = q \circ c$ and $A_i' = q(A_i)$. Let us consider the case that both A_0 and A_1 are fixed points, the other cases are similar but easier. Since $f' \circ c' \sim c' : I, 0, 1 \to F, A_0', A_1'$, by Lemma 2.5, there is a path γ' in Fix(f') such that $\gamma' \sim c' : I, 0, 1 \to F, A_0', A_1'$.

Since Fix(f') is doubly covered by Fix(f), there is a path γ in Fix(f) that covers γ', starting from A_0 and ending at some $x_1 \in q^{-1}(A_1')$. Since $\gamma' \sim c' : I, 0, 1 \to F, A_0', A_1'$, the loop $\gamma'^{-1} c'$ is contractible in F. So the path $\gamma^{-1} c$ is homotopic in M to a path τ in the fiber $q^{-1}(A_1')$, i.e. $c \sim \gamma\tau$. Now $f \circ c \sim c$ implies $f \circ (\gamma\tau) = (f \circ \gamma)(f \circ \tau) = \gamma(f \circ \tau) \sim \gamma\tau$, hence $f \circ \tau \sim \tau$ in M. By Lemma 2.6, we see $f \circ \tau \sim \tau$ in the fiber $q^{-1}(A_1')$. But f reverses the orientation of the invariant fiber $q^{-1}(A_1')$, so τ must be a contractible loop. Therefore $c \sim \gamma : I, 0, 1 \to M, A_0, A_1$. □

Lemma 2.9. *Suppose $q : M \to F$ is an orientable circle bundle over an orientable compact surface F where $\chi(F) < 0$ or F is the torus. Suppose $f : M \to M$ is a fiber preserving homeomorphism that reverses the fiber orientation, and the induced map $f' : F \to F$ is in standard form. Let C' be a component of Fix(f'), and let $C = q^{-1}(C') \cap$ Fix(f). Then either*

(a) *C is a connected fixed point class of f and index$(f, C) = 2$ index$(f', C') \leq 0$, or*

(b) *C splits into two components C_1 and C_2, each a fixed point classes of f, and index$(f, C_1) = $ index$(f, C_2) = $ index$(f', C') \leq 1$.*

Proof. The conclusions follow easily from the fact that C' is doubly covered by C and the fact that the fixed point classes of f are connected (by Lemma 2.8). The index inequalities follow from Lemma 2.4 where only a C' of types (7)–(12) can admit connected double cover. □

Proof of Proposition 2.3. First suppose f preserves the fiber orientation. By perturbing the induced map $f' : F \to F$ and then lifting it, we deform f so that it has only a finite number of invariant fibers. Since f is orientation preserving on every invariant fiber, it is easily deformed along these fibers to a fixed point free map. Hence $L(f) = N(f) = 0$.

Now suppose f reverses the orientations of all its invariant fibers. By Lemma 2.7, we may assume that the induced map $f' : F \to F$ is in standard form. Now the conclusion follows from Lemma 2.9, and the observation (by Lemma 2.4) that every fixed point class of f' with index 1 must be a point or an arc. □

We have proved Proposition 2.3, hence also Theorem 2.1.

The above discussion also prepares us toward a "nice form" to be used in Section 4.

Proposition 2.10. *Suppose $q : M \to F$ is an orientable circle bundle over an orientable surface with boundary, $\chi(F) < 0$. Then M is a product bundle $F \times S^1$. Suppose $f : M \to M$ is a fiber preserving homeomorphism such that on each invariant boundary torus T of M, the restriction $f|_T : T \to T$ is isotopic to a periodic map. Then f can be deformed in a fiber preserving way so that*

(1) *when f preserves the fiber orientation, f is fixed point free;*
(2) *when f reverses the fiber orientation, the induced map $f' : F \to F$ is in standard form; and*
(3) *on each invariant boundary torus T of M, the restriction $f|_T : T \to T$ is a periodic map.*

Proof. The circle bundle is trivial because the base is homotopy equivalent to a bouqué of circles, and every orientable circle bundle over the circle is trivial.

By Lemma 2.7, we may assume that the induced map $f' : F \to F$ is in standard form, and on every invariant boundary torus T the restriction $f|_T$ is periodic. When f reverses the fiber orientation, our goal is achieved.

When f preserves the fiber orientation, by the Remark after Lemma 2.7, we may assume that each $f|_T$ is fixed point free. By perturbing the induced map $f' : F \to F$ rel ∂F and then lifting it, we deform f rel ∂M so that it has only a finite number of invariant fibers. Then get rid of all fixed points by sliding along these invariant fibers. □

SECTION 3. NICE FORM FOR HOMEOMORPHISMS ON $T^2 \times D^1$

The analysis in the next section also needs a nice form for homeomorphisms in the neighborhood of the decomposition tori. So in this section we study homeomorphisms on $T^2 \times I$. Although the argument is elementary, we shall do it carefully.

Let D^1 be the closed interval $[-1, 1]$ and $\varphi : T^2 \times D^1 \to T^2 \times D^1$ be a homeomorphism such that $\varphi|\partial(T^2 \times D^1)$ is periodic. We now consider a nice form of φ under isotopies relative to $\partial(T^2 \times D^1)$.

As preparation we introduce some notation and recall some facts concerning homeomorphisms of the torus.

A homeomorphism $f : \mathbb{R}^2 \to \mathbb{R}^2$ is said to be affine if it is of the form

(*) $f(x_1, x_2) = (x_1, x_2)A + (b_1, b_2) \qquad (x_1, x_2) \in \mathbb{R}^2$,

where $A \in GL_2(R)$ and (b_1, b_2) is a real vector.

Represent T^2 as $\mathbb{R}^2/\mathbb{Z}^2$. Then points of T^2 are written as $x = (x_1, x_2)$, where x_1, x_2 are real numbers mod 1. A homeomorphism $T^2 \to T^2$ is said to be affine if it lifts to an affine transformation $\mathbb{R}^2 \to \mathbb{R}^2$. Its matrix A is then in $GL_2(\mathbb{Z})$. We shall use liftings to specify maps.

Let H_1 be the identity component of the group H of homeomorphisms of T^2. Let $L \subset H$ be the subgroup of all affine homeomorphisms. We know $H = LH_1$. It follows that if $k \in H$ is conjugate to an affine map, it can be conjugated into L via some $h_1 \in H_1$. In fact, suppose $k = h^{-1}fh$ with $f \in L$, we have $h = gh_1$ with $g \in L$ and $h_1 \in H_1$. Then $k = h_1^{-1}(g^{-1}fg)h_1$ and $g^{-1}fg \in L$. The "mapping class group" $\pi_0(H) = H/H_1 \cong GL_2(\mathbb{Z})$, each component of H corresponding to the matrix A of its action on the first homology group of T^2.

For later reference we define some special matrices:

$$A_1 = \begin{pmatrix} 1 & 0 \\ 0 & 1 \end{pmatrix}, \qquad A_2 = \begin{pmatrix} 0 & 1 \\ -1 & 0 \end{pmatrix}, \qquad A_3 = \begin{pmatrix} 0 & 1 \\ -1 & -1 \end{pmatrix},$$

$$A_4 = \begin{pmatrix} 0 & 1 \\ -1 & 1 \end{pmatrix}, \qquad A_5 = \begin{pmatrix} 1 & 0 \\ 0 & -1 \end{pmatrix}, \qquad A_6 = \begin{pmatrix} 1 & 1 \\ 0 & -1 \end{pmatrix}.$$

It is well known that every periodic map f on T^2 is conjugate to an affine map. (T^2 admits an f-invariant Riemannian metric, hence also an f-invariant conformal structure. By the Uniformization Theorem in two-dimensional conformal geometry, f is conformal, hence isometric, with respect to a representation of T^2 as the complex plane modulo some lattice. Therefore f is affine with respect to the representation of T^2 as $\mathbb{R}^2/\mathbb{Z}^2$.) Thus up to conjugation f has the form (*). The matrix A is, up to conjugation in $GL_2(\mathbb{Z})$, one of the following seven matrices (cf. [H, p.123]): $A_1, -A_1, A_2, A_3, A_4, A_5, A_6$. Periodicity implies both b_1, b_2 are rational, and by translation we can make $b_1 = b_2 = 0$ if $A = -A_1, A_2, A_3, A_4$, and $b_2 = 0$ if $A = A_5, A_6$. When $A = A_1$, we can make $b_1 = 0$ by conjugation in $GL_2(\mathbb{Z})$.

A homeomorphism $f : \mathbb{R}^2 \times D^1 \to \mathbb{R}^2 \times D^1$ is said to be *affine* if it is the restriction to $\mathbb{R}^2 \times D^1$ of some affine map on $\mathbb{R}^3$, i.e. of the form

(**) $f(x, s) = (f_s(x), \pm s), \qquad f_s(x_1, x_2) = (x_1, x_2)A + (b_1, b_2) + s(c_1, c_2),$

where $A \in GL_2(R)$, $b = (b_1, b_2)$ and $c = (c_1, c_2) \in \mathbb{R}^2$. A homeomorphism $N \to N$ is said to be affine if it lifts to an affine map $\mathbb{R}^2 \times D^1 \to \mathbb{R}^2 \times D^1$. Its matrix A is then in $GL_2(\mathbb{Z})$.

Proposition 3.1. Let $\varphi : T^2 \times D^1 \to T^2 \times D^1$ be a homeomorphism. Then φ is isotopic rel $\partial(T^2 \times D^1)$ to a homeomorphism $\psi : T^2 \times D^1 \to T^2 \times D^1$ with the following properties. When φ does not interchange the boundary components and $\varphi|\partial(T^2 \times D^1)$ is periodic, then either

(1) every fixed point class of ψ is inessential and is a circle or a torus in $\partial(T^2 \times D^1)$, in particular there are no fixed points in $\text{int}(T^2 \times D^1)$; or

(2) every fixed point class of ψ is an arc intersecting both components of $\partial(T^2 \times D^1)$ transversely, index $= 1$.

When φ interchanges the boundary components, then

(3) every fixed point class of ψ is a point in $\text{int}(T^2 \times D^1)$, index $= \pm 1$.

In any case, $\text{Fix}(\psi) \cap \text{int}(T^2 \times D^1)$ is a disjoint union of points and open arcs.

Proof. Since $T^2 \times D^1$ is a Haken manifold, homotopic homeomorphisms are isotopic.

First suppose $\varphi : T^2 \times D^1 \to T^2 \times D^1$ preserves the boundary components, and $\varphi|\partial(T^2 \times D^1)$ is periodic. We set out to simplify φ by conjugation on $T^2 \times D^1$ and by homotopy rel $\partial(T^2 \times D^1)$. We claim that φ can be so reduced to an affine map of the form (**), where A is one of $A_1, -A_1, A_2, A_3, A_4, A_5, A_6$.

Since φ_{-1} and φ_1 are periodic, there exist h_{-1} and $h_1 \in H_1$ such that $h_{-1}\varphi_{-1}h_{-1}^{-1}$ and $h_1\varphi_1 h_1^{-1} \in L$. Let $\{h_s\}_{s \in D^1}$ be a path in H_1 from h_{-1} to h_1. Via the homeomorphism $h : T^2 \times D^1 \to T^2 \times D^1$, $(x, s) \mapsto (h_s(x), s)$, φ is conjugated (to $h\varphi h^{-1}$ which we still call φ) so that both φ_{-1} and φ_1 are affine maps of T^2. Since φ_{-1} and φ_1 are homotopic, they have the same matrix A. By conjugation in $GL_2(\mathbb{Z})$ we can assume A is one of $A_1, -A_1, A_2, A_3, A_4, A_5, A_6$. Lift $\varphi : T^2 \times D^1 \to T^2 \times D^1$ to $\tilde{\varphi} : \mathbb{R}^2 \times D^1 \to \mathbb{R}^2 \times D^1$. Define $\tilde{\psi} : \mathbb{R}^2 \times D^1 \to \mathbb{R}^2 \times D^1$ by

$$\tilde{\psi}(x, s) = \frac{1-s}{2}\tilde{\varphi}(x, -1) + \frac{1+s}{2}\tilde{\varphi}(x, 1) \qquad x \in \mathbb{R}^2, s \in D^1.$$

It is easy to see that the homotopy $\{(1-t)\tilde{\varphi} + t\tilde{\psi}\}_{t \in I}$ projects to a homotopy $\varphi \simeq \psi$ on $T^2 \times D^1$ rel $\partial(T^2 \times D^1)$, where ψ is clearly of the form (**).

Consider two distinct cases:

(1) $A = A_1, A_5$ or A_6. Since $T^2 \times D^1$ is homotopy equivalent to T^2, we have $N(\psi) = L(\psi) = 0$. Hence every fixed point class of ψ is inessential. Deform ψ rel $\partial(T^2 \times D^1)$ to a new ψ of the form

$$\psi(x, s) = (xA + b'(s), s^3),$$

where $b'(s)$ is a vector function such that $b'(0) = (\frac{1}{2}, 0) \in \mathbb{R}^2$. This ψ is of type (1).

(2) $A = -A_1, A_2, A_3$ or A_4. Via affine conjugation we can replace ψ with a new ψ of the form $\psi(x, s) = (xA, s)$. For example, when $A = -A_1$, define $h : T^2 \times D^1 \to T^2 \times D^1$ by $h(x, s) = (x - \frac{1}{2}b - \frac{s}{2}c, s)$, then $h\varphi h^{-1}$ will be in the desired form with $b = c = 0$. The fixed point behavior of this ψ is of type (2).

Next suppose φ interchanges the boundary components. Then obviously φ is isotopic rel boundary to a homeomorphism ψ of the form $\psi(x, s) = (\psi_s(x), -s)$, where $\psi_0 : T^2 \to T^2$ is an affine map. Moreover, ψ_0 can be made fixed point free if the matrix A has eigenvalue 1. This ψ is of type (3). $\quad\square$

SECTION 4. 3-MANIFOLDS WITH NON-TRIVIAL TORUS DECOMPOSITION

Recall that a 3-manifold M admits a non-trivial torus decomposition, if M can be cut along a non-empty minimal collection Γ of two-sided incompressible tori so that each piece is either a Seifert manifold or a hyperbolic manifold. The minimal collection Γ of decomposition tori is unique up to isotopy. Hence we can assume that the homeomorphism $f : M \to M$ leaves an open regular neighborhood $\mathcal{N}(\Gamma)$ of Γ invariant, i.e. $f(\mathcal{N}(\Gamma)) = \mathcal{N}(\Gamma)$. When we refer to a *piece* of M, we always mean a component of $M - \mathcal{N}(\Gamma)$.

Theorem 4.1. *Suppose M is a closed 3-manifold finitely covered by a 3-manifold with non-trivial torus decomposition and $f : M \to M$ is a homeomorphism. Then the index of each fixed point class of f is not greater than 1, hence $L(f) \leq N(f)$. Furthermore, if f is orientation preserving, then the index of each essential fixed point class of f is ± 1, hence $|L(f)| \leq N(f)$.*

A Seifert piece of M will be called a product piece if it is homeomorphic to $F \times S^1$ for some compact orientable surface F.

Proposition 4.2. *If M is a closed 3-manifold with non-trivial torus decomposition, then there is an orientable finite cover $\tilde{M}$ of M such that each Seifert piece of $\tilde{M}$ is a product piece.*

Proof. See [W] for a short proof based on the residual finiteness of the fundamental group of Haken 3-manifolds. □

When M is an orientable 3-manifold with non-trivial torus decomposition such that each Seifert piece of M is a product piece, every finite cover of M is also a manifold of this type. Therefore, by Lemmas 0.1 and 0.2, to prove Theorem 4.1 it suffices to prove the following

Proposition 4.3. *Suppose M is a closed orientable 3-manifold with non-trivial torus decomposition such that each Seifert piece of M is a product piece. Suppose $f : M \to M$ is homeomorphism. Then the index of each fixed point class of f is not greater than 1, hence $L(f) \leq N(f)$. If furthermore, f is orientation preserving, then the index of each essential fixed point class of f is ± 1, hence $|L(f)| \leq N(f)$.*

The remaining part of this section is devoted to the proof of Proposition 4.3.

Lemma 4.4. *Suppose M is a closed manifold with non-trivial torus decomposition such that each Seifert piece of M is a product piece. Then each product piece has a unique Seifert fibration, hence is of the form $F \times S^1$ with $\chi(F) < 0$, except when M is a tours bundle over the circle and Γ is a single fiber.*

Proof. Since each decomposition torus is incompressible, there is no product piece homeomorphic to $S^1 \times D^2$. Since Γ is minimal, there is no product piece homeomorphic to $T^2 \times I$ except when the two boundary components of $T^2 \times I$ arise from cutting M along the same torus. In this exceptional case M is a torus bundle over the circle. In all other cases, it follows from [S, Theorem 3.9] that each product piece has a unique Seifert fibration. □

The exceptional case is simple and will be treated first.

Proof of Proposition 4.3 for the Exceptional Case. M is a torus bundle over the circle and Γ is a single fiber. Since the homeomorphism f makes Γ invariant, up to isotopy we can assume that f is fiber preserving and induces a homeomorphism $f' : S^1 \to S^1$ on the base.

If f' is of degree 1, then we can deform f so that f is fixed point free. Hence $L(f) = N(f) = 0$.

If f' is of degree -1, then f has two invariant fibers T_1 and T_2. Obviously the fixed points in different fibers are in different fixed point classes of f. We claim that every fixed point class of f intersects T_i in a fixed point class of $f|_{T_i}$. In fact, suppose $x, y \in \text{Fix}(f|_{T_i})$ are in the same fixed point class of f, then there is a path $c : I, 0, 1 \to M, x, y$ such that $f \circ c \sim c : I, 0, 1 \to M, x, y$. Let $c' = q \circ c$, then c' is a loop at $q(x)$ and $f' \circ c' \sim c'$. However f' is orientation reversing, this implies that $c' \sim 0$. So c can be deformed into T_i. Now $\pi_1(T_i)$ injects into $\pi_1(M)$, hence x and y are in the same fixed point class of $f|_{T_i}$, as we claimed. Since the index of each essential fixed point class of $f|_{T_i}$ is ± 1, the index of each essential fixed point class of f is also ± 1. $\square$

From now on, we assume the generic case that every product piece of M is of the form $F \times S^1$ with $\chi(F) < 0$, and $f : M \to M$ is fiber preserving on product pieces.

Lemma 4.5. *Let T be an f-invariant component of $\partial \mathcal{N}(\Gamma)$. Then the restriction f_T is isotopic to a periodic map.*

Proof. Suppose T is a boundary component of a component N of $\overline{\mathcal{N}(\Gamma)}$. N is homeomorphic to $T^2 \times I$. Let its another boundary component be T'. Suppose T and T' are boundary components of decomposition pieces P and P' respectively.

Since f leaves T invariant, T', P and P' are also f-invariant. $f|_T$ is isotopic to a periodic map if and only if $f|_{T'}$ is isotopic to a periodic map. If P is a hyperbolic piece, then $f|_T$ is a periodic map. Similarly for P'. Now suppose both P and P' are product pieces. Let c and c' be a fiber of P and P' respectively. Since f is fiber preserving, both c and c' are invariant under $f|_N$ up to isotopy. Since Γ is minimal, P and P' can not be combined into one big product piece, so c and c' form a basis of $H_1(N)$. It follows that $f|_N$, therefore $f|_T$ also, is of order 1 or 2 up to isotopy. $\square$

Now we deform f to a nice form that facilitates the analysis of fixed point classes.

Let P be the generic notation for an f-invariant component of $M - \mathcal{N}(\Gamma)$, N be the generic notation for an f-invariant component of $\overline{\mathcal{N}(\Gamma)}$, and T be the generic notation for an f-invariant component of $\partial \mathcal{N}(\Gamma)$.

We isotope f so that f restricted to each P takes the nice forms specified at the end of Sections 1 and 2. Then further isotope rel $M - \mathcal{N}(\Gamma)$ so that f restricted to each N takes the nice forms of Proposition 3.1.

Let C be a component of $\text{Fix}(f)$. Observe from the construction above that every component of $C \cap \partial \mathcal{N}(\Gamma)$ is either a point or a circle. Also observe that if one component of $C \cap \partial \mathcal{N}(\Gamma)$ is a circle, then every component of $C \cap \partial \mathcal{N}(\Gamma)$ is a circle. Because C is a union of components of $\mathrm{Fix}(f|_P)$ and $\mathrm{Fix}(f|_N)$ and we know from Sections 1–3 that the latter components have the desired property.

Definition. Define B to be the union of the components C of $\mathrm{Fix}(f)$ such that one (hence every) component of $C \cap \partial \mathcal{N}(\Gamma)$ is a circle. Define W to be $\mathrm{Fix}(f) - B$. Alternatively, W is the union of the components C of $\mathrm{Fix}(f)$ such that either $C \cap \partial \mathcal{N}(\Gamma)$ is empty or every component of $C \cap \partial \mathcal{N}(\Gamma)$ is a point.

We need a 3-dimension analogue of [JG, Lemma 3.3].

Lemma 4.6. *Let $T = T_1 \cup \cdots \cup T_n$ be a disjoint union of 2-sided incompressible surfaces in* $\mathrm{int} M$. *Let $l : S^1 \to M$ be a loop homotopically trivial in M. Assume that l is transverse to T.*

(i) If l is in $M - T$, then l is homotopically trivial in $M - T$.

(ii) If l does cross T, then it must have at least two "compressible segments".

Here by a "segment" we mean a component E of $S^1 - l^{-1}(T)$; it is "compressible" if $l|A$ can be deformed through $M - T$ into T rel endpoints, i.e. if $l|E$ can be extended to $h : D \to M$ such that $h^{-1}(T) = \overline{S^1 - E}$, D being the closed unit disc bounded by S^1.

Proof. The same as the proof of [JG, Lemma 3.3]. □

Lemma 4.7. *Let A_0, A_1 be two fixed points in W. Suppose A_0 and A_1 are f-related via a path $c : I, 0, 1 \to M, A_0, A_1$. Then there is a path γ in $\mathrm{Fix}(f)$ such that $\gamma \sim c : I, 0, 1 \to M, A_0, A_1$.*

Proof. Suppose the path c is transverse to $\partial \mathcal{N}(\Gamma)$ and crosses $\partial \mathcal{N}(\Gamma)$ k times. We prove the lemma by induction on k.

The case $k = 0$: The path is either contained in a component P of $M - \mathcal{N}(\Gamma)$, or contained in a component N of $\overline{\mathcal{N}(\Gamma)}$. In the former case, $f(P) = P$ and the conclusion follows from Lemmas 1.3 and 2.8. In the latter case, $f(N) = N$. The conclusion follows from Proposition 3.1 and the definition of W.

The inductive step: Assume inductively that the conclusion is true if the number of crossings is less than k.

Suppose now the path c crosses $\partial \mathcal{N}(\Gamma)$ at k points $z_1, \cdots, z_k$ that cut c into $k+1$ subpaths $c_0, c_1, \cdots, c_k$. The condition $c \sim f \circ c : I, 0, 1 \to M, A_0, A_1$ implies

$$\ell := c_1 \cdots c_k (f \circ c_k)^{-1} ... (f \circ c_0)^{-1} c_0$$

is null homotopic in M. Let $e_0 := (f \circ c_0)^{-1} c_0$ and $e_1 := c_k (f \circ c_k)^{-1}$.

Without loss we may assume that none of the subpaths $c_0, c_1, \cdots, c_k$ is a compressible segment, because otherwise c is homotopic rel endpoint to a path crossing $\mathcal{N}(\Gamma)$ less than k times so the desired conclusion follows from the inductive hypothesis. Then none of $(f \circ c_0), \cdots, (f \circ c_k)$ can be compressible either, since $f \circ c_i$ is compressible iff c_i is compressible. Now Lemma 4.6 guarantees at least two compressible segments of ℓ. So the only possible segment decomposition of ℓ is $\ell = c_1 \cdots c_{k-1} e_1 (f \circ c_{k-1})^{-1} \cdots (f \circ c_1)^{-1} e_0$, and both e_0 and e_1 are compressible segments.

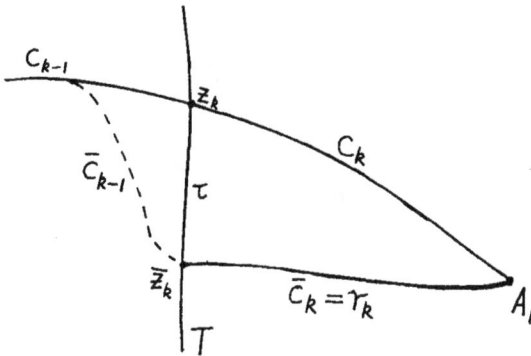

FIGURE 3

If e_1 is in a component P of $M - \mathcal{N}(\Gamma)$, then either P is a hyperbolic piece, or P is a product piece and f reverses the orientation of S^1. Suppose z_k lies in a boundary component T of P. Then $f(P) = P$ and $f(T) = T$. The compressibility of e_1 into T implies that T is $(f|_P)$-related to A_1 via the path c_k. It follows from Lemma 1.3 or Lemma 2.8 that there is a path γ_k in $\text{Fix}(f)$ and a path τ in T so that $c_k \sim \tau\gamma_k$ rel endpoints.

If e_1 is in a component N of $\overline{\mathcal{N}(\Gamma)}$ and z_k lies in a boundary component T of N, the compressibility of e_1 implies that $f(T) = T$. A_1 lies in W implies that the restriction of f on N is of type (2) of Proposition 3.1. So, as in the previous paragraph, there is a also path γ_k in $\text{Fix}(f)$ and a path τ in T so that $c_k \sim \tau\gamma_k$ rel endpoints.

Adjusting c_{k-1}, c_k as indicated in Figure 3 (and then removing the bar from the notation) we may assume that $c_k = \gamma_k$ and z_k is a fixed point.

Consider the path $c' = c_0 c_1 ... c_{k-1}$ from A_0 to z_k. From $\ell \sim 0$ and $e_1 \sim 0$, we have $c' \sim f \circ c' : I, 0, 1 \to M, A_0, z_k$. This c' crosses $\partial \mathcal{N}(\Gamma)$ only $k - 1$ times (cf. Fig.3). By the inductive hypothesis, there is a path γ' in $\text{Fix}(f)$ such that $\gamma \sim c : I, 0, 1 \to M, A_0, z_k$. It follows that $\gamma := \gamma'\gamma_k$ is the desired path of the conclusion.

The inductive step is now complete. $\square$

Corollary 4.8. *Every fixed point class of f, as a disjoint union of components of B and W, contains at most one component of W.* $\square$

For the calculation of the indices, we quote the following Mayer-Vietoris type formula ([JG, Lemma 3.7]).

Lemma 4.9. *Let $f : X \to X$ be a self-map of a compact polyhedron X. Suppose X_0, X_1, X_2 are subpolyhedra of X such that $X = X_1 \cup X_2$, $X_0 = X_1 \cap X_2$. Suppose $f(X_i) \subset X_i$ and write $f_i : X_i \to X_i$ for the restriction of f, for $i = 0, 1, 2$.*

Let $A \subset \text{Fix}(f)$ be both open and closed in $\text{Fix}(f)$, and let $A_i = A \cap X_i$, $i = 0, 1, 2$. Then

$$\text{index}(f, A) = \text{index}(f_1, A_1) + \text{index}(f_2, A_2) - \text{index}(f_0, A_0). \quad \square$$

Lemma 4.10. *The possible values of* $\text{index}(f, C)$, *for all types of components* C *of* $\text{Fix}(f)$, *are listed in the following table, where the subscript* $+$ *or* $-$ *means* $f : M \to M$ *is orientation preserving or reversing,* $\varnothing$ *means no such* C *is available.*

type of C	$C_+ \subset W$	$C_+ \subset B$	$C_- \subset W$	$C_- \subset B$
a point	$\pm 1, 0$	$\varnothing$	≤ 1	$\varnothing$
an arc	$\pm 1, 0$	$\varnothing$	$\varnothing$	$\varnothing$
a circle	0	0	≤ 0	≤ 0
a surface	$\varnothing$	$\varnothing$	≤ 0	≤ 0

Proof. From Sections 1–3 we collect the following three tables of indices for all types of components E of $\text{Fix}(f|_P)$ or $\text{Fix}(f|_N)$.

(1) $\text{index}(f|_P, E)$ for a hyperbolic piece P.

type of E	$E_+ \subset W$	$E_+ \subset B$	$E_- \subset W$	$E_- \subset B$
a point	$\varnothing$	$\varnothing$	1	$\varnothing$
an arc	1	$\varnothing$	$\varnothing$	$\varnothing$
a circle	0	$\varnothing$	$\varnothing$	$\varnothing$
a surface	$\varnothing$	$\varnothing$	≤ 0	≤ 0

(2) $\text{index}(f|_P, E)$ for a Seifert piece P where f reverses the fiber orientation.

type of E	$E_+ \subset W$	$E_+ \subset B$	$E_- \subset W$	$E_- \subset B$
a point	$\pm 1, 0$	$\varnothing$	≤ 1	$\varnothing$
an arc	$\pm 1, 0$	$\varnothing$	$\varnothing$	$\varnothing$
a circle	0	0	≤ 0	≤ 0
a surface	$\varnothing$	$\varnothing$	≤ 0	≤ 0

(3) $\text{index}(f|_N, E)$ for a component N of $\overline{\mathcal{N}(\Gamma)}$.

type of E	$E_+ \subset W$	$E_+ \subset B$	$E_- \subset W$	$E_- \subset B$
a point	$\varnothing$	$\varnothing$	≤ 1	$\varnothing$
an arc	1	$\varnothing$	$\varnothing$	$\varnothing$
a circle	$\varnothing$	0	$\varnothing$	0
a surface	$\varnothing$	$\varnothing$	$\varnothing$	$\varnothing$

The table of the Lemma follows from these three tables by examining how these E's join together at $\partial \mathcal{N}(\Gamma)$ to form C. Use the previous lemma for index calculation. The interesting case is to see how arcs join together to form an arc or a circle. $\quad \square$

Proof of Proposition 4.3 for the Generic Case. The desired conclusion follows from Corollary 4.8 and Lemma 4.10. $\quad \square$

SECTION 5. EXAMPLES

Let T^0 be a point and T^n be the product of n copies of S^1's. Let h_0 be the identity on T^0, h_1 be the orientation reversing involution on $T^1 = S^1$. Let h_n be defined on T^n as the product of n copies of h_1's. Then h_n is a homeomorphism on T^n with 2^n fixed point classes of index 1.

Example 1. Let ϕ be an Anosov map on T^2 with more than two fixed points such that each fixed point of ϕ is a fixed point class of ϕ of index -1. Let F_2 be a closed orientable surface of genus 2 and $p : F_2 \to T^2$ be a double cover branched over the two fixed points of ϕ. Then there is a lift $\tilde{\phi}$ of ϕ such that (1) two of the fixed points of $\tilde{\phi}$ have index -3 and the other fixed points of $\tilde{\phi}$ have index -1, (2) $\tilde{\phi}$ is pseudo Anosov; hence two fixed point classes of $\tilde{\phi}$ have index -3 and other fixed point classes of $\tilde{\phi}$ have index -1.

For any $n > 3$ we define the homeomorphism f as

$$\tilde{\phi} \times \phi \times h_{n-4} : F_2 \times T^2 \times T^{n-4} \to F_2 \times T^2 \times T^{n-4}$$

Then $2 \times 2^{n-4}$ fixed point classes of f has index 3 and other fixed point classes of f have index 1; Hence $L(f) > N(f)$. Here $F_2 \times T^2 \times T^{n-4}$ is aspherical.

The following example is simpler but less interesting.

Example 2. Let id be the identity map on $M^4 = F_2 \times F_2$. Let

$$f = id \times h_{n-4} : M^4 \times T^{n-4} \to M^4 \times T^{n-4}$$

Then all the fixed point classes of f have index 4; hence $L(f) > N(f)$. Here $M^4 \times T^{n-4}$ is aspherical.

Example 3. Consider S^1 as the unit circle in the complex plane and define $g : S^1 \to S^1$ to be $g(z) = z^2$. Then

$$f = \tilde{\phi} \times g : F_2 \times S^1 \to F_2 \times S^1$$

is a map on an aspherical 3-manifold such that $L(f) > N(f)$.

Example 4. Let id be the identity map on S^2. Then

$$f = id \times h_1 : S^2 \times S^1 \to S^2 \times S^1$$

is a homeomorphism on a closed 3-manifold such that $4 = L(f) > N(f) = 2$.

Acknowledgement. We would like to express our gratitude to the referee for his comments which led to extensive revision of the exposition. We also wish to thank Professor Chengye You for helpful conversations.

136   BOJU JIANG AND SHICHENG WANG

REFERENCES

[B]   Brown, R.F., *The Lefschetz Fixed Point Theorem*, Scott, Foresman and Co., Chicago, 1971.
[H]   Hempel, J., *3-manifolds*, Ann. Math. Studies vol. 86, Princeton Univ. Press, Princeton, 1976.
[J]   Jaco, W., *Lectures on Three-manifold Topology*, CBMS Regional Conference Series in Math. vol. 43, AMS, Providence, 1980.
[J1] Jiang, B., *Lectures on Nielsen Fixed Point Theory*, Contemporary Math. vol. 14, AMS, Providence, 1983.
[J2] Jiang, B., *Fixed points of surface homeomorphisms*, Bull. AMS **5** (1981), 176–178.
[JG] Jiang, B., Guo, J., *Fixed points of surface diffeomorphisms*, Research Report No.13, Dept. of Math., Peking Univ., 1990.
[M]   Morgan, J.W., *On Thurston's uniformization theorem for three-dimensional manifolds*, The Smith Conjecture (Morgan, J. et al., eds.), Academic Press, New York, 1984, pp. 37–125.
[S]   Scott, P., *The geometries of 3-manifolds*, Bull. London Math. Soc. **15** (1983), 401–487.
[St] Steenrod, N.E., *The Topology of Fibre Bundles*, Princeton Univ. Press, Princeton, 1951.
[T1] Thurston, W., *The Geometry and Topology of 3-manifolds*, preprint.
[T2] Thurston, W., *Three dimensional manifolds, Kleinian groups and hyperbolic geometry*, Bull. AMS **6** (1982), 357–381.
[T3] Thurston, W., *On the geometry and dynamics of diffeomorphisms of surfaces*, Bull. AMS **19** (1988), 417–431.
[W]   Wang, S., *The existence of nonzero degree maps between closed aspherical 3-manifolds*, Research Report No.23, Dept. of Math., Peking Univ., 1990 (to appear in Math. Zeit.).

DEPARTMENT OF MATHEMATICS, PEKING UNIVERSITY, BEIJING 100871, CHINA

CYCLIC HOMOLOGY OF TRIANGULAR MATRIX ALGEBRAS

LARS KADISON

Roskilde University

ABSTRACT. We compute the cyclic homology of general upper triangular matrix algebras. The computation is interesting because it is valid over the integers or any other commutative ring, and gives an outcome that algebraic K-theory also satisfies. The method of computation is to use relative homological algebra by defining new cyclic homology groups for subalgebras containing unity. One then reduces the problem to relative cyclic chains where the canonical idempotents in the triangular algebra are free to move across the tensors. This method also simplifies previous presentations of Morita invariance of cyclic homology, as well as a special case of the Künneth formula of C. Kassel. We end with a survey of other applications of relative cyclic homology to group rings and the Bass conjecture, to cyclic homology of quiver algebras, to norm continuous cyclic cohomology of nest algebras, and to algebraic K-theory.

1. INTRODUCTION

The purpose of this paper is to give an account of the methods and proof of a property of cyclic homology shared by algebraic K-theory [1, 7, 20, 26]:

Theorem 1.1. $HC_n \begin{pmatrix} R & M \\ 0 & T \end{pmatrix} = HC_n(R) \oplus HC_n(T) \quad (n = 0, 1, 2, \dots)$[1]

R and T are unital k-algebras where k is an arbitrary unital commutative ring, M is a unitary R–T bimodule, and the left-hand side is given matrix addition and multiplication.

The method of proof, oddly enough, is to *generalize* cyclic homology of a k-algebra A, $HC_*(A)$, to noncommutative or non-central scalars in a subalgebra S of A, which contains the unity element 1_A. We get homology groups $HC_n(A, S)$ which we call relative cyclic homology groups because they are related by the Connes long exact sequence to the relative Hochschild homology groups of [13]. When $S = k1_A$ we recover ordinary cyclic homology $HC_n(A)$. For any unital subalgebra there is a natural homomorphism $\phi : HC_n(A) \to HC_n(A, S)$. If S is a separable k-algebra, then ϕ is an isomorphism. Then choosing the right scalars, we give direct proofs at the level of cyclic module for Theorem 1.1 and Morita invariance of cyclic homology, which we revisit in Section 6.

---

[1]announced in [14] and independently in [31].

Cyclic homology of algebras is a noncommutative generalization of de Rham cohomology of manifolds [5, 23] that is related to K-theory via a generalized Chern character [18]. Many similarities of cyclic theory with algebraic K-theory were known by 1983: e.g., Morita invariance of both theories, the Loday-Quillen-Tsygan theorem which provides an additive version of the Milnor-Moore-Quillen theorem in rational algebraic K-theory [23], and relations with algebraic K-theory of spaces [21] and rational algebraic K-theory relativized by a nilpotent ideal [10]. It became urgent and interesting to re-compute, and in many cases extend, the well-known computations of K-theory for the new cyclic homology functor. Thus, cyclic homology of many rings of functions, filtered algebras, group algebras, and self-adjoint operator algebras were computed: the last two computations led, respectively, to much progress on the Bass conjecture in group theory [8], and the settling via H-unitality of the Karoubi conjecture stating the equality of topological and algebraic K-theories of stable C^*-algebras [30]. The methods of computation that evolved from the most basic one of the Connes long exact sequence and projective resolution of homological algebra include: mixed complexes and strongly homotopy linear maps [19], homotopy-theoretic computations with fibrations associated to a cyclic set [3], an assortment of spectral sequences [2, 5, 27, 31], and H-unitality of C^*-algebras [31].

In particular, Theorem 1.1 and its analogue in algebraic K-theory together indicate some extension of Goodwillie's theorem to rings not containing the rationals [22]. This paper is organized as follows. Two preliminary sections cover the theory of cyclic modules in the relative case, and separable k-algebras from the homological viewpoint of [12]. In Section 4 we prove our main theorem, which is already known and applied in Hochschild cohomology with general coefficients [9], though we give a new simplicial proof for the cyclic theory. In Section 5 we prove Theorem 1.1. In Section 6 we give a new, simplified proof of Morita invariance of cyclic homology [29] and a special case of the Künneth theorem [19].[2] We conclude with Section 7 in which is discussed a key role played by relative cyclic homology in getting further results on the Bass conjecture [28], continuous cyclic cohomology of non-self adjoint operator algebras, and other computations that use a relative homological approach.

2. The Cyclic Module $Z(A, S)$

Fix the notation $k1_A \subseteq S \subseteq A$ of Section 1. Let A^e denote $A \otimes_k A^{op}$, the device that turns bimodules into one-sided modules.

Define a cyclic k-module $Z(A, S)$, i.e., a simplicial object in k-**Mod** with actions from all finite cyclic groups, [5, 10] by

$$Z_n(A, S) = \underbrace{A \otimes_S \cdots \otimes_S A}_{n+1} \otimes_{S^e} S.$$

---

[2]We thank Loday for pointing out this approach.

Note that $Z_0(A, S) = A/[A, S]$, and elements of $Z_n(A, S)$ can be written as linear combinations of $a_0 \otimes_S \cdots \otimes a_n$ with the extra condition

$$sa_0 \otimes_S \cdots \otimes a_n = a_0 \otimes_S \cdots \otimes a_n s \quad \forall s \in S$$

The $n + 1$ face maps $d_i : Z_n(A, S) \to Z_{n-1}(A, S)$ are defined by

$$d_i(a_0 \otimes_S \cdots \otimes a_n) = a_0 \otimes_S \cdots \otimes a_i a_{i+1} \otimes \cdots \otimes a_n \quad (i = 0, 1, \ldots, n - 1)$$
$$d_n(a_0 \otimes_S \cdots \otimes a_n) = a_n a_0 \otimes \cdots \otimes a_{n-1}.$$

The action of the cyclic group $Z/n + 1$ on $Z_{n+1}(A, S)$ is given by

$$t_{n+1}(a_0 \otimes_S \cdots \otimes a_n) = a_n \otimes_S a_0 \otimes \cdots \otimes a_{n-1}.$$

Even in the presence of non-central scalars S, the face and cyclic action maps are well-defined thanks to the extra circularity relation.

Degeneracy maps can also be defined:

$$s_i(a_0 \otimes_S \cdots \otimes a_n) = a_0 \otimes_S \cdots \otimes a_i \otimes 1 \otimes a_{i+1} \otimes \cdots \otimes a_n \quad (i = 0, 1, \cdots, n).$$

Cyclic modules satisfy the usual simplicial relations [24] together with three additional relations involving t_{n+1} [10], the most obvious being $t_{n+1}^{n+1} = 1$.

Lemma 2.1. $Z(A, S)$ is a cyclic module.

Proof. If S is the unit subalgebra $k1_A$ then $Z(A, S)$ is the usual cyclic module ZA associated to an algebra [10]. There is a canonical surjection $\phi : ZA \to Z(A, S)$ given by

$$a_0 \otimes_k \cdots \otimes a_n \longmapsto a_0 \otimes_S \cdots \otimes a_n,$$

which commutes with face, degeneracy, and cyclic action maps. Then $Z(A, S)$ satisfies all the relations of ZA; in particular, it is a cyclic module.□

For the convenience of the reader we review the theory of cyclic modules [10] in order to show how relative cyclic and Hochschild homologies are defined and related by the Connes long exact sequence.

The first step is to form the Tsygan first quadrant bicomplex $C_{**}(A, S)$:

$$
\begin{array}{ccccccc}
\downarrow b & & \downarrow -b' & & \downarrow b & & \\
Z_{n+1}(A, S) & \xleftarrow{1-T} & Z_{n+1}(A, S) & \xleftarrow{N} & Z_{n+1}(A, S) & \xleftarrow{1-T} & \\
\downarrow b & & \downarrow -b' & & \downarrow b & & \\
Z_n(A, S) & \xleftarrow{1-T} & Z_n(A, S) & \xleftarrow{N} & Z_n(A, S) & \xleftarrow{1-T} & \\
\downarrow b & & \downarrow -b' & & \downarrow b & & \\
\end{array}
$$

where $b, b' : Z_n(A, S) \to Z_{n-1}(A, S)$ $(n \geq 1)$ are defined by

$$b = \sum_{i=0}^{n}(-1)^i d_i \quad b' = \sum_{i=0}^{n-1}(-1)^i d_i = b - (-1)^n d_n$$

and $N, T : Z_n(A, S) \to Z_n(A, S)$ are defined by

$$T = (-1)^n t_{n+1} \quad N = 1 + T + \cdots + T^n$$

The rows and columns are indeed chain complexes, and the squares commute [23]. The even columns are identically $(Z_n(A, S), b)$, which we call the relative Hochschild complex with groups $HH_n(A, S)$ $(= \mathrm{Tor}_n^{(A^e, S \otimes A^{op})}(A, A))$, a functor of relative homological algebra [11]). The odd rows are identically $(Z_n(A, S), -b')$, an acyclic complex since $b's + sb' = 1$ where $s = (-1)^n s_n$.

Definition 2.1. *Relative cyclic homology is defined as*

$$HC_n(A, S) = H_n(\mathrm{Tor}\, C_{**}(A, S), d' + d''),$$

i.e., as the homology of the total complex with $d' + d''$ the sum of vertical and horizontal differentials.

Lemma 2.2. *If k is a field of characteristic zero, then*

$$HC_*(A, S) = H_*(Z_n(A, S)/\mathrm{im}(1 - T), b).$$

Proof. One checks that the rows of $C_{**}(A, S)$ are exact under the hypothesis on k. □

The Connes long exact sequence. It is easy to check that the cokernel sequence for the inclusion I of the first two columns of $C_{**}(A, S)$ yields a homology long exact sequence

$$\cdots \to HC_{n-1}(A, S) \xrightarrow{B} HH_n(A, S) \xrightarrow{I} HC_n(A, S) \xrightarrow{S} HC_{n-2}(A, S) \to \cdots$$

where S is the homology map induced by projection (the cokernel of I) and B is the connecting homomorphism induced by $(1 - T)sN$.

Remark 2.1. Fix the notation in Lemma 2.1 for the surjection

$$\phi : ZA \to Z(A, S).$$

Certainly a morphism of cyclic modules induces a morphism of Tsygan bicomplexes, their total complexes, their Hochschild and cyclic homologies and the Connes long exact sequences. We emphasize here that ϕ very much depends on the presence of 1_A in S. We study the map ϕ because its image has the advantage of its reduced size due to more tensorial relations.

The variants of cyclic homology, HC^-_* and HP_*, are derived from an obvious extension to the second quadrant of the Tsygan bicomplex. The next lemma is valid also for HC^-_* and in the forward implication for HP_*.

Lemma 2.3. *If $\psi : Z \to W$ is a morphism of cyclic modules, then the induced map in Hochschild homology is an isomorphism iff the induced map in cyclic homology is an isomorphism.*

Proof. ($\Rightarrow$) The induced morphism on Connes long exact sequences is an isomorphism on every third arrow. But $HC_0 = HH_0$.[3] Then apply induction and the five lemma. ($\Leftarrow$) Apply only the five lemma. $\square$

3. Separable k-Algebras

We recall here some facts and examples in this classical subject. Again, k is *any* commutative ground ring with unity. In this section we briefly use S in a different context from its role as subalgebra.

Definition 3.1. *S is a separable k-algebra (or k-separable) iff there exists an element $e = \sum_{i=1}^{N} u_i \otimes v_i \in S^e$, called a separability idempotent, such that $\sum_{i=1}^{N} u_i v_i = 1$ and $se = es \ \forall s \in S$.*

Example 3.2. $S = M_n(k)$, the full matrix algebra over k, is k-separable with $e = \sum_{i=1}^{n} E_{i1} \otimes_k E_{1i}$ where the E_{ij} are the matrix units.

Example 3.3. $S = C[G]$ where $o(G) < \infty$ is C-separable with

$$e = \frac{1}{o(G)} \sum_{g \in G} g \otimes_k g^{-1}$$

Example 3.4. $S = H$, the real quaternions, is a separable R-algebra with $e = \frac{1}{4}(1 \otimes 1 - i \otimes i - j \otimes j - k \otimes k)$.

Example 3.5. Any separable finite extension of a field k is a separable k-algebra, as is any finite dimensional semi-simple algebra over a perfect field k.

Proposition 3.1 (Hochschild [12]). *S is a separable k-algebra iff the Hochschild (co)homology with any coefficient bimodule vanishes in positive degrees. Every extension of a separable k-algebra by a bimodule is trivial, therefore (Wedderburn's principal theorem) every finite dimensional algebra over a perfect field is a direct sum of a separable subalgebra and its radical ideal.*

Separable extensions of algebras. Consistent with the idea that separable k-algebras S satisfy a homology vanishing condition, one defines a separable extension of k-algebras $1_A \in S \subseteq A$ as precisely those satisfying $H_n(A, S; M) = 0 \ \forall M_{A^e}$ and $n > 0$. These relative homology groups with coefficients are obtained as the homology of the relative Hochschild complex with M replacing the first tensor factor A in the first column of $C_{**}(A, S)$. Three non-trivial examples are $(A \otimes_k S, A)$ where S is a separable k-algebra, $(k[G], k[H])$ where k is a field of characteristic p, H a Sylow p-subgroup of a finite group G, and (M, N) where M and N are type II_1 or type III factors, N a subfactor of M of finite Jones index [17]. This approach yields straightforward generalizations of the theorems below [16].

---

[3] $HC_0(A, S) = A/[A, A] =$ the cotrace group

4. Main Theorem

Theorem 4.1. *The canonical map* $\phi : HC_*(A) \to HC_*(A, S)$ *is an isomorphism if S is k-separable.*

Proof. The method of proof will be to find a section ψ for the epimorphism of simplicial modules underlying ϕ such that $\psi \circ \phi$ is chain homotopic to the identity in the Hochschild complex. Then $\phi : HH_*(A) \to HH_*(A, S)$ is an isomorphism, and we finish the proof by invoking Lemma 2.3.

Let $e = \sum_{i=1}^{m} u_i \otimes_k v_i$ be a separability idempotent in S^e. Define the simplicial map $\psi : Z(A, S) \to ZA$ by

$$\psi(a_0 \otimes_S \cdots \otimes a_n) = \sum_{i_0, \dots, i_n = 1}^{m} v_{i_n} a_0 u_{i_0} \otimes_k v_{i_0} a_1 u_{i_1} \otimes \cdots \otimes v_{i_{n-1}} a_n u_{i_n}.$$

ψ can be shown to be well-defined since $se = es$ $\forall s \in S$. Clearly ψ commutes with all face maps and $\phi \circ \psi = 1$ since $\sum u_i v_i = 1$.

A simplicial homotopy [24] between $\psi \circ \phi$ and 1 is given by $n + 1$ maps $h_i : Z_n A \to Z_{n+1} A$ where

$$h_i(a_0 \otimes_k \cdots \otimes a_n)$$
$$= \sum_{j_0, \dots, j_i = 1}^{m} a_0 u_{j_0} \otimes_k v_{j_0} a_1 u_{j_1} \otimes \cdots \otimes v_{j_{i-1}} a_i u_{j_i} \otimes v_{j_i} \otimes a_{i+1} \otimes \cdots \otimes a_n$$

Then $h'_n = \sum_{i=0}^{n} (-1)^i h_i$ satisfies $b h'_n + h'_{n-1} b = 1 - \psi \circ \phi$. $\square$

An analogue of Theorem 4.1 is valid for Hochschild cohomology with any coefficients (see [9] where this technique is used to prove that Hochschild cohomology $H^n(A, A)$ is simplicial cohomology of a finite simplicial complex if A is an associated poset algebra).

5. Triangular Algebras

Let R and T be unital k-algebras, and let M be a unitary R–T bimodule. Then

$$A = \begin{pmatrix} R & M \\ 0 & T \end{pmatrix}$$

is a k-algebra with 2×2 matrix multiplication and k-module structure. Many non-singular rings are iterations of this basic construction [1] and their cyclic homology is computed from the following

Theorem 5.1. *(see Theorem 1.1)* $HC_n \begin{pmatrix} R & M \\ 0 & T \end{pmatrix} = HC_n(R) \oplus HC_n(T)$ *for* $(n = 0, 1, 2, \dots)$

Proof. Consider the following idempotents:

$$e_1 = \begin{pmatrix} 1_R & 0 \\ 0 & 0 \end{pmatrix}, \quad e_2 = \begin{pmatrix} 0 & 0 \\ 0 & 1_T \end{pmatrix}$$

and the subalgebra $S = ke_1 + ke_2$. S is k-separable since $e_1 \otimes_k e_1 + e_2 \otimes_k e_2$ is a separability idempotent in S^e.

We next show that $Z(A, S) \cong ZR \oplus ZT$ as cyclic modules (clearly an abelian category), whence $HC_*(A) \cong HC_*(R) \oplus HC_*(T)$ by Theorem 4.1. It follows readily from our proof that the split surjection $\left(\begin{smallmatrix} r & m \\ 0 & t \end{smallmatrix}\right) \mapsto \left(\begin{smallmatrix} r & 0 \\ 0 & t \end{smallmatrix}\right)$ induces an isomorphism in cyclic homology.

Let $a_i = \left(\begin{smallmatrix} r_i & m_i \\ 0 & t_i \end{smallmatrix}\right)$, $a_i' = \left(\begin{smallmatrix} r_i & 0 \\ 0 & t_i \end{smallmatrix}\right)$, $m_i' = \left(\begin{smallmatrix} 0 & m_i \\ 0 & 0 \end{smallmatrix}\right)$, $r_i' = \left(\begin{smallmatrix} r_i & 0 \\ 0 & 0 \end{smallmatrix}\right)$, and $t_i' = \left(\begin{smallmatrix} 0 & 0 \\ 0 & t_i \end{smallmatrix}\right)$ for $i = 0, 1, \ldots, n$. Define a morphism of cyclic modules $G : Z(A, S) \to ZR \oplus ZT$ by

$$G(a_0 \otimes_S \cdots \otimes a_n) = (r_0 \otimes_k \cdots \otimes r_n, t_0 \otimes_k \cdots \otimes t_n).$$

Now define a right inverse to G, $F : ZR \oplus ZT \to Z(A, S)$ by

$$F(r_0 \otimes_k \cdots \otimes r_n, t_0 \otimes_k \cdots \otimes t_n) = a_0' \otimes \cdots \otimes a_n'.$$

F is in fact also a left inverse of G by the following computation with S-relative chains:

If $a_i = m_i'$ for some $i = 0, 1, \ldots, n$ then

$$
\begin{aligned}
a_0 \otimes_S \cdots \otimes a_i \otimes \cdots \otimes a_n &= a_0 \otimes \cdots \otimes e_1 m_i' e_2 \otimes \cdots \otimes a_n \\
&= a_0 \otimes \cdots \otimes e_1 m_i' \otimes e_2 a_{i+1} \otimes \cdots \otimes a_n \\
&= a_0 \otimes \cdots \otimes e_1 m_i' \otimes t_{i+1}' e_2 \otimes \cdots \otimes a_n \\
&\ \ \vdots \\
&= e_2 a_0 \otimes \cdots \otimes e_1 m_i' \otimes t_{i+1}' \otimes \cdots \otimes t_n' \\
&\ \ \vdots \\
&= t_0' \otimes \cdots \otimes e_2 e_1 m_i' \otimes t_{i+1}' \otimes \cdots \otimes t_n' \\
&= 0.
\end{aligned}
$$

In general, $a_0 \otimes_S \cdots \otimes a_n = (a_0' + m_0') \otimes_S \cdots \otimes (a_n' + m_n') = a_0' \otimes_S \cdots \otimes a_n'$. $\square$

Remark. Let $I = \left(\begin{smallmatrix} 0 & M \\ 0 & 0 \end{smallmatrix}\right)$ denote the evident nilpotent ideal in A. Since the projection $A \to A/I$ induces isomorphism in both algebraic K-theory [20] and cyclic theory it follows that in the relative algebraic K-theory and cyclic theory of [10] we have $K_n(A, I) = HC_{n-1}(A, I) = 0$ for $(n = 1, 2, \ldots)$. This would provide evidence together with [22] for some improvement of Goodwillie's theorem [11] $(K_n(A, I) \otimes Q = HC_{n-1}(A, I) \otimes Q)$ to the non-rational case. Of course, one can turn this around and give an easy proof from the known result in algebraic K-theory that $A \to A/I$ gives an isomorphism in cyclic homology if A contains a copy of the rationals.

6. MORITA INVARIANCE REVISITED

Theorem 6.1. *If S is a separable k-algebra, A is any unital k-algebra, and $S/[S,S]$ is a flat k-module, then*

$$HC_*(A \otimes_k S) = HC_*(A) \otimes S/[S,S].$$

Proof. $1_A \otimes S$ is k-separable since it is a homomorphic image of S. It will suffice to show by 4.1 that $Z(A \otimes S, 1 \otimes S) \cong ZA \otimes S/[S,S]$ as cyclic modules (where the right side is tensor product of all modules and maps by constant module and identity).

Define $F : Z(A \otimes_k S, 1 \otimes S) \to ZA \otimes_k S/[S,S]$ by

$$F(a_0 \otimes_k s_0 \otimes_S \cdots \otimes a_n \otimes s_n) = a_0 \otimes_k \cdots \otimes a_n \otimes_k (s_0 \cdots s_n + [S,S]),$$

$(n = 0, 1, 2, \dots)$, a well-defined cyclic module morphism because we map into S modulo commutators.

Its inverse is given by

$$a_0 \otimes_k \cdots \otimes a_n \otimes_k (s + [S,S]) \mapsto a_0 \otimes_k s \otimes_S a_1 \otimes 1 \otimes \cdots \otimes a_n \otimes 1.$$

$\square$

Corollary 6.1. *For any $n \geq 1$, $HC_*(M_n(A)) = HC_*(A)$.*

Proof. $S = M_n(k)$ is k-separable as noted in example 3.2. $\square$

Remark 6.1. Note that $F \circ \phi$ is the Dennis trace map. More generally, the Dennis trace map descends to an isomorphism of cyclic modules $Z(M_n(A), M_n(S)) \xrightarrow{\cong} Z(A, S)$ for any subalgebra S [16].

7. DISCUSSION

With just slightly more effort the computations in Sections 4–6 could be carried out only with relative Hochschild homology. However, J. Schafer has made a strong case in [28] for relative cyclic homology as a legitimate abstract entity. We recapitulate the main argument in this light.

Remark 7.1. Let R be any unital ring and P a finitely generated (f.g.) projective R-module. The Hattori-Stallings rank r_P of P is an element of the cotrace group $R/[R,R] = HC_0(R)$; e.g., if $P = eR^n$ for some idempotent $n \times n$ matrix e, then $r_P = trace(e) = \sum_{i=1}^{n} e_{ii} + [R,R]$.

Now fix $R = Q[G]$, any rational group algebra, and $S = Q[H]$, where H is any normal subgroup of G. It is easy to check that $HC_0(R)$ is a rational vector space with the conjugacy classes of G as basis. The Hattori-Stallings rank then has components in each conjugacy class denoted by $r_P(x)$ for a representative x.

Conjecture 7.1 (Bass over $\mathbb{Q}$). *For arbitrary G and P as above, $r_P(x) = 0$ if x has infinite order in G. (A group G satisfying this hypothesis on $r_P(x)$ we will say satisfies the Bass conjecture.)*

Remark 7.2. Burghelea computes in [3] $HC_*(R) = \bigoplus_{[x]} HC_*^{[x]}(G)$, a direct sum over all conjugacy classes where the summands are the group homology of certain subquotients of G that depend on $[x]$. The Connes long exact sequence undergoes a similar direct sum decomposition. Schafer computes relative cyclic homology to have the similar form $HC_*(R, S) = \bigoplus_{[x]} HC_*^{[x]}(G/H)$, also indexed by the conjugacy classes of G, but now the summands are group homology of analogous subquotients of G/H. The map ϕ behaves naturally with respect to these decompositions (see diagram below).

Theorem 7.1 (Eckmann [8]). *G satisfies the Bass conjecture if G is both a group of finite rational homology dimension and is either (i) a nilpotent group, (ii) a torsion-free solvable group, (iii) or a group of rational cohomological dimension ≤ 2.*[4]

Remark 7.3. There are three conceptual steps to the proof. First, $[P] \mapsto r_P \in HC_0(R)$ defines a homomorphism $K_0(R) \mapsto HC_0(R)$, in fact the Chern character Ch_0 of Connes [C1]. Second, Karoubi shows the existence of lifts to higher cyclic homology $HC_{2n}(R)$ via the "higher Chern characters" $\mathrm{Ch}_0^n : K_0(R) \to HC_{2n}(R)$ and the periodicity operator S (see the diagram below). Third, given G satisfying the hypothesis in Theorem 7.1 one shows using group homology theory that $HC_{2n}^{[x]}(R) = 0$ for n large enough if x has infinite order in G. Then the left part of the commutative diagram below gives $r_P(x) = 0$. This argument, the computation for relative cyclic homology for group algebras, and the commutative diagram below lead to

Theorem 7.2 (Schafer [28]). *Let $x \in G$ with H a normal subgroup such that G/H satisfies the hypotheses of Theorem 7.1. If Hx has infinite order in G/H, then $r_P(x) = 0$ for arbitrary f.g. projective G-module P. In particular, G satisfies the Bass conjecture if there exists a finite normal subgroup H for which G/H satisfies the hypothesis of Theorem 7.1.*

$$
\begin{array}{ccc}
& HC_{2n}(R) & \xrightarrow{\phi} & HC_{2n}(R, S) \\
\mathrm{Ch}_0^n \nearrow & \downarrow{\scriptstyle S^n} & & \downarrow{\scriptstyle S^n} \\
K_0(R) \xrightarrow{\mathrm{Ch}_0} & HC_0(R) & \xrightarrow{\cong} & HC_0(R, S)
\end{array}
$$

Remark 7.4. The calculation in Section 5 is basic to any calculation of cyclic homology for triangular algebras. But consider a fiber product of an upper and lower

---

[4]Eckmann reproves a result by Bass for a fourth class of groups

triangular algebra over the diagonal, $A = \begin{pmatrix} R & M \\ N & T \end{pmatrix}$ where R and T are unital k-algebras, M is an R–T bimodule, N a T–R bimodule, and $N \cdot M = 0 = M \cdot N$. The idempotent trick in Section 5 is rendered useless in this situation. A step in the right direction is taken in [4] which computes cyclic homology of 2-nilpotent algebra A over a field. It is shown that if $A = S \oplus J$ where S is k-separable and J is an ideal, then $HC_n(A) = HC_n(S) \oplus HC_n^S(J)$, where the last summand is cyclic homology of the cyclic δ-module $Z(J, S)$ (i.e., no degeneracy maps). If $J^2 = 0$, $HC_n^S(J)$ decomposes into a direct sum of group homologies. It turns out that $HC_n(A)$ depends on the number of proper oriented cycles of length dividing $n + 1$ in the quiver graph representing A. For example, if $A = \begin{pmatrix} k & k \\ k & k \end{pmatrix}$ as a special case of the fiber product above, the underlying quiver is $\cdot \leftrightarrow \cdot$ and if k has characteristic 0 then

$$HC_n(A) = \begin{cases} k \oplus k & \text{for } n \text{ even} \\ k & \text{for } n \text{ odd} \end{cases}$$

Remark 7.5. The non-self adjoint operator algebras that provide an infinite dimensional generalization of the complex algebra of block-upper triangular matrices is the nest algebra N_F of bounded linear operators on a separable Hilbert space H that leave invariant a totally ordered set F of orthogonal projections including 0 and 1. The core C of N_F is the commutative C^*-algebra generated by the projections. C is *amenable*, and plays the role (with respect to norm continuous Hochschild cohomology) of the separable algebra $S = ke_1 + ke_2$ in Section 5. If F is discrete, then a characterization of the Jacobson radical J of N_F by Ringrose gives a continuous version of $\begin{pmatrix} 0 & M \\ 0 & 0 \end{pmatrix} = \{a \in A : e_1 a e_1 + e_2 a e_2 = 0\}$ in Section 5. A computation like 5.1 gives the following calculation in continuous cyclic cohomology

Theorem 7.3 ([15]). $H_\lambda^n(N_F) \cong H_\lambda^n(N_F/J)$

Remark 7.6. It seems quite probable that a result can be obtained like that for K_0 in terms only of the diagonal algebra [25]. In passing we note the easy theorem below, which gives hope to compute H_λ^n of any nest algebra from only its atomic part, as for K_0 in [25].

Theorem 7.4. *If F is a continuous nest, then $H_\lambda^n(N_F) = 0$ $(n \geq 0)$.*

Proof. It is well-known that with this hypothesis on F, N_F is isomorphic to $N_F \otimes B(H)$ [25]. Now a theorem of Wodzicki [31, p. 49] states that $H_\lambda^n(A \otimes B(H)) = 0$ for any H-unital complex algebra A, which easily adapts to the continuous case. $\square$

Remark 7.7. A K-theoretic version of relative cyclic homology $HC_n(A, S)$ exists [LK3]. The simplicial space underlying $BGLA$ may be quotiented out by relations of the form

$$[\cdots \mid g_i h \mid g_{i+1} \mid \cdots] = [\cdots \mid g_i \mid h g_{i+1} \mid \cdots]$$

where the $g_i \in GLA$ and $h \in GLS$, to obtain a spectrum $K(A, S)$ with $\pi_1 = GLA/\text{normalclosure}(GLS)$. Define relative K-groups $K_n(A, S) = \pi_n(K(A, S))$ for

$n \geq 1$. There exist Chern character homomorphisms c_n that make the square below commute (where ν is the natural map)

$$
\begin{array}{ccc}
K_n(A) & \xrightarrow{\;\nu\;} & K_n(A,S) \\
Ch_n^0 \downarrow & & \downarrow c_n \\
HC_n(A) & \xrightarrow{\;\phi\;} & HC_n(A,S)
\end{array}
$$

REFERENCES

1. A. J. Berrick and M. E. Keating, *The K-theory of triangular matrix rings*, Contemporary Math. **55 part 1** (1986), 69–74.
2. J. Block, *Cyclic homology of filtered algebras*, K-theory **1** (1988).
3. D. Burghelea, *Cyclic homology of group rings*, Comment. Helv. Math. **91** (1984), 305-317.
4. C. Cibils, *Cyclic and Hochschild homologies of 2-nilpotent algebras*, K-theory **4** (1990), 131–141.
5. A. Connes, *Non commutative differential geometry*, Publ. Math. I.H.E.S. **62** (1986), 257–360.
6. _____, *Cohomologie cyclique et foncteurs Ext^n*, C. R. Acad. Sci. Paris **290** (1983), 953–958.
7. R. K. Dennis and S. C. Geller, *K_i of upper triangular matrix rings*, Proc. A. M. S. **56** (1976), 73–78.
8. B. Eckmann, *Cyclic homology of groups and the Bass conjecture*, Comment. Math. Helv. **61** (1986), 193–202.
9. M. Gerstenhaber and S. D. Schack, *Algebraic cohomology and deformation theory* (1988), S. U. N. Y. Buffalo (preprint).
10. T. Goodwillie, *Derivations, cyclic homology, and the free loopspace*, Topology **35** (1985), 501–569.
11. _____, *Relative algebraic K-theory and cyclic homology*, Ann. Math. **124** (1986), 347–402.
12. G. Hochschild, *On the cohomology groups of an associative algebra*, Ann. Math. **46** (1945), 58–67.
13. _____, *Relative homological algebra*, Trans. A. M. S. **82** (1956), 246–269.
14. L. Kadison, *A relative cyclic cohomology theory useful for computations*, C. R. Acad. Sci. Paris **308** (1989), 569–573.
15. _____, *On the cyclic cohomology of nest algebras and a spectral sequence induced by a subalgebra in Hochschild cohomology*, C. R. Acad. Sci. Paris **311** (1990), 247–252.
16. _____, *Cyclic homology of extension algebras with applications*, Aarhus Universitet, Matematisk Institut Preprint Series **23** (March) (1989/90), 1–116.
17. L. Kadison and D. Kastler, *Cohomological aspects and relative separability of finite Jones index subfactors*, Proc. Goettingen Acad. Sci. (to appear).
18. M. Karoubi, *Homologie cyclique et K-théorie*, Astérisque **149** (1987).
19. C. Kassel, *Cyclic homology, comodules, and mixed complexes*, J. Alg. **107** (1987), 195–216.
20. M.E. Keating, *The K-theory of triangular matrix rings, II*, Proc. A. M. S. **100**, no. 2, (1987), 235–236.
21. J.-L. Loday, *Cyclic homology: a survey*, Geometric and Algebraic Topology, Banach Center Publ. **18** (1986), Warsaw, 281–303.
22. _____, *Comparaison des homologies du groupe linéaire et de son algèbre de Lie*, Ann. Inst. Fourier **37**, 4 (1987), Grenoble, 167–190.
23. J.-L. Loday and D. Quillen, *Cyclic homology and the Lie algebra of matrices*, Comment.Math. Helv. **59** (1984), 565–591.
24. J. P. May, *Simplicial objects in algebraic topology*, Van Nostrand, Princeton, 1967.
25. D. Pitts, *On the K_0 groups of nest algebras*, K-theory **2** (1989), 737–752.
26. D. Quillen, *Lecture at conf. on alg. K-theory*, June 23–28, 1974, Oberwolfach, F. R. Germany.

27. _____, *Cyclic cohomology and algebra extensions*, K-theory **3** (1989), 205–246.
28. J. A. Schafer, *Relative cyclic homology and the Bass conjecture*, Comment. Math. Helv. (to appear).
29. P. Seibt, *Cyclic homology of algebras*, World Scientific, Singapore, 1987.
30. A. A. Suslin and M. Wodzicki, *Excision in algebraic K-theory and Karoubi's conjecture*, Proc. Nat. Acad. Sci. U.S.A. (to appear).
31. M. Wodzicki, *Excision in cyclic cohomology and rational algebraic K-theory*, Ann. Math. **129** (1989), 591–639.

INSTITUTE FOR MATHEMATICS AND PHYSICS, ROSKILDE UNIVERSITY, POSTBOX 260, 4000 ROSKILDE, DENMARK
E-mail address: kadison@fatou.ruc.dk

TOPOLOGICAL PROBLEMS ARISING
FROM REAL ALGEBRAIC GEOMETRY

HENRY C. KING

University of Maryland

ABSTRACT. We discuss some theorems about putting real algebraic structures on topological spaces. In the course of doing so a number of topological questions arise which we make explicit.

In this talk I will investigate some aspects of the problem of when a topological entity has an algebraic structure. To illustrate this, we start out with the problem of making smooth manifolds algebraic. Take a smooth manifold M. We wish to find a real algebraic variety V diffeomorphic to M. (A real algebraic variety $V \subset \mathbf{R}^n$ is the solution set $V = p^{-1}(0)$ for a polynomial function $p: \mathbf{R}^n \to \mathbf{R}^k$.)

The naive attack would be:

(1) Imbed M in some $\mathbf{R}^n$.
(2) Write $M = f^{-1}(0)$ for some smooth function $f: \mathbf{R}^n \to \mathbf{R}^k$.
(3) Approximate f by a polynomial map p.
(4) Hope that $p^{-1}(0)$ is diffeomorphic to $f^{-1}(0)$.

But there are problems with this. You can do steps 1 and 2 (in fact any closed subset of $\mathbf{R}^n$ is the zeroes of some smooth function). For step 3 however, you can only approximate a smooth function by a polynomial on a compact set K. You have no control over what happens to p outside K and in fact what tends to happen is that the polynomial approximation p oscillates wildly as soon as you leave K. Consequently $p^{-1}(0)$ tends to contain extra stuff outside K that you do not want. But even if you agree to ignore this extra stuff, you still have problems with step 4. Just because p approximates f, there is no reason $p^{-1}(0)$ and $f^{-1}(0)$ should be diffeomorphic. You need some sort of stability of f to conclude this.

But not all is lost, there are situations where this naive attack works. In 1936, H. Seifert showed that if M is compact and has trivial normal bundle in $\mathbf{R}^n$ then M is diffeomorphic to a component of a real algebraic variety [S]. His proof was just the above naive attack where we let k be the codimension of M, we let K be a tubular neighborhood of M, we make sure 0 is a regular value of $f|_K$ and we ignore the extra stuff coming from step 3. Furthermore, by using a trick in step 3 to get rid of the extra stuff, he could show that if M has codimension one then M is diffeomorphic to a real algebraic variety, not just a component of one.

The reason Seifert's proof worked was transversality, since the map f is transverse to 0 we know that step 4 works. But we needed M to have trivial normal bundle to get this. For a general compact manifold M we must modify the above argument to make it go through.

In particular, if $M \subset \mathbf{R}^n$ has codimension k, let G be the Grassmannian of k planes in $\mathbf{R}^n$ and let $\pi: E \to G$ be the canonical k bundle. Then there is a smooth map $f: K \to E$ from a neighborhood K of M so that f is transverse to the zero section G and $f^{-1}(G) = M$. The spaces E and G can be given an algebraic structure. We then approximate f by a polynomial p and let $V = p^{-1}(G)$. Then V is a real algebraic variety and it has a component $V \cap K$ which is diffeomorphic to M. This is the proof which J. Nash gave in 1952 [N].

There is a point in Nash's proof which bears a closer look however. Suppose E is given as a real algebraic variety in $\mathbf{R}^m$. Then it is easy to approximate f by a polynomial $p: \mathbf{R}^n \to \mathbf{R}^m$ but it is not clear how to ensure that the image of p will lie inside E. A moments reflection on the above proof shows that this is necessary to make the proof work. Nash got around this point by a device which involved increasing n. Thus a statement of Nash's theorem is as follows:

Theorem (Nash). *Let $M \subset \mathbf{R}^n$ be a smooth compact submanifold. Then there is a real algebraic variety $V \subset \mathbf{R}^{n+m}$ and a component V_0 of V so that V_0 is diffeomorphic to M.*

Nash conjectured in [N] that it was not actually necessary to increase n, i.e.,

Conjecture (Nash). *Let $M \subset \mathbf{R}^n$ be a smooth compact submanifold. Then there is a real algebraic variety $V \subset \mathbf{R}^n$ and a component V_0 of V so that V_0 is diffeomorphic to M.*

In 1974, A. Tognoli was able to get rid of all the extra components [T]. By refining Nash's proof he showed that in fact any smooth compact manifold is diffeomorphic to a real algebraic variety, but again it may be imbedded in a much bigger Euclidean space.

We can now ask the first question (to which we do not know the answer).

Question 1. *Given a compact manifold M and an integer n, what does the following set look like? ($\approx$ denotes diffeomorphism).*

$$\{\, V \subset \mathbf{R}^n \mid V \text{ is a real algebraic variety and } V \approx M \,\}$$

The above gives some invariants of the manifold M, namely spaces of algebraic structures on M. We can modify the above question to a similar question.

Question 2. *What does the set below look like?*

$$\{\, p: \mathbf{R}^n \to \mathbf{R}^k \mid p \text{ is a polynomial of degree } \leq d \text{ and } p^{-1}(0) \approx M \,\}$$

The above is a subset of $\mathbf{R}^N$ for some big N, since polynomials of degree $\leq d$ are just determined by their coefficients. In fact it is a semialgebraic set, it is given by polynomial equations and inequalities.

Restricting the degree is useful because of the following, easily proven using Thom's isotopy Lemmas.

Theorem. *Given n, k and d there are only a finite number of topological types of real algebraic varieties $p^{-1}(0)$ for polynomials $p: \mathbf{R}^n \to \mathbf{R}^k$ of degree $\leq d$.*

This brings to mind the finiteness theorems of Grove, Peterson, Wu, Gromov, Cheeger, etc., which say that when you put certain curvature, diameter and volume bounds on a Riemannian manifold then only finitely many diffeomorphism types of manifolds are possible [GP].

It raises the question of whether similar types of theorems come out of real algebraic geometry. The idea is that curvature bounds should mean that the map $f: M \to G$ to the Grassmannian is not too wiggly and thus can be approximated by a polynomial with bounds on the degree. This then gives you finiteness theorems.

Question 3. *Do bounds on the geometry of a compact Riemannian manifold M give bounds on the degree of a rational function $\varphi: M \to G$ needed to induce the tangent bundle of M? (A positive answer gives finiteness theorems.)*

I will now mention some recent results proven by S. Akbulut and me. The first includes a solution of Nash's 1952 conjecture mentioned above.

Theorem 1. *If $M \hookrightarrow \mathbf{R}^n$ is a smooth compact immersed submanifold then M is regularly homotopic to a component of a real algebraic variety $V \subset \mathbf{R}^n$. If M is imbedded then M is ϵ-isotopic to a real algebraic variety $V' \subset \mathbf{R}^{n+1}$.*

Theorem 2. *If a smooth compact immersed submanifold $M \hookrightarrow \mathbf{R}^n$ is immersion cobordant to an algebraic immersed submanifold, then M is regularly homotopic to a real algebraic variety $V \subset \mathbf{R}^n$.*

Note that these two theorems have some consequences. Since immersion cobordism classes of m manifolds in $\mathbf{R}^n$ are in one to one correspondence with

$$\pi_n(\Omega^\infty S^\infty MO(n-m))$$

we have a subgroup

$$\pi_n^{alg}(\Omega^\infty S^\infty MO(n-m)) \subset \pi_n(\Omega^\infty S^\infty MO(n-m))$$

which determines when an immersion is regularly homotopic to a real algebraic variety. We also have a subgroup

$$\pi_n^{alg}(MO(n-m)) \subset \pi_n(MO(n-m))$$

which determines when a closed m dimensional submanifold is ϵ-isotopic to a smooth real algebraic variety in $\mathbf{R}^n$. We have

$$(*) \qquad \pi_n^{alg}(MO(n-m)) = \theta^{-1}(\pi_n^{alg}(\Omega^\infty S^\infty MO(n-m)))$$

where

$$\theta: \pi_n(MO(n-m)) \to \pi_n(\Omega^\infty S^\infty MO(n-m))$$

is the canonical map. Theorem 1 then says that

$$(**) \qquad Image(\pi_{n-1}(MO(n-m)) \to \pi_n(MO(n-m))) \subset \pi_n^{alg}(MO(n-m)).$$

To summarize, these cobordism subgroups determine whether an imbedded or immersed submanifold of $\mathbf{R}^n$ can be approximated by a real algebraic variety in $\mathbf{R}^n$. They might be everything, but meanwhile:

Question 4. *What do (*) and (**) above imply about $\pi_n^{alg}(\Omega^\infty S^\infty MO(n-m))$ and $\pi_n^{alg}(MO(n-m))$?*

There are similar theorems where $\mathbf{R}^n$ is replaced by a nonsingular real algebraic variety V. It is necessary to have an extra condition though, that the bordism class of $M \to V$ is algebraic.

I now indicate a proof of Theorem 1 which appears in [AK1]. First you factor the immersion into the composition of an imbedding $M \subset \mathbf{R}^N$ and projection $\pi : \mathbf{R}^N \to \mathbf{R}^n$. By old results (Tognoli) we can approximate M by a real algebraic variety $W \subset \mathbf{R}^N$. So you have an algebraic subset $W \subset \mathbf{R}^N$ and $\pi|_W : W \to \mathbf{R}^n$ is regularly homotopic to your immersion. But the image $\pi(W)$ might not be a real algebraic variety.

For example consider

$$W = \{ (x, y, z) \in \mathbf{R}^3 \mid z^2 = (x-1)(2-x), \; y = zx \}$$

which is a circle and let π be projection to the x, y plane. Then $\pi(W)$ is not a real algebraic variety. However $\pi(W) \cup \{(0,0)\}$ is the real algebraic variety $\{ y^2 = x^2(x-1)(2-x) \}$. The thing is that the extra point $(0,0)$ is $\pi(0, 0, \sqrt{-2})$ and $(0, 0, \sqrt{-2})$ is a complex solution to the polynomials defining W.

In the complex case, you know that under mild conditions (for example properness) the projection of a complex variety is a variety. But there is no analogous result for the reals. So as in the example above, we turn to the complexification $W_{\mathbf{C}}$ of W. $W_{\mathbf{C}}$ is the set of all complex solutions of the polynomials defining W.

By algebraic geometrical generic projection arguments we may assume that the projections $\pi' : W_{\mathbf{C}} \subset \mathbf{C}^N \to \mathbf{C}^{n-1}$ and $\pi : W_{\mathbf{C}} \subset \mathbf{C}^N \to \mathbf{C}^n$ are finite maps. This is a technical algebraic condition which implies the maps are proper and have finite fibers.

Now $\pi(W_{\mathbf{C}}) \cap \mathbf{R}^n$ is a real algebraic variety but it consists of two parts, $\pi(W) \approx M$ and $\pi(\text{complex points})$. If those two parts are disjoint then we are done, the problem is that they might intersect. The map π is (π', f) where f is projection to the n-th coordinate. Letting $Q = \pi'^{-1}(\mathbf{R}^{n-1}) \cap W_{\mathbf{C}} - W$ we see the nonreal points with real image are contained in Q. See Figure 1.

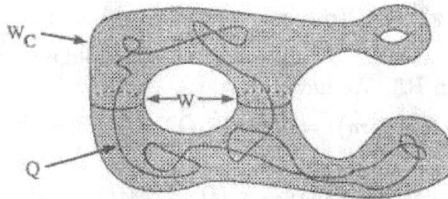

FIGURE 1: $W_{\mathbf{C}}$ AND ITS POINTS WITH POSSIBLY REAL IMAGE

So we want to replace f by a polynomial p which approximates f near W, but so (π', p) takes $Q \cap p^{-1}(\mathbf{R})$ far away from $\pi(W)$. Then we would have

$$(\pi', p)(W_{\mathbf{C}}) \cap \mathbf{R}^n = (\pi', p)(W) \cup Z$$

where $Z = (\pi', p)(Q \cap p^{-1}(\mathbf{R}))$ is far away from $(\pi', p)(W)$. We would then be done since $(\pi', p)(W)$ is a component of the real algebraic variety $V = (\pi', p)(W_{\mathbf{C}}) \cap \mathbf{R}^n$ regularly homotopic to our original immersion. See Figure 2.

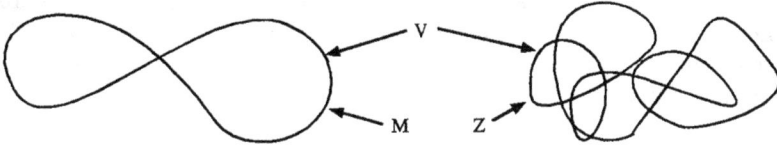

FIGURE 2: $(\pi', p)(W) \approx M$ AS A COMPONENT OF V

We first find a continuous version g of p, a continuous function $g: Q \cup W \to \mathbf{C}$ so that g agrees with f near W and $g(y)$ is very big if $g(y) \in \mathbf{R}$ and $y \notin W$. We must then approximate g by a complex polynomial. Ordinarily there is no hope, one can approximate continuous functions on compact subsets of $\mathbf{R}^n$ by real polynomials, but one has no hope in general of approximating continuous functions on subsets of $\mathbf{C}^n$ by complex polynomials. Just think of the Cauchy-Riemann equations for example. However, this particular situation is special enough that one can do the approximation.

So we have indicated a proof of the first part of Theorem 1. For the second part, we note that by algebraic degree theory, $\dim Z < \dim M$. (This is because the mod 2 number of real points in $(\pi', p)^{-1}(z)$ is 1 for a generic point point $z \in V$. But $(\pi', p)^{-1}(z)$ has no real points for $z \in Z$.) One can then show that if M is imbedded then Z is the set of singular points of V, so $Z = q^{-1}(0)$ for some polynomial q. We may then let $V' = \{ (z, t) \in V \times \mathbf{R} \mid tq(z) = 1 \}$ and see that V' is isotopic to M.

We can now prove Theorem 2, details appear in [AK2]. The immersion cobordism hypothesis means there is a properly immersed compact submanifold $W \looparrowright \mathbf{R}^n \times [0, 1]$ with $\partial W = M \cup V$ which restricts to $M \looparrowright \mathbf{R}^n \times 0$ on M and which restricts to an immersion of V to an almost nonsingular real algebraic variety $K' \times 1 \subset \mathbf{R}^n \times 1$. Pick an affine hyperplane L in $\mathbf{R}^n$ which does not intersect K'. Let K be the reflection of K' through L. By adding on $K \times [0, 1]$ and the half of the surface of revolution of K' in $\mathbf{R}^n \times [1, \infty)$, we actually get a proper immersion of W in $\mathbf{R}^n \times [0, \infty)$. See Figure 3.

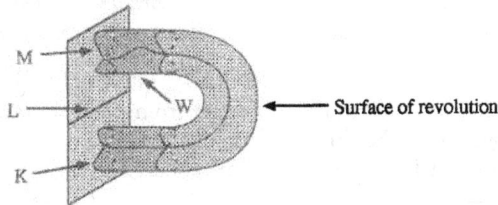

FIGURE 3: A PROPER IMMERSION OF W IN $\mathbf{R}^n \times [0, \infty)$

Over $\mathbf{R}^n \times 0$ the immersion of W is the disjoint union of the immersion of M and an immersion of V to the almost nonsingular real algebraic variety K. We now double W to get an immersion of a closed manifold into $\mathbf{R}^n \times \mathbf{R}$ whose intersection with $\mathbf{R}^n \times 0$ is K' union the image of M. We now go through the above proof of Theorem 1 to make this immersion a component of a real algebraic variety Z, only we are careful to make sure it still contains K and the extra points are far from $\mathbf{R}^n \times 0$. See Figure 4.

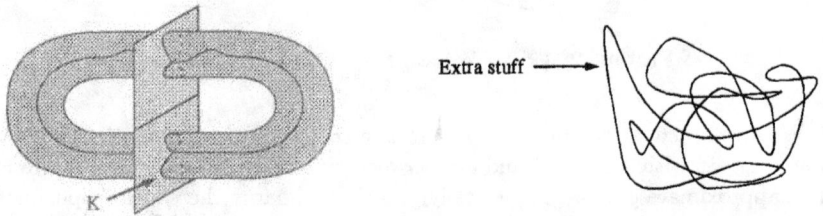

FIGURE 4: THE DOUBLE OF THE IMMERSION MADE ALGEBRAIC

Then $Z \cap \mathbf{R}^n \times 0 = K \cup Z'$ where Z' is an immersed manifold regularly homotopic to M. But by being careful you can insure that Z' is itself a real algebraic variety. Thus Theorem 2 is proven.

The above proof of Theorem 2 also illustrates the the general principle:

General Principle. *If a topological situation is bordant to an algebraic situation, then it is isomorphic to an algebraic situation (if you have the right notion of bordism and do a lot of work).*

We have another example of this principle in the works of Dovermann, Masuda, Petrie and Suh where they show that if G is a compact Lie group and M is a smooth G-manifold then M is G-diffeomorphic to a nonsingular real algebraic G-variety if and only if M is G-cobordant to a nonsingular real algebraic G-variety [DMP].

Not all real algebraic varietiesare manifolds, so the next question is what a general real algebraic variety looks like. Akbulut and I explore this question in [AK3]. We state the following theorem rather loosely. It is proven by careful use of resolution of singularities.

Theorem 3. *Every (real or complex) algebraic variety is obtained from nonsingular algebraic varieties by gluing them together with monomial functions.*

For example, the real algebraic variety $\{ (x,y,x) \in \mathbf{R}^3 \mid z^2 = x^2(1 - x^2 - y^2) \}$ which looks as in Figure 5, is obtained from a sphere and a line by collapsing the sphere's equator to the line. We could use the collapsing map $(x,y,z) \mapsto (x,y,xz)$ from the standard unit sphere. This collapsing map is monomial because on the equator $\{ x = 0 \}$ it locally looks like either $t \mapsto t$ or $t \mapsto t^2$.

The next question is whether we can go the other way, and the next Theorem says we can.

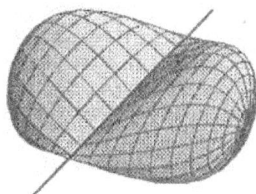

FIGURE 5: A SAMPLE SINGULAR REAL ALGEBRAIC VARIETY

Theorem 4. *Suppose a topological space X can be obtained from smooth manifolds by gluing together with monomial functions. Then X is homeomorphic to a real algebraic variety.*

Thus for example in dimensions ≤ 3 one can give a complete and easily computable topological characterization of real algebraic varieties. In fact in order to decide whether or not a compact 3 dimensional polyhedron is homeomorphic to a real algebraic variety, it is only necessary to do some counting of the number of simplices adhering to each simplex.

In higher dimensions the gluing conditions needed for Theorem 4 are a little stronger than those obtained in Theorem 3. Thus the topological characterization of varieties in dimensions > 3 is not yet complete. It depends on the solution of a question on resolution of singularities of rational functions. We state a simplified version, which is a natural conjecture for what resolution of singularities of rational functions should be. We can prove this in dimensions ≤ 2 (where it does us no good) and T. C. Kuo has also given an independent proof in these dimensions.

We say a map $f: M \to N$ between smooth manifolds M and N is *nice* if there are codimension one submanifolds $M_i \subset M$, $i = 1, \ldots, k$ and $N_i \subset N$, $i = 1, \ldots, \ell$ in general position so that f restricts to a submersion on the resulting strata. In other words, for each subset $A \subset \{1, 2, \ldots, k\}$ there is a subset $B \subset \{1, 2, \ldots, \ell\}$ so that if $M_A = \bigcap_{i \in A} M_i - \bigcup_{i \notin A} M_i$ and $N_B = \bigcap_{i \in B} N_i - \bigcup_{i \notin B} N_i$ then $f(M_A) \subset N_B$ and in fact f restricts to a submersion of M_A to N_B.

Then in rough form, the resolution conjecture is as follows:

Conjecture. *Let $f: M \to N$ be an entire rational function between real or complex algebraic varieties. Then there are sequences of algebraic blowups $\pi: M' \to M$ and $\rho: N' \to N$ and a nice map $g: M' \to N'$ so that $\rho g = f \pi$.*

REFERENCES

[**AK1**]   S. Akbulut and H. King, *On approximating submanifolds by algebraic sets and a solution to the Nash conjecture*, to appear, Inventiones Math.

[AK2]   S. Akbulut and H. King, *Algebraicity of immersions in* R^n, (preprint).

[AK3]   S. Akbulut and H. King, *The topology of real algebraic sets*, to appear, M.S.R.I. Publications.

[DMP]   H. Dovermann, M. Masuda and T. Petrie, *Fixed point free algebraic actions on varieties diffeomorphic to* R^n, Topological Methods in Algebraic Transformation Groups, Birkhäuser, Boston, 1989, pp. 49–80.

[GP]   K. Grove and P. Peterson, Annals of Math. **128** (1988), 195–206.

[N]   J. Nash, *Real algebraic manifolds*, Annals of Math. **56** (1952), 405–421.

[S]   H. Seifert, *Algebraische Approximation von Mannigfaltigkeiten*, Math. Zeit. **41** (1936), 1–17.

[T]   A. Tognoli, *Su una congettura di Nash*, Ann. Sc. Norm. Sup. Pisa **27** (1973), 167–185.

DEPARTMENT OF MATHEMATICS, UNIVERSITY OF MARYLAND, COLLEGE PARK, MD 20742
E-mail address: hck@ frances.umd.edu

COMPLEMENTS OF 2-SPHERES IN 4-MANIFOLDS

Vo Thanh Liem and Gerald A. Venema

ABSTRACT. Complements of topologically embedded 2-spheres in compact 4-manifolds are investigated. The main technical result gives a necessary condition for such a complement to be homotopically dominated by a compact polyhedron. Corollaries give conditions under which the end of the complement will have a collar or a weak collar.

INTRODUCTION

Let $h : S^2 \to M^4$ be a topological embedding of the 2-sphere into the interior of the compact 4-manifold M^4. We use Σ to denote $h(S^2)$ and W to denote $M - \Sigma$; then W is a non-compact 4-manifold with one end which we call ϵ. We wish to investigate properties of ϵ that are related to the way in which Σ is embedded in M. In particular, we give a condition which implies that neighborhoods of ϵ are homotopically dominated by finite complexes and a second condition which implies that the end is collared (or weakly collared).

Let us begin with a definition. A compact set X in the interior of a manifold M is said to be *globally 1-alg* if for every neighborhood U of X in M there is a neighborhood V of X in U such that the image of $\pi_1(V - X)$ in $\pi_1(U - X)$ is abelian. We are most interested in the case in which $X = \Sigma$, as in the paragraph above. In that case duality shows that the image of $\pi_1(V - \Sigma)$ in $\pi_1(U - \Sigma)$ will be cyclic. The order of this cyclic group is determined by $\Sigma \cdot \Sigma$, the integer-valued self-intersection number of $[\Sigma] \in H_2(M; \mathbb{Z})$. If Σ is globally 1-alg and $\Sigma \cdot \Sigma = 0$, then $\pi_1(\epsilon)$ is infinite cyclic. Otherwise $\pi_1(\epsilon)$ is a finite cyclic group whose order is $\Sigma \cdot \Sigma$. This is explained in Lemma 1, below.

We can now state our main results.

Theorem. *If $h : S^2 \to M^4$ is a topological embedding into the interior of the 4-manifold M and $\Sigma = h(S^2)$ is globally 1-alg, then any 0-neighborhood of the end of $M - \Sigma$ is finitely dominated.*

In the infinite cyclic case we can obtain a collar on the end.

The second author was partially supported by N. S. F. Grant DMS-8900822

Corollary 1. *If Σ is a 2-sphere topologically embedded in the interior of the 4-manifold M and π_1 of the end of $M - \Sigma$ is infinite cyclic, then $M - \Sigma \cong M - K$ for some locally flat 2-sphere K in M. Moreover, there exists a neighborhood N of Σ such that $N \cong S^2 \times B^2$ and $N - \Sigma \cong \partial N \times [0,1)$.*

We do not know whether the end of the complement of every globally 1-alg 2-sphere can be collared. The Theorem does show that the end is always tame. In §3 we prove that the finiteness obstruction $\sigma(\epsilon) \in \tilde{K}_0(\mathbb{Z}[\pi_1(\epsilon)])$ is trivial and so we are at least able to obtain a weak collar. (See [2, §11.9B] for the definition of a weak collar.)

Corollary 2. *If Σ is a 2-sphere topologically embedded in the interior of the 4-manifold M and π_1 of the end of $M - \Sigma$ is a finite cyclic group, then the end of $M - \Sigma$ has a weak collar.*

The Theorem and its Corollaries generalize results in [3], where such theorems are proved for the special case in which the ambient manifold is S^4. This generalization is interesting in its own right, but is also motivated by an application to a different problem. That problem is the problem of giving a homotopy characterization of local flatness for surfaces embedded in 4-manifolds. The conjecture is that a local version of the 1-alg property will imply local flatness for a surface N topologically embedded in the interior of the 4-manifold M. The surprising fact is that the global structure of the end of $M - N$ plays a role in this local problem. In [4], Corollary 1 above is used to prove the following result: *If Σ is a 2-sphere topologically embedded in the interior of the 4-manifold M^4 so that Σ is locally 1-alg at x for every $x \in \Sigma$ and $\Sigma \cdot \Sigma = 0$, then Σ is locally flat at x for every $x \in \Sigma$.*

It should be noted that the Theorem only asserts that the global 1-alg condition is a sufficient condition to guarantee that the end of the complement of Σ is finitely dominated. In fact it is not a necessary condition, even in case Σ is a 2-sphere in S^4. In [4] we show that a necessary condition would involve $\pi_2(M - \Sigma)$. We also show by example that conditions on $\pi_1(M - \Sigma)$ alone will not suffice; in [4] we construct an example of a topological 2-sphere $\Sigma \subset S^4$ such that $\pi_1(S^4 - \Sigma) \cong \mathbb{Z}$ but $\pi_2(S^4 - \Sigma)$ is nontrivial. Such an example must necessarily be wildly embedded with $\pi_1(\epsilon)$ unstable.

The results of this paper apply only in dimension 4. In high dimensions the global 1-alg condition by itself is not strong enough to imply that that the end of the complement of a topologically embedded codimension 2 sphere is finitely dominated. Examples are discussed in [4].

Notation. Throughout the remainder of this paper, M will denote a compact 4-manifold, $h : S^2 \to \text{int}\, M$ will denote a topological embedding and $\Sigma = h(S^2)$. Then $M - \Sigma$ is a noncompact manifold with one end which we call ϵ. If V is a closed neighborhood of the end ϵ, we write $\overline{V} = V \cup \Sigma$.

Remark. All the results above hold true if Σ is a compactum having the shape of S^2; i.e., the topological embedding h can be replaced with a shape equivalence from S^2 to a compact subset Σ of M^4.

1. Proof of the Theorem

There are some technical results needed for the proof and we will state them as lemmas. The first lemma explains the structure of $\pi_1(\epsilon)$.

Lemma 1. *If Σ is globally 1-alg embedded in int M, then the fundamental group of the end ϵ is stable and cyclic. More precisely, $\pi_1(\epsilon) \cong \mathbb{Z}/n\mathbb{Z}$, where $n = \Sigma \cdot \Sigma$, the integer-valued self-intersection number of $[\Sigma] \in H_2(M; \mathbb{Z})$.*

Proof. Since Σ is an ANR, we can choose a nested sequence $\{V_n\}_{n=1}^{\infty}$ of closed, connected, PL manifold neighborhoods of ϵ such that $\cap V_n = \emptyset$ and each $\overline{V}_{n+1} \hookrightarrow \overline{V}_n$ factors, up to homotopy, through a retraction $r_{n+1} : \overline{V}_{n+1} \to \Sigma$. Consider the following diagram.

$$
\begin{array}{ccccccc}
H_2(\overline{V}_{n+1}; \mathbb{Z}) & \xrightarrow{\alpha_1} & H_2(\overline{V}_{n+1}, V_{n+1}; \mathbb{Z}) & \xrightarrow{\alpha_2} & H_1(V_{n+1}; \mathbb{Z}) & \xrightarrow{\alpha_3} & H_1(\overline{V}_{n+1}; \mathbb{Z}) \\
\downarrow{\beta_1} & & \downarrow{\beta_2} & & \downarrow{\beta_3} & & \downarrow{\beta_4} \\
H_2(\overline{V}_n; \mathbb{Z}) & \xrightarrow{\gamma_1} & H_2(\overline{V}_n, V_n; \mathbb{Z}) & \xrightarrow{\gamma_2} & H_1(V_n; \mathbb{Z}) & \xrightarrow{\gamma_3} & H_1(\overline{V}_n; \mathbb{Z})
\end{array}
$$

Notice that β_2 is an isomorphism (excision) and $\beta_4 = 0$ (by the choice of V_n), so a simple diagram chasing argument shows that $\operatorname{im}\beta_3 = \operatorname{im}\gamma_2$. Now duality gives $H_2(\overline{V}_n, V_n; \mathbb{Z}) \cong \check{H}^2(\Sigma) \cong \mathbb{Z}$, and the duality isomorphism takes $x \in H_2(\overline{V}_n, V_n; \mathbb{Z})$ to $x \cdot [\Sigma] \in \mathbb{Z}$. Furthermore, the images of $H_2(\overline{V}_n; \mathbb{Z})$ and $H_2(\overline{V}_{n+1}; \mathbb{Z})$ in $H_2(\overline{V}_n, V_n; \mathbb{Z})$ are equal, so $\operatorname{im}\gamma_1$ is generated by $\gamma_1([\Sigma])$, where $[\Sigma] \in H_2(V_n; \mathbb{Z})$. Thus $\operatorname{im}\beta_3 = \operatorname{im}\gamma_2 \cong \mathbb{Z}/n\mathbb{Z}$ where $n = \Sigma \cdot \Sigma$. It follows that the inverse sequence $\{H_1(V_1; \mathbb{Z}) \leftarrow H_1(V_2; \mathbb{Z}) \leftarrow \cdots\}$ is stable and

$$
\varprojlim_n H_1(V_n; \mathbb{Z}) \cong \mathbb{Z}/n\mathbb{Z}.
$$

Finally, the fact that Σ is globally 1-alg implies that the inverse sequence $\{\pi_1(V_1) \leftarrow \pi_1(V_2) \leftarrow \cdots\}$ is also stable and $\pi_1(\epsilon) \cong \varprojlim_n \{\pi_1(V_n)\} \cong \varprojlim_n \{H_1(V_n; \mathbb{Z})\}$. $\square$

Using Lemma 1, take $V \subset V_0$ to be a pair of connected neighborhoods of ϵ such that

(1) ∂V and ∂V_0 are connected,
(2) the image of $\pi_1(V)$ in $\pi_1(V_0)$ is isomorphic to $\pi_1(\epsilon)$, and
(3) the image of $\pi_1(\partial V)$ in $\pi_1(V)$ contains the image of $\pi_1(\epsilon)$ in $\pi_1(V)$.

Observe that the inclusion induced map $\pi_1(V) \to \pi_1(V_0)$ factors through $H_1(V)$, so $\ker[\pi_1(V) \to \pi_1(V_0)]$ is a normally generated by a finite set. We can therefore do 1-dimensional surgery on the interior of V and arrange that $i_* : \pi_1(\epsilon) \to \pi_1(V)$ is an isomorphism. Consequently, $\pi_1(\partial V) \to \pi_1(V)$ will be surjective by (3) above. We shall prove that V is dominated by a finite complex. Then every 0-neighborhood is dominated by a finite complex by [6, Proposition 4.3].

Lemma 2. *If $\widetilde{V}$ is the universal covering space of V and $\{E_n\}$ is a nested sequence of connected neighborhoods of the end of $\widetilde{V}$ with $\bigcap E_n = \emptyset$, then the compact support cohomology group $H_c^1(\widetilde{V}; \mathbb{Z}) = \varinjlim H^1(\widetilde{V}, E_n) = 0$.*

Proof. See the proof of [3, Lemma 3]. $\square$

Lemma 3. *$H_2(V, \partial V; \mathbb{Z})$ is finitely generated.*

Proof. From the exact sequence of the pair $(\overline{V}, V)$, it follows that $H_2(V; \mathbb{Z})$ is finitely generated because $H_3(\overline{V}, V; \mathbb{Z}) \cong \check{H}^1(\Sigma) = 0$. Thus the Lemma follows from the exact sequence of the pair $(V, \partial V)$. $\square$

We now define a subpolyhedron W of V as follows. Let $\partial V(3)$ denote the collection of all closed 3-simplices of ∂V. For each $\sigma \in \partial V(3)$, let L_σ be a proper PL ray in V starting at the barycenter of σ, let N_σ be a regular neighborhood of L_σ in V and let $\mathrm{int}_V N_\sigma$ be the topological interior of N_σ in V. Without loss of generality we may assume that

(4) each ∂N_σ is PL homeomorphic to $\mathbb{R}^3$ (the union of an increasing sequence of PL 3-balls),

(5) $N_\sigma \cap \partial V = \sigma$, and

(6) the family $\{\mathrm{int}_V N_\sigma | \sigma \in \partial V(3)\}$ is pairwise disjoint.

Define $W = V - \cup\{\mathrm{int}_V N_\sigma | \sigma \in \partial V(3)\}$. Then

(7) $W \cap \partial V = \partial V^{(2)}$, the 2-skeleton of ∂V, and

(8) $\pi_1(\epsilon) \cong \pi_1(V) \cong \pi_1(W)$.

Let Λ denote the group ring $\mathbb{Z}[\pi_1(\epsilon)]$.

Lemma 4. *If H is the homology or cohomology functor and $\mathcal{B}$ is any coefficient bundle of Λ-modules, then $H(V, \partial V; \mathcal{B}) \cong H(W, \partial V^{(2)}; \mathcal{B})$.*

Proof. Let $A = \cup\{N_\sigma | \sigma \in \partial V(3)\}$. By (4), each N_σ strong deformation retracts onto ∂N_σ; we abbreviate this $N_\sigma \searrow \partial N_\sigma$. It follows that $A \searrow \partial A$ and consequently $V \searrow W \cup \partial V$. Hence,

$$H(V, \partial V; \mathcal{B}) \cong H(W \cup \partial V, \partial V; \mathcal{B})$$
$$\cong H(W, \partial V^{(2)}; \mathcal{B}),$$

the last isomorphism being an excision isomorphism. $\square$

Since V is homotopy equivalent to $W \cup \partial V = W \cup (\cup\{\sigma | \sigma \in \partial V(3)\})$ and $\partial V(3)$ is a finite collection, we need only show that W is dominated by a finite complex. Since Λ is Noetherian, Theorems B and F in [9] imply that it is sufficient to show that W satisfies conditions D_2 and NF_2 in [9]. Refer to [10, Theorem 5] for an equivalent statement of D_2. We need only consider the case of countably generated free Λ-modules since we will only be involved with countably generated free chain complexes.

Lemma 5. *W satisfies condition D_2; i.e.,*

   (i) *$H_i(W; \Lambda) = 0$ for all $i > 2$, and*

   (ii) *$H^3(W; \mathcal{B}) = 0$ for every coefficient bundle $\mathcal{B}$ of countably generated free Λ-modules.*

Proof. First, for $j = 3, 4$ we have

$$
\begin{aligned}
H_j(W, \partial V^{(2)}; \Lambda) &\cong H_j(V, \partial V; \Lambda) &\text{(Lemma 4)}\\
&\cong H_j(\widetilde{V}, \partial\widetilde{V}; \mathbb{Z})\\
&\cong H_c^{4-j}(\widetilde{V}; \mathbb{Z}) &\text{(duality)}\\
&= 0 &\text{(Lemma 2)}.
\end{aligned}
$$

From the homology sequence of the pair $(W, \partial V^{(2)})$, we can infer that $H_j(W; \Lambda) = 0$. So (i) is proved.

Next, from the cohomology exact sequence of the pair $(W, \partial V^{(2)})$ we can infer that $H^3(W; \mathcal{B}) = 0$ since $H^3(\partial V^{(2)}; \mathcal{B}) = 0$ and since $H^3(W, \partial V^{(2)}; \mathcal{B}) \cong H^3(V, \partial V; \mathcal{B}) \cong H_1^\infty(V; \mathcal{B}) = 0$ (by Lemma 4 and the relative version of [3, Lemma 5], respectively). Therefore (ii) is proved. $\square$

Lemma 6. *W satisfies condition NF_2; i.e., $H_2(W; \Lambda)$ is a finitely generated Λ-module.*

Proof. Since the sequence

$$
\cdots \to H_2(\partial V^{(2)}; \Lambda) \to H_2(W; \Lambda) \to H_2(W, \partial V^{(2)}; \Lambda) \to \cdots
$$

is exact and Λ is Noetherian, we need only prove that $H_2(W, \partial V^{(2)}; \Lambda)$ is finitely generated [9, Lemma 1.5]. From the exact sequence

$$
H_2(W, \partial V^{(1)}; \Lambda) \to H_2(W, \partial V^{(2)}; \Lambda) \to H_1(\partial V^{(2)}, \partial V^{(1)}; \Lambda) = 0
$$

of the triple $(W, \partial V^{(2)}, \partial V^{(1)})$, we see that we will only need to prove that the Λ-module $H_2(W, \partial V^{(1)}; \Lambda)$ is finitely generated.

Now let $i : \partial V^{(1)} \to W$ be the inclusion map. Since $i_* : \pi_1(\partial V^{(1)}) \to \pi_1(W) \cong \pi_1(V)$ is onto (by (3)) and W satisfies the condition D_2 (by Lemma 5), it follows from the proof of Lemma 2.1 of [9] that $H_2(W, \partial V^{(1)}; \Lambda)$ is a projective Λ-module. Suppose, on the contrary, that $H_2(W, \partial V^{(1)}; \Lambda)$ is not a finitely generated Λ-module. Then it will be a free Λ-module by [7, Theorem 2.3] since $\pi_1(\epsilon)$ is cyclic. Consequently, the abelianization $\pi_2(i)_{\mathrm{ab}} \cong H_2(W, \partial V^{(1)}; \Lambda)$ of the homotopy group $\pi_2(i)$ is a free Λ-module. (Refer to [5, p. 222] for notation and terminology.) It follows from Theorem 4.3 of [5] that W has the homotopy type of an almost finite 2-complex X; i.e., X is the wedge of a finite complex Y (containing $\partial V^{(1)}$) and infinitely many copies of the 2-sphere. Consequently, $H_2(X, Y; \mathbb{Z})$ is free and of infinite rank and it follows from the exact sequence of the triple $(X, Y, \partial V^{(1)})$ that

$H_2(W, \partial V^{(1)}; \mathbb{Z}) \cong H_2(X, \partial V^{(1)}; \mathbb{Z})$ is of infinite rank. Finally, the exact sequence of the triple $(W, \partial V^{(2)}, \partial V^{(1)})$

$$H_2(\partial V^{(2)}, \partial V^{(1)}; \mathbb{Z}) \to H_2(W, \partial V^{(1)}; \mathbb{Z})$$
$$\to H_2(W, \partial V^{(2)}; \mathbb{Z}) \to H_1(\partial V^{(2)}, \partial V^{(1)}; \mathbb{Z}) = 0$$

will imply that $H_2(W, \partial V^{(2)}; \mathbb{Z})$ is of infinite rank. Thus Lemma 4 implies that $H_2(V, \partial V; \mathbb{Z}) \cong H_2(W, \partial V^{(2)}; \mathbb{Z})$ is of infinite rank. This contradicts Lemma 3. Hence $H_2(W, \partial V^{(1)}; \Lambda)$ is a finitely generated Λ-module and so is $H_2(W; \Lambda)$. $\square$

Remark. In case $\pi_1(\epsilon) \cong \mathbb{Z}$ we can use Lemma 5 (*ii*) and the proof of [2, Lemma 5] to show that $H_2(W; \mathbb{Z})$ is a projective Λ-module.

2. Infinite cyclic $\pi_1(\epsilon)$

The proof of Corollary 1 is essentially the same as the proof of [3, Theorem 1], but for the sake of completeness we provide an outline. We first note that, since $\widetilde{K}_0(\mathbb{Z}[\pi_1(\epsilon)]) = 0$, the end of $M - \Sigma$ has a weak collar U by [2, Theorem 11.9B] (see [3, Lemma 8]). A *weak collar* is a closed manifold neighborhood U of the end which homotopically looks like a collar. In particular, $\pi_1(U) \cong \mathbb{Z}$ and $\ker[\pi_1(\partial U) \to \pi_1(U)]$ is perfect. Second, it follows from 4-dimensional surgery theory [2, §11.6] that there is a compact 4-manifold V such that $\partial V \cong \partial U$ and V has the homotopy type of S^1 (see [3, Lemma 9]). As before, we use $\overline{U}$ to denote $U \cup \Sigma$. Next, $\overline{U} \cup_\partial V$ is homeomorphic to S^4 by the Poincaré conjecture (see [3, Lemma 10]). We have thus succeeded in embedding a neighborhood of Σ in S^4, and we can therefore use the theorems of [3] which apply in case the ambient manifold is S^4. Let $W' = U \cup V \subset S^4$. Then $\pi_1(W') \cong \mathbb{Z}$ by Van Kampen's Theorem. By [3, Theorem 5'], there is a compact manifold $N \subset S^4$ such that

$$\phi : S^2 \times B^2 \xrightarrow{\cong} N, \Sigma = S^4 - W' \subset \operatorname{int} N, \text{ and } N \cap W' = N - \Sigma \cong \partial N \times [0, 1).$$

Pushing along the radial structure of $\partial N \times [0, 1)$, we can assume that $N \subset U \subset M$ and define $K = \phi(S^2 \times \{0\}) \subset N \subset M$. The proof of the Corollary is complete.

3. Finite cyclic $\pi_1(\epsilon)$

A *Swan space* for a group G is a space with the homotopy type of a CW-complex, fundamental group G, and universal cover of the homotopy type of some sphere S^k (refer to [2, p. 228]). Let $\mathbb{Z}_n$ denote the cyclic group of order $n, n < \infty$.

Lemma 7. *Let Σ be a 2-sphere embedded in the interior of the compact 4-manifold M and let ϵ denote the end of $M - \Sigma$. If $\pi_1(\epsilon) \cong \mathbb{Z}_n$, then ϵ has neighborhoods that are Swan spaces for $\pi_1(\epsilon)$.*

Proof. Let E denote the homotopy collar of the end of $M - \Sigma$ (as defined on p. 214 of [2]). By [2, Theorem 11.9E] there is an open neighborhood U of the end ϵ

such that U is homeomorphic to the infinite cyclic cover of a compact manifold N with N homotopy equivalent to $E \times S^1$. Thus $\pi_1(U) \cong \pi_1(E) \cong \mathbb{Z}_n$ and U has two ends. One of the two ends is ϵ, which is finitely dominated by the Theorem. But both ends are proper homotopy equivalent to $E \times [0, \infty)$, so the other end of U must also be finitely dominated and have fundamental group $\mathbb{Z}_n$. Moreover, since $\pi_1(U)$ is finite, the universal cover $\tilde{U}$ of U also has two finitely dominated ends, both simply connected. Therefore each end of $\tilde{U}$ has a simply connected weak collar whose boundary is a homology 3-sphere [2, Theorem 11.9C]. Thus $\tilde{U} \cong S^3 \times \mathbb{R}$ [1, Corollary 1.3]. In other words, U is a Swan space for $\pi_1(\epsilon) \cong \mathbb{Z}_n$. $\square$

Proof of Corollary 2. Let U be an open neighborhood of the end ϵ. By the lemma above we may assume that U is a Swan space for $\pi_1(\epsilon)$. The finiteness obstruction

$$\sigma(U) \in \tilde{K}_0(\mathbb{Z}[\pi_1(\epsilon)])$$

is trivial by [8, Corollary 2.4(i)] and so ϵ has a weak collar [2, Theorem 11.9B]. $\square$

REFERENCES

1. M. H. Freedman, *The topology of four-dimensional manifolds*, J. Diff. Geom. **17** (1982), 357–453.
2. M. H. Freedman and F. Quinn, *Topology of 4-manifolds*, Princeton University Press, Princeton, 1990.
3. Vo Thanh Liem and G. A. Venema, *Characterization of knot complements in the 4-sphere*, Topology and its Appl. (to appear).
4. _____, *On the asphericity of knot complements*, preprint.
5. J. G. Ratcliffe, *On complexes dominated by a 2-complex*, Combinatorial group theory and topology, Annals of Math. Studies, vol. 111, 1987, pp. 221–259.
6. L. C. Siebenmann, *The obstruction to finding a boundary for an open manifold of dimension greater than five*, PhD dissertation, Princeton University, Princeton, New Jersey, 1965.
7. P. F. Smith, *A note on idempotent ideals in group rings*, Arch. Math. **27** (1976), 22–27.
8. C. B. Thomas and C. T. C. Wall, *The topological spherical space form problem I*, Compositio Math. **23** (1971), 101–114.
9. C. T. C. Wall, *Finiteness conditions for CW-complexes*, Annals of Math. **81** (1965), 56–69.
10. _____, *Finiteness conditions for CW complexes, II*, Proc. Roy. Soc., Ser. A **295** (1966), 129–139.
11. _____, *Surgery on Compact Manifolds*, Academic Press, London, 1970.

PRINCETON UNIVERSITY, PRINCETON, NEW JERSEY 08544 AND UNIVERSITY OF ALABAMA, TUSCALOOSA, ALABAMA 35487-0350

CALVIN COLLEGE, GRAND RAPIDS, MICHIGAN 49546

AN INVARIANT OF 3–VALENT SPATIAL GRAPHS

Kenneth C. Millett

University of California at Santa Barbara

ABSTRACT. An invariant of 3–valent spatial graphs is presented in this note. It is a one variable finite Laurent polynomial which is unchanged under topological isotopy of the embedding of the graph. The development of polynomial invariants of classical knots and links is briefly reviewed and the Yamada invariant theory for graphs presented. This provides the foundation for the extension of the theory to the category of regular 3–valent graphs. This extension, its basic properties, and several applications to the study of molecular graphs arising in the study of DNA are described. This note concludes with a brief discussion of some open problems.

1. AN INTRODUCTION TO THE POLYNOMIAL THEORIES

Topologists have associated, for some time, algebraic equations or invariants to knots and links in 3–space in order to provide a means to distinguish between spatially distinct configurations. Until recently, the most familiar of these has been the **Alexander polynomial**, $\Delta(L)(t) = \det(V - tV^T)$, where L denotes an oriented knot or link, where V denotes the linking matrix associated to a connected oriented surface whose boundary is this oriented link, and where V^T denotes its transpose [A, R]. J. H. Conway, [C], noticed that one could as easily work with a recursive definition of the Alexander polynomial. This definition, which does not require the calculation of the determinant (c.f., Kauffman, [K0], for discussion), is:

$$\Delta(L)(t) = \det(t^{1/2}V - t^{-1/2}V^T).$$

By means of this symmetric formulation, Conway noted the following facts, already implicit in the original paper of Alexander:

Theorem 1.1. *The **Alexander polynomial** has the following properties:*
  (i) *$\Delta(U) = 1$, where "U" denotes the trivial knot, and*
  (ii) *if L_+, L_-, and L_0 are planar pictures of oriented links in each of which is identified a small circular region of the picture containing either a single crossing or, in the last case, no crossing at all, as shown in Figure 1.2, and such that outside these small circular regions shown in the figure, the planar pictures are <u>exactly</u> the same, then the Alexander polynomial satisfies the formula:*

$$\Delta(L_+)(t) - \Delta(L_-)(t) - (t^{-1/2} - t^{1/2})\Delta(L_0)(t) = 0.$$

165

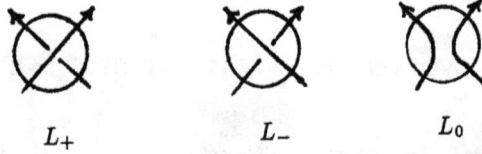

$$L_+ \qquad\qquad L_- \qquad\qquad L_0$$

Figure 1.2

If one replaces the quantity, $t^{-1/2} - t^{1/2}$, with "z" in the above formula, one has the definition of the Conway potential function. From this formula one can recursively calculate the Alexander polynomial since any knot or link can be changed to an unknotted presentation by changing crossings. The Alexander polynomial of these unknotted presentations is one, for one component, and zero, for more than one component. Thus, the Alexander polynomial is the unique finite Laurent polynomial in $t^{\pm 1/2}$ satisfying conditions (i) and (ii) of Theorem 1.1.

In the spring of 1984 the **Jones polynomial** $V(L)(t)$, a completely new finite integral Laurent polynomial spatial invariant of an oriented link, L, in a single variable, also called "t", was discovered [J1]. Although Vaughan Jones defined it by means of a braid presentation of the link, a representation of the braid group into a certain Hecke algebra over the field of fractions of $Z[t^{1/2}]$, and an appropriately normalized trace function, it was soon noted that there was a relationship between $V(L_+)$, $V(L_-)$ and, $V(L_0)$ which resembled the earlier relationship enjoyed by the Alexander polynomial. This is expressed in the following theorem:

Theorem 1.3. *The* **Jones polynomial**, $V(L)(t)$, *of an oriented knot or link, L, is the unique Laurent polynomial in "t", which satisfies the following fundamental formulae:*

(i) *if U denotes the standard unknotted circle in the plane, then $V(U)(t) = 1$, and*

(ii) *if L_+, L_-, and L_0 are planar pictures of oriented links in each of which we have identified a small circular region of the picture containing either a single crossing or, in the last case, no crossing at all, and such that outside these small circular regions shown in Figure 1.2, the planar pictures are* <u>exactly</u> *the same, then the* **Jones polynomial** *satisfies the formula:*

$$t^{-1}V(L_+)(t) - tV(L_-)(t) + (t^{-1/2} - t^{1/2})V(L_0)(t) = 0.$$

Jones noticed immediately that his polynomial must be functionally independent from the Alexander polynomial since his polynomial was able to distinguish between the right and left handed trefoil knots and was not zero for distant unions of links whereas neither of these properties is true for the Alexander polynomial. The observed similarity between the recursion formulae for the Alexander and Jones polynomials was too great to be accidental: a number of people believed that they must be both instances of a more general polynomial invariant for isotopy classes of oriented links. Indeed this is the case as was announced independently and simultaneously, in August 1984, by Lickorish and myself, Freyd and Yetter, Ocneanu, and by Hoste, [F4]; and also, in January 1985, by Przytycki and Traczyk, [P]. The theorem is as follows:

Theorem 1.4. *The* **oriented polynomial** $P(L)(\ell, m)$ *of an oriented knot or link, L is the unique Laurent polynomial in the variables, "ℓ" and "m", which satisfies the following fundamental formulae:*

(i) *if U denotes the standard unknotted circle in the plane, then $P(U)(\ell, m) = 1$, and*

(ii) *if L_+, L_-, and L_0 are planar pictures of oriented links in each of which we have identified a small circular region of the picture containing either a single crossing or, in the last case, no crossing at all, and such that outside these small circular regions shown in Figure 1.2, the planar pictures are* exactly *the same, then the* **oriented polynomial** *satisfies the formula:*

$$\ell P(L_+)(\ell, m) + \ell^{-1} P(L_-)(\ell, m) + m P(L_0)(\ell, m) = 0.$$

Stimulated by the desire to extend this theory of spatial invariants of knots and links to a theory for spatial graphs, another polynomial was discovered by Brandt, Lickorish, and myself, [B2], and, independently, by Ho, [H], which required no orientations of any sort (hence the name 'absolute polynomial' was given to it by Conway) but which involved the consideration the other possible way of removing the crossing, denoted by L_∞ in the Figure 1.6.

Theorem 1.5. *The* **absolute polynomial**, $Q(L)(z)$, *is the unique Laurent polynomial in the variable "z" associated to the isotopy classes of unoriented knots and links, which satisfies the following fundamental formulae:*

(i) *if U denotes the standard unknotted circle in the plane, then $Q(U)(z) = 1$, and*

(ii) *if L_+, L_-, L_0, and L_∞ are planar pictures of oriented links in each of which we have identified a small circular region of the picture in the figure above containing either a single crossing or, in the last cases, no crossing at all, according to the convention shown in Figure 1.6 (with orientations ignored), and such that outside these small circular regions the planar pictures are* exactly *the same, then the* **absolute polynomial** *satisfies the formula:*

$$Q(L_+)(z) + Q(L_-)(z) = z[Q(L_0)(z) + Q(L_\infty)(z)].$$

The absolute polynomial was quickly extended by Kauffman, [K2], who explained how to introduce a second variable, "a", into the absolute polynomial. As a result, one has still another polynomial, distinct from the previous ones and which satisfies

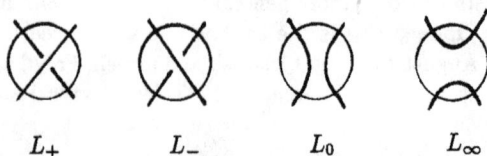

$$L_+ \qquad L_- \qquad L_0 \qquad L_\infty$$

Figure 1.6

a somewhat more complicated recursive formulation involving two versions of the crossing relation depending upon the number of distinct strands involved in the crossings. Recall that the number of strands, or components, in a link L is denoted by $c(L)$. Let $\langle X, Y \rangle$ where X and Y mutually disjoint links, denote the algebraic linking number of X and Y. This number can be calculated from a presentation of the links as one–half the sum of $+1$'s and -1's for each of the crossings between X and Y, depending upon whether the crossing is positive or negative as indicated in the figure. The orientation conventions that are to be employed in the recursive calculation depend upon whether the crossing in L_+ involves the same or distinct strands of the link. These are shown in Figure 1.7. The result is the following theorem:

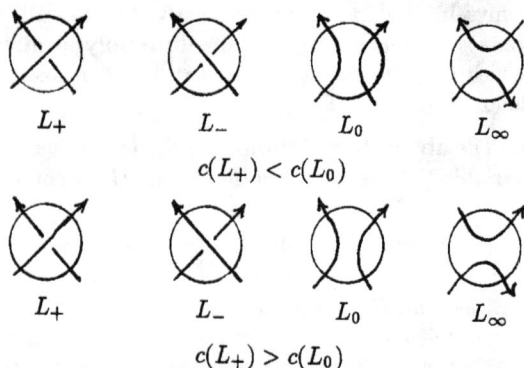

$$L_+ \qquad L_- \qquad L_0 \qquad L_\infty$$
$$c(L_+) < c(L_0)$$

$$L_+ \qquad L_- \qquad L_0 \qquad L_\infty$$
$$c(L_+) > c(L_0)$$

Figure 1.7

Theorem 1.8. *The **semioriented polynomial** $F(L)(a, x)$, of an oriented knot or link, L, is the unique Laurent polynomial in the variables, "a" an "x", which satisfies the following fundamental formulae:*

(i) *if U denotes the standard unknotted circle in the plane, then $F(U)(a, x) = 1$, and*

(ii) *if L_+, L_-, and L_0 are planar presentations of oriented links in each of which we have identified a small circular region of the picture containing either a* *single crossing or, in the last cases, no crossing at all, and such that outside these small circular regions shown in Figure 1.7, the planar pictures are* <u>exactly</u> *the same, then the* **semioriented polynomial** *satisfies one of the formulae, depending upon whether the crossing in L_+ involves the same or distinct components:*

    (I) If $c(L_+) < c(L_0)$, let $\lambda = \langle X, L_0 - X \rangle$ and
$$aF(L_+)(a,x) + a^{-1}F(L_-)(a,x) = x[F(L_0)(a,x) + a^{-4\lambda}F(L_\infty)(a,x)]$$
    (II) If $c(L_+) > c(L_0)$, let $\mu = \langle X, L_+ - X \rangle$ and
$$aF(L_+)(a,x) + a^{-1}F(L_-)(a,x) = x[F(L_0)(a,x) + a^{-4\mu+2}F(L_\infty)(a,x)].$$

Since Kauffman's semioriented polynomial [K2, LM2], distinguishes different pairs of knots and links than does the oriented polynomial, [LM1], the oriented and semioriented polynomials provide independent spatial invariants of these objects. The discovery of the absolute polynomial provoked, however, another very important advance; the discovery of a state model for the Jones polynomial, also by Kauffman, [K3]. The state model approach provides an extraordinarily simple proof of the existence of the Jones polynomial and underlies many efforts to extend the polynomial invariants to the category of graphs as well as hinting at profound connections with certain elements of theoretical physics. Thus, stimulated largely by questions of chirality of spatial graphs, i.e. topological invariance under mirror reflection, arising in chemistry, a number of mathematicians have sought algebraic invariants which would apply to these spatial graphs.

By way of a conceptual introduction to the search for spatial invariants of graphs, we first consider another approach to the development of both topological and chimerical invariants, [JM2], of graphs. By a chimerical graph and chimerical equivalence of such graphs, we mean the specification of a rigid structure on the edges in a neighborhood of each vertex of the graph and the preservation of these structures near each of the vertices under all allowable spatial movements of the graph as provided with this enhanced structure. Kauffman's idea, [K1], was to associate to the spatial realization of a graph a collection of knots and links which would remain unchanged under either the topological or the chimerical equivalence, as appropriate. For example, if the spatial graph contains a nontrivial knot as a subgraph, then the graph cannot be equivalent to a planar graph since all subgraphs in a planar graph are unknotted. Furthermore, if the graph contains a subgraph which is a chiral knot but the graph does not contain the mirror reflection of this knot one can conclude that this spatial embedding is topologically chiral. Since this method is not always sufficient, one is led to seek other means to detect chirality by the exploitation of the richer aspects of the graph structure. One can, if desired, associate the polynomial invariants discussed above to this collection and, thereby, extract an algebraic invariant for the graph.

Thus, stimulated by the potential applications to the natural sciences and fundamental connections with deep issues in mathematical physics, similar polynomial invariants were developed for various categories of graphs subject to a variety of assumptions on the nature of allowable spatial movements under which these equations were to be unchanged. Among the developers of such theories are Yamada [Y1], Kauffman [K1], Jonish–Millett [JM2], Kauffman–Vogel [K4], Jaeger [Jae], and Witten [W1,2&3]. In this paper, we shall focus our attention upon those aspects of these theories which are most directly applicable to our molecular ladder models of the structure of DNA and their knotting, linking and, supercoiling.

The method is a generalization of the polynomial invariant of Yamada, [Y1] which was, in turn, inspired by the Kauffman state model method. We shall first describe the Yamada approach. He defines a finite one variable Laurent polynomial which is a chimerical invariant of unoriented graphs up to multiplication by plus or minus a power of the variable. In the special case that the maximum degree of the vertices of the graph is three, the chimerical invariance is the equivalent to topological invariance, i.e. the preservation of a vertex template for the vertices of degree three does not introduce any additional restriction. In addition, if one restricts to the case of regular graphs of degree 2, i.e., unoriented knots or links, Yamada notes that one has a specialization of the Kauffman's semioriented polynomial.

We shall quickly review the definition of the Yamada invariant: Let G denote a graph having vertex set V and edge set E and let $\mu(G)$ and $\beta(G)$ denote the number of components of G and the first Betti number of G, respectively. For $F \subset E$ let $[F]$ denote the number of elements of F and let $G - F$ denote the graph $(V, E - F)$. Yamada defines a preliminary polynomial as follows:

$$h(G)(x,y) = \sum_{F \subset G} (-x)^{-[F]+\mu(G-F)} y^{\beta(G-F)}$$

and introduces a state model as follows: For any representation of a spatial graph, **G**, one considers the collection of functions, S, from the set of crossings of the

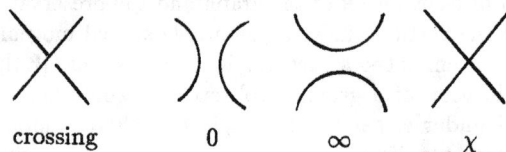

crossing   0   ∞   χ

Figure 1.9

representation from the set of crossings of the representation into the set $\{0, \infty, \chi\}$ and for each $s \in S$ one defines the associated state of **G**, $s(\mathbf{G})$, replacing each crossing by the configuration determined by the value of s according to Figure 1.9. Let $[s] = p - q$, where p is the number of times that s takes the value 0 and q denotes the number of times that s takes the value ∞. A one variable finite Laurent polynomial is then defined by

$$R(\mathbf{G})(A) = \sum_{s \in S} A^{[s]} h(s(\mathbf{G}))(-1, -A^{-1} - 2 - A).$$

Yamada shows, by direct calculation, that this expression is invariant under generalized type II and type III Reidemeister moves, Figures 2.2 and 2.3, and that it is unchanged, up to multiplication by plus or minus a power of A, under the chimerical type I move in which the specific rigid structure is preserved. By eliminating the χ state from the expansion formula and using the perspective given for the crossing given in Figure 1.9 as the definition of the "+" crossing, one can give the following formula relating the "+" and "-" crossing positions:

$$R(\mathbf{G}_+)(A) - R(\mathbf{G}_-)(A) = (A - A^{-1})[R(\mathbf{G}_0) - R(\mathbf{G}_\infty)]$$

which resembles the recursion relation used to define the absolute polynomial given in Theorem 1.5. If one introduces an appropriate factor to insure invariance under the desired type I moves one has, in the case of classical knots and links, a version of the semioriented polynomial. This version has been called the Dubrovnik polynomial by Kauffman. In the development of these polynomial theories there are two critical issues that must be addressed. The first of these is the evaluation of the base states, i.e. the trivial knots and links or the planar graphs, in the case of Yamada. Thus Yamada utilizes the "h" function which has been noted, by Jaeger, to be a version of the flow polynomial or, equivalently, the chromatic polynomial of the planar dual of the graph. In the case of Kauffman, the choice is determined by the invariance under the type II move. The second of these critical issues is the choice of normalization to achieve type I invariance. The Kauffman choice leads to the definition of the semioriented polynomial.

For certain graphs, such as those which determine classical knots and links or those which give spatial realizations of the θ^n graphs which are shown in Figure 1.10, Yamada shows how one is able to remove the indeterminacy in the polynomial equation, associated to the chimerical type I move, by another normalization. Because this normalization problem is the key obstruction to the definition of a spatial invariant for 3–valent graphs via Yamada base state evaluation, we shall review his method. For a link presentation, L, Yamada defines the twisting number of the presentation, $t(L)$, to be the sum of the self crossing numbers associated to each of the individual components, or cycles, of the presentation. For $\theta^n = (\{u, v\}, \{e_1, \ldots, e_n\})$ let c_{ij} denote the cycle consisting of the two vertices and the ith edges and j^{th} edges. Yamada defines the twisting number of a spatial realization of θ^n, θ, by

$$t(\boldsymbol{\theta}) = \sum_{i<j} t(\mathbf{c}_{ij})/(n-1)$$

and shows that $(-A)^{-2t(\boldsymbol{\theta})}R(\boldsymbol{\theta})$ is chimerical invariant of θ and, in the 3–valent case, it is a spatial invariant of $\boldsymbol{\theta}$.

By way of examples, consider the case of the trivial θ^3, shown in Figure 1.11 (a), for which one has the polynomial $-A^{-2} - A^{-1} - 2 - A - A^2$. The second embedding, Figure 1.11 (b), has the polynomial $A^{-8} + A^{-7} - A^{-6} + A^{-5} + A^{-4} - A^{-3} + A^{-1} + 2A + A^2 + 2A^3 - A^5 + A^6 - 2A^7 - A^8 + A^9$ and associated twisting number equal to $-3/2$. As a consequence, the second embedding is topologically distinct

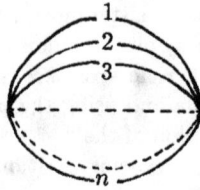

Figure 1.10

from the first. It is topologically chiral because $(-A)^{-2t(\theta)}R(\theta)$ of an achiral embedding, θ, must be invariant under the change of variables taking A to A^{-1}, which is the case for the invariants associated to the classical knots and links. Therefore 1.11(b) and 1.11(c) are spatially distinct conformations.

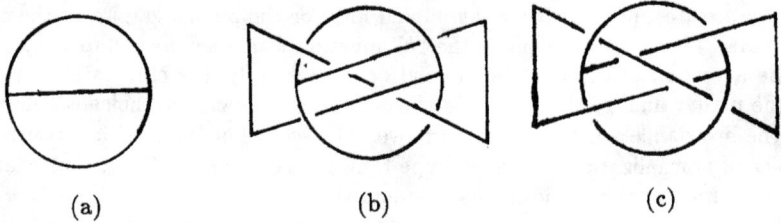

(a)   (b)   (c)

Figure 1.11 The Kinoshita Examples

Further analysis of the Yamada invariant shows that it is an extension of the semioriented polynomial invariant of Kauffman [K2]. For example, if one restricts attention to the case of the classical knots and links, then

$$(-1)^{c(L)-1}A^{4\lambda(L)}(A^{-1}+1+A)F_G(L)(iA^2, i(A^{-1}-A)) = A^{-2t(L)}R(L)(A),$$

where $\lambda(L)$ denotes the total linking number of L, i.e., the sum of the linking numbers of all pairs of distinct components of L. Thus, in this fundamental way, the Yamada polynomial is an extension of the semioriented polynomial to the special case of these 3–valent graphs.

We recall a couple of elementary relations described by Yamada, [Y], which can be profitably employed to simplify calculations of these invariants. The most important of these relations shows that the Yamada expression associated to a graph is equal to the sum of those associated to the result of removing and collapsing a non loop edge. In addition, the expression associated to the distant union is the product of the expressions and the expression associated to a vertex union is the negative of the product. If there is a "cut edge", the expression is zero. In this situation, a **cut edge** in a spatial graph is defined to be an edge for which there exists an embedded 2–sphere meeting the spatial graph in precisely one interior point of the edge. Finally, if we define $R(U) = \sigma = A^{-1} + 1 + A$ and let $\mathbf{G_1}$ and $\mathbf{G_2}$ denote two graphs whose connected sum is denoted by $\mathbf{G_1}^\#\mathbf{G_2}$, then $R(\mathbf{G_1}^\#\mathbf{G_2}) = \sigma^{-1}R(\mathbf{G_1})R(\mathbf{G_2})$. We note that a planar graph is a connected sum if there is a circle in the plane meeting the graph in exactly two points in such a manner that the graphs in the two complementary region are non–trivial. If the graph cannot be decomposed as a non–trivial connected sum we say that it is irreducible. For spatial graphs, the requirement is that there is a two sphere in 3–space meeting the graph in exactly two points in such a manner that the graphs in the two complementary regions are non–trivial spatial graphs. As is the case for classical links, the notion of a connected sum is not well defined because the construction of the connected sum can take place on any pair of edges. Nevertheless, the associated invariants are equal and therefore the invariant of the connected sum is well defined.

2. An invariant of spatial graphs

In this section we shall show that it is possible to extend the Yamada polynomial, defined in the previous section, to encompass a class of 3–valent graphs which contains those often encountered in the study of the molecular models of DNA. This is accomplished by choosing an appropriate normalization so as to insure invariance of the polynomial under all of the elementary movements. We shall briefly develop the fundamental aspects of this new polynomial theory before describing its specific applications to the theory of the molecular ladders associated to the DNA structure problems in the next section.

As is the case for the classical knots and links, c.f. [LM1], or for the Yamada invariant discussed in the previous section, one can demonstrate the existence of a topological invariant of a spatial graph by defining an expression associated to each representation of the spatial graph and showing that this expression is unchanged by each of three elementary movements with which each spatial equivalence can be expressed. In the case of the spatial graphs, we call these the generalized Reidemeister moves. They are shown in Figures 2.1 through 2.3.

Yamada has shown that the expression $R(G)(A)$ is unchanged by type II and type III moves. The awkward point is the type I move. We shall consider the case of 3–valent graphs, i.e. every vertex has valence 3. (N.B. Any vertices of valence

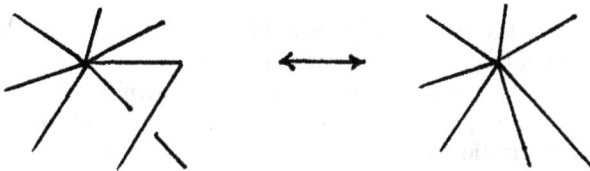

Figure 2.1 Generalized type I Reidemeister move

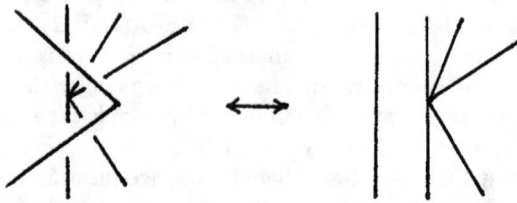

Figure 2.2 Generalized type II Reidemeister move

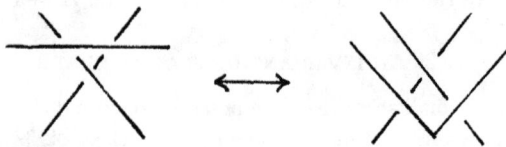

Figure 2.3 Type III Reidemeister move

two are removed by amalgamating the contiguous edges and thereby defining a topologically equivalent graph.) Any specific planar realization determines a collection of circuits in the graph given by the boundaries of the complementary regions, e.g., Figure 2.4. Note that each edge of the graph is a member of precisely two such

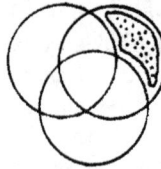

Figure 2.4

circuits. This is also the case for any regular 3–valent graph, i.e., any such can be embedded in an orientable surface of minimal genus whose complementary regions determine a family of circuits. These collections of circuits will not be uniquely associated to the isomorphism class of the graph but, rather, depend upon the specific choice of realization as a subset of the minimal genus surface.

If Γ is the planar realization of a connected irreducible planar graph without cut edges or, more generally, a regular 3–valent graph together with a specific realization of the graph in a minimal genus oriented surface, let $C(\Gamma)$ denote the collection of circuits associated to this realization. This collection is strongly dependent upon

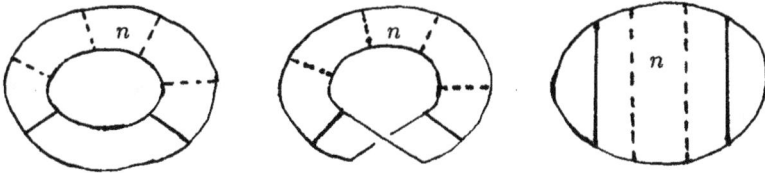

Figure 2.5 $\mathcal{O}_n$, μ_n, and θ^n, respectively

the specific realization of the graph. As a consequence, when we speak of a graph, Γ, we shall intend by that a specific fixed embedding of the abstract graph in a minimal genus orientable surface. For example, there are two cases which arise in which this distinction will be particularly important: the cases of $\mathcal{O}_n$ and θ^n, shown in Figure 2.5. Although $\mathcal{O}_1$ has a cut edge, the complementary regions determine a unique collection of "circuits", one of which has a "double edge", i.e. the cut edge. In general, if there is a cut edge it has the corresponding property of occurring "twice" in the circuit. Since the invariant associated to a planar graph with a cut edge is zero we shall not encounter any difficulties due to this point. Note that all the θ^n and $\mathcal{O}_2 = \theta^2$ are connected sums. Indeed, θ^n is the connected sum of n copies of θ. In the case of the Mobius ladder graphs, they are not planar graphs for $n \geq 3$, as a consequence of the Kuratowski theorem, since each then contains a copy of $\mu_3 = K_{3,3}$. The other cases are μ_0, a spatially unknotted circle; μ_1, a spatially unknotted embedding of θ; and μ_2, spatially equivalent to a planar realization of the edge graph of a tetrahedron. The three classes of graphs, $\mathcal{O}_n$, μ_n, and θ^n play a fundamental role in the discussion of molecular models for closed circular DNA in the next section.

In as much as the connected sum is not well defined, there is the corresponding problem with the definition of the family of circuits associated to the graph. It is for this reason that a fixed realization is attached to the graph and distinct planar realizations of the same abstractly planar graph will be considered as being distinct.

If Γ denotes some spatial embedding of the graph Γ, we define the **reduced writhe** of the embedding $\hat{\omega}(\Gamma)$, to be the sum of the writhe's of each of the individual circuits. Since the writhe of a knot is independent of the orientation of the knot this quantity does not require nor depend upon any choice of orientation of the edges of the graph or of the various circuits. By considering the cases which can occur, one easily demonstrates:

Proposition (2.6). *The reduced writhe is unchanged by the elementary spatial movements of types II and III.*

As in the case of the $\boldsymbol{\theta^n}$ graphs discussed in the previous section, one can choose an appropriate normalization so as to insure invariance under the type I moves and, thereby, define a topological invariant of the spatial embedding.

Proposition (2.7). *If Γ is the spatial realization of a connected regular 3–valent graph, Γ, then*

$$S(\Gamma) = (-A)^{\hat{\omega}(\Gamma)} R(\Gamma)$$

is invariant under the generalized Reidemeister moves and, therefore, defines a spatial invariant of Γ.

Proof. Each crossing occurring in the type I Reidemeister move makes a contribution of ± 1, depending on the sign of the crossing, to each circuit in $C(L)$ in which the two edges occur. If the crossing is associated to a 3–valent vertex of the graph there is only one circuit in which the crossing makes a contribution, therefore corresponding to the contribution of a factor of $(-A)^{\pm 1}$ arising from the Yamada calculation of R. If the crossing is associated to a single edge, as in the case of a classical knot or link, there are precisely two circuits (or a double contribution from a single circuit in the case of a cut edge) in which the crossing makes a contribution, for a total of ± 2, therefore corresponding to the contribution of a factor of $(-A)^{\pm 2}$. Therefore $S(\Gamma)$ is invariant under the type I move. Since the type II and III moves do not change the reduced writhe and $R(\Gamma)$ is invariant under these moves, $S(\Gamma)$ is a spatial invariant of Γ.

Proposition (2.8). *If $\Gamma = \theta^3$, $S(\Gamma) = (-A)^{-2t(\Gamma)}R(\Gamma)$, i.e. $S(\Gamma)$ is a generalization of the Yamada topological invariant.*

Proof. Recall that Yamada defines the twisting number of a spatial realization of θ^n, $\boldsymbol{\theta}^n$, by

$$t(\boldsymbol{\theta}) = \sum_{i<j} t(c_{ij})/(n-1)$$

In the case $n = 3$, the three circuits that arise are the same. Since the divisor "$n-1$" is equal to 2, one has the same normalization factor in this case and, therefore, the same invariant equation.

3. INVARIANTS FOR MOLECULAR GRAPHS ASSOCIATED TO DNA

The fundamental properties of this new spatial invariant applied to a class of molecular graphs that arise in the study of the conformational structure of DNA will be discussed in this section. We shall first concern ourselves with the evaluations of the basic configurations, $\mathcal{O}_n$, θ^n and τ_n, described in Figure 2.5 with the latter being described in Figure 3.1.

Proposition (3.2). *Let $\sigma = A^{-1} + 1 + A$, then*

    (i) $S(\mathcal{O}_1) = 0$, $S(\theta^0) = \sigma$, $S(\tau_0) = -\sigma^2$

    (ii) $S(\theta^n) = \sigma(1-\sigma)^n$

    (iii) $S(\tau_n) = -\sigma^2(2-\sigma)^n + \sigma \sum_{j=0}^{n-1}(2-\sigma)^j(1-\sigma)^{n-1-j}$
             $= \sigma(1-\sigma)[(2-\sigma)^n - (1-\sigma)^{n-1}]$

(iv) $S(\mathcal{O}_n) = \sum_{j=1}^{n-1}[-\sigma^2(2-\sigma)^j + \sigma\sum_{k=0}^{j-1}(2-\sigma)^k(1-\sigma)^{j-1-k}]$

$\qquad = -\sigma(2-\sigma)(1-(2-\sigma)^{n-1}) - (1-\sigma)(1-(1-\sigma)^{n-1})$

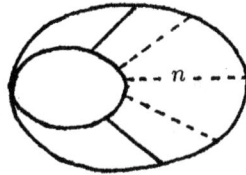

Figure 3.1 τ_n

Proof. Parts (i) and (ii) are by direct calculation while (iii) uses the connected sum formula. For (iv) one proceeds by induction of n using the fact that $S(\mathcal{O}_2) = S(\theta^2) = (A^{-1}+A)^2(A^{-1}+1+A)$. Using the recursion formula on the edges, one finds that

$$S(\mathcal{O}_n) = S(\mathcal{O}_{n-1}) + S(\tau_{n-1}) \quad \text{and} \quad S(\tau_n) = (2-\sigma)S(\tau_{n-1}) + S(\theta^{n-1}),$$

where $\sigma = A^{-1}+1+A$ and $S(\tau_0) = -\sigma^2$. The proof that

$$S(\tau_n) = -\sigma^2(2-\sigma)^n + \sigma\Sigma(2-\sigma)^j(1-\sigma)^{n-1-j}$$

is by induction on n:

$$S(\tau_{n+1}) = (2-\sigma)S(\tau_n) + S(\tau_n) + S(\theta^n)$$

$$= (2-\sigma[-\sigma^2(2-\sigma)^n + \sigma\sum_{j=0}^{n-1}(2-\sigma)^j(1-\sigma)^{n-1-j}] + \sigma(1-\sigma)^n$$

$$= -\sigma^2(2-\sigma)^{n+1} + \sigma\sum_{j=0}^{n-1}(2-\sigma)^{j+1}(1-\sigma)^{n-1-j} + \sigma(1-\sigma)^n$$

$$= -\sigma^2(2-\sigma)^n + \sigma\sum_{j=0}^{n}(2-\sigma)^j(1-\sigma)^{n-1-j}$$

This allows one to complete the proof of the proposition by induction on n:

$$S(\mathcal{O}_n) = S(\mathcal{O}_{n-1} + S(\tau_{n-1})$$

$$= \sum_{j=1}^{n-2}[-\sigma^2(2-\sigma)^j + \sigma\sum_{k=0}^{j-1}(2-\sigma)^k(1-\sigma)^{j-1-k}]$$

$$+ [-\sigma^2(2-\sigma)^{n-1} + \sigma\sum_{j=0}^{n-2}(2-\sigma)^j(1-\sigma)^{n-1-j}]$$

$$= \sum_{j=1}^{n-1}[-\sigma^2(2-\sigma)^j + \sigma\sum_{k=0}^{j-1}(2-\sigma)^k(1-\sigma)^{j-1-k}].$$

Because the graphs whose invariants were just calculated are planar graphs, the formulae are invariant under the change of variables A goes to A^{-1}. Since $\sigma = A^{-1} + 1 + A$ is invariant under the involution taking A to A^{-1}, all equations in the expression σ such as those in the proposition are invariant under the involution. If, on the other hand, one considers the simplest regular 3–valent graph for which no planar embedding is possible, such as $K_{3,3}$ shown in Figure 3.3, it is important to observe that the invariant detects the nonplanarity of the graph. Indeed this is the case as the invariant associated to the embedding of $K_{3,3}$, denoted Γ, is calculated showing that the total twisting is zero, with respect of the circuits arising from the genus one surface constructed by making the over crossing pass over by means of the one handle. One then takes the expansion given by the projection to find $S(\Gamma) = (1 - A)(A^{-1} + A)(A^{-1} + 1 + A)(2A^{-1} - a + 2A)$. Noting that $S(\Gamma)$ is changed under the change of variables taking A to A^{-1} demonstrates that Γ is not deformable to a plane embedding.

That the analysis of the spatial properties is a subtile matter is indicated by the fact that the spatial realization of $K_{3,3}$ given in Figure 3.3 is spatially equivalent to its mirror image! How does this fact not violate the previous analysis demonstrating that the spatial realization is not equivalent to a planar realization? The key issue

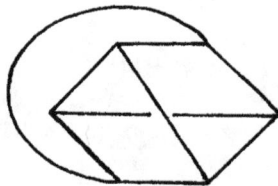

Figure 3.3 A spatial embedding of $K_{3,3}$

is the definition that associated to Γ is the collection of circuits which are utilized to determine the reduced writhe of the spatial graph. Thus, although we calculate that

$\hat{\omega}(\Gamma) = 0$, when one follows the spatial equivalence between Γ and $\overline{\Gamma}$ to establish the correspondence between circuits one discovers that $\omega(\overline{\Gamma}) = 1$. Thus, although $R(\Gamma) = (1-A)(A^{-1}+A)(A^{-1}+1+A)(2A^{-1}-1+2A)$ and $R(\overline{\Gamma}) = (1-A^{-1})(A^{-1}+A)(A^{-1}+1+A)(2A^{-1}-1+2A)$, the reduced writhe change explains the fact that $S(\Gamma) = R(\Gamma) = (-A)^{\omega(\overline{\Gamma})}R(\overline{\Gamma}) = S(\overline{\Gamma})$ as is required by the fact that they are spatially equivalent.

The example of $K_{3,3}$ is a particular case of a Mobius ladder, Figure 2.5, having 3 rungs. The Mobius ladders, μ_n, have planar embeddings for $n = 0$, 1, and 2; the standard model is achiral for $n = 3$; and the standard embeddings are chiral for all $n > 3$. This latter statement is a result of J. Simon, [S1,S2], and can be demonstrated using the above invariants.

Proposition 3.4 (Simon). *The Mobius ladders, μ_n, are chiral for all $n > 3$.*

Proof. Since $\mu_n \subset \mu_{n+1}$, it is sufficient to show that μ_4 is chiral. Let Γ denote the standard spatial embedding of μ_4. We calculate that

$$R(\Gamma) = (A^{-1} + A)(A^{-1} + 1 + A)(A^{-1} - 1 + A)(-A^{-1} + 4 - 3A + 3A^2) \quad \text{and}$$
$$R(\overline{\Gamma}) = (A^{-1} + A)(A^{-1} + 1 + A)(A^{-1} - 1 + A)(3A^{-2} - 3A^{-1} + 4 - A).$$

If it were the case that Γ was spatially equivalent to $\overline{\Gamma}$, $R(\Gamma)/R(\overline{\Gamma})$ would be a power of $(-A)$ determined by the change in the reduced writhe associated to the two configurations necessary to insure that S is a spatial invariant. Because this is not the case, Γ is not spatially equivalent to $\overline{\Gamma}$, thereby proving the proposition.

These examples of applications of the spatial invariant to regular 3-valent graphs have a secondary but, nevertheless, important role in the development and study of invariants associated to those spatial graphs serving as models for closed circular DNA. Let K denote a molecular model for closed circular DNA given as a spatial embedding which is a knotted $\mathcal{O}_n$, such as that which is illustrated in Figure 3.5.

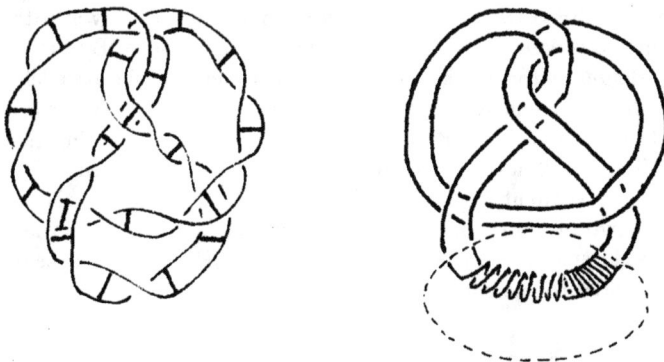

Figure 3.5 A knotted molecular ladder

This would correspond to a typical spatial configuration of the knotted closed circular DNA that has been identified by electron microscopy in the course of an experimental procedure. While respecting the topological structure and for mathematical (as opposed to physical) calculational purposes, the rungs and supercoils (i.e. twists) of the molecular ladder can be deformed along the spatial graph so that they are finally concentrated in a small region, such as that shown at the right in Figure 3.5.

The importance of this mathematical manipulation is that it allows one to exploit the localization principle in the skein theory of these invariants thereby facilitating their calculation and study, c.f. [LM1]. This localization can be used in the detection of knotting, linking and, supercoiling by means of the polynomial invariants. In the model shown in Figure 3.5, one can identify and isolate the two types of contributions, denoted **L** and **M** respectively, within the regions shown in Figure 3.6. There is shown the configuration in which we are interested and which will be denoted $N(\mathbf{L}\#\mathbf{M})$, the 'numerator' of the sum of **L** and **M** in the Conway vocabulary [C].

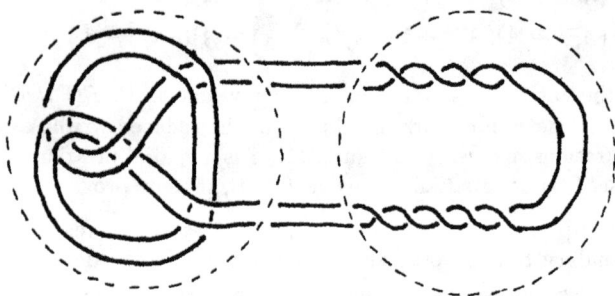

Figure 3.6 Localization

This sum is defined by making the lower connections as shown in the figure while the numerator is defined by virtue of the upper connections. By way of contrast, the Conway 'denominator' of the sum is defined by joining the parallel upper strands to each other as they leave the circular regions, as shown in Figure 3.7. In this case, $D(\mathbf{L}\#\mathbf{M})$ is seen to be spatially equivalent to θ^n by retracting the 'band' along itself until all the 'knotting and twisting' have been eliminated.

For any constituent, e.g. **L**, we let $N(\mathbf{L})$ and $D(\mathbf{L})$ denote the respective completions shown in Figure 3.8. The localization of the skein theory allows one to determine the spatial invariant associated to $N(\mathbf{L}\#\mathbf{M})$ by reference to basic expressions associated to the constituent elements.

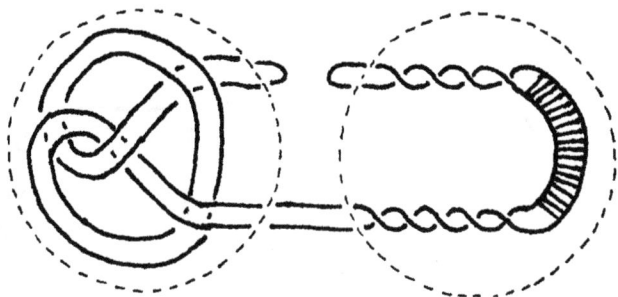

Figure 3.7 The denominator construction

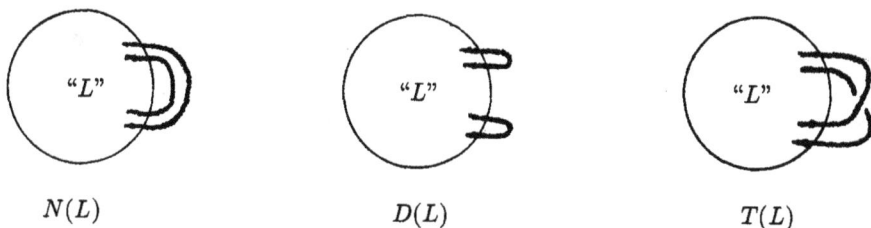

$N(L)$ $\qquad\qquad\qquad\qquad$ $D(L)$ $\qquad\qquad\qquad\qquad$ $T(L)$

Figure 3.8 The numerator and denominator, and the twist

Proposition 3.9. *The following expression, in which* $V = A^{-1} - A - A^2$ *and* $W = A^{-2} + A^{-1} + A$, *gives the invariant for the type of configuration shown in Figure 3.10:*

$$
\begin{aligned}
&S(N(\mathbf{L}\#\theta^N)) \\
&= \frac{[(A^{-1}+1)S(N(\mathbf{L})) + \sigma S(T(\mathbf{L})) - A^{2\omega(K)+2}(A+1)]}{[\sigma^2(\sigma-1)]} S(\mathcal{O}_n) \\
&+ \frac{[-A\,S(N(\mathbf{L})) + \sigma S(T(\mathbf{L})) + A^{2\omega(K)}(A+1)\sigma]}{[\sigma^2(\sigma-1)]} S(\theta^n) \\
&+ \frac{[VS(N(\mathbf{L})) + A^{-1}(A+1)^2\sigma S(T(\mathbf{L})) + A^{2\omega(K)}W\sigma]}{[\sigma^3(\sigma-1)]} S(\tau_n).
\end{aligned}
$$

Proof. The fundamental principle of the skein theory is the use of multilinearity and the expansion of the invariant in terms of the generators of each skein room. In the present setting we shall use the generators whose symmetric representatives are shown in Figure 3.11. One has the expansion

$$
R(N(\mathbf{L}\#\theta^n)) = X\ R(N(\zeta\#\theta^n)) + Y\ R(N(\xi\#\theta^n)) + Z\ R(N(\chi\#\theta^n))
$$

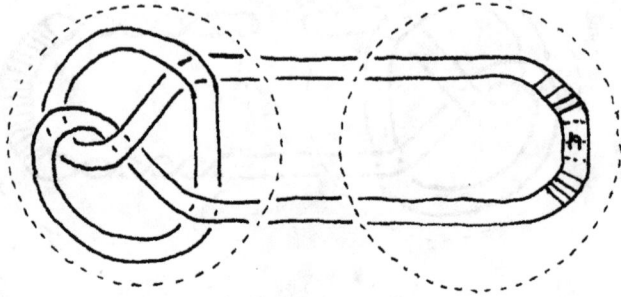

Figure 3.10

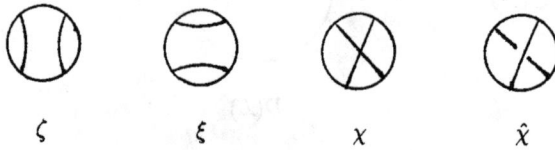

$$\zeta \qquad \xi \qquad \chi \qquad \hat{\chi}$$

Figure 3.11

in which terms X, Y, and Z are finite integral Laurent polynomials determined by the Yamada state expansion.

By observing that $R(N(\zeta\#\theta^n)) = R(\mathcal{O}_n)$, $R(N(\xi\#\theta^n)) = R(\theta^n)$, and that $R(N(\chi\#\theta^n)) = R(\tau_n)$ we see that the calculation is reduced to the determination of polynomials X, Y, and Z. This is accomplished by the family of equations:

$$R(N(\mathbf{L})) = X\ R(N(\zeta\#\zeta)) + Y\ R(N(\xi\#\zeta)) + Z\ R(N(\chi\#\zeta))$$
$$R(D(\mathbf{L})) = X\ R(N(\zeta\#\xi)) + Y\ R(N(\xi\#\xi)) + Z\ R(N(\chi\#\xi))$$
$$R(T(\mathbf{L})) = X\ R(N(\zeta\#\hat{\chi})) + Y\ R(N(\xi\#\hat{\chi})) + Z\ R(N(\chi\#\hat{\chi})).$$

In our case, $N(\mathbf{L})$ is the double of the diagram of a knot K thereby giving a certain twisted double of the knot, the amount of twisting determined by the writhe of the diagram of K. $D(\mathbf{L})$ is a diagram of the trivial knot and $T(\mathbf{L})$ is the (2,1) cable of the diagram of K giving a certain twisted double of the knot, the amount of twisting again being determined by the writhe of the diagram for K. The values

$$R(N(\zeta \# \zeta)) = R(N(\xi \# \xi)) = \sigma^2,$$
$$R(N(\zeta \# \xi)) = R(N(\xi \# \zeta)) = \sigma,$$
$$R(N(\chi \# \zeta)) = R(N(\chi \# \xi)) = -\sigma^2,$$
$$R(N(\zeta \# \hat{\chi})) = A^2,$$
$$R(N(\xi \# \hat{\chi})) = A^{-2}, \quad \text{and}$$
$$R(N(\chi \# \hat{\chi})) = \sigma$$

give the system of equations:

$$R(N(\mathbf{L})) = \sigma^2 X + \sigma Y - \sigma^2 Z$$
$$\sigma = \sigma X + \sigma^2 Y - \sigma^2 Z$$
$$R(T(\mathbf{L})) = A^2 X + A^{-2} Y + \sigma Z,$$

whose solution is:

$$X = \frac{[(A^{-1}+1)R(N(\mathbf{L})) + \sigma R(T(\mathbf{L})) - A^2(A+1)]}{[\sigma^2(\sigma - 1)]}$$

$$Y = \frac{[-AR(N(\mathbf{L})) + \sigma R(T(\mathbf{L})) + (A+1)\sigma]}{[\sigma^2(\sigma - 1)]}$$

$$Z = \frac{[VR(N(\mathbf{L})) + A^{-1}(A+1)^2\sigma R(T(\mathbf{L})) + W\sigma]}{[\sigma^3(\sigma - 1)]}$$

Substitution in the first equation completes the calculation of $R(N(\mathbf{L}\#\boldsymbol{\theta}^n))$:

$$R(N(\mathbf{L}\#\boldsymbol{\theta}^n))$$
$$= \frac{[(A^{-1}+1)R(N(\mathbf{L})) + \sigma R(T(\mathbf{L})) - A^2(A+1)]}{[\sigma^2(\sigma - 1)]} R(\mathcal{O}_n)$$
$$+ \frac{[-AR(N(\mathbf{L})) + \sigma R(T(\mathbf{L})) + (A+1)\sigma]}{[\sigma^2(\sigma - 1)]} R(\theta^n)$$
$$+ \frac{[VR(N(\mathbf{L})) + A^{-1}(A+1)^2\sigma R(T(\mathbf{L})) + W\sigma]}{[\sigma^3(\sigma - 1)]} R(\tau_n).$$

To complete the proof of the proposition we determine the contribution of the reduced writhe, $\hat{\omega}(N(\mathbf{L}\#\boldsymbol{\theta}^n))$, in terms of the writhe of the knot K. Specifically, we find that $\hat{\omega}(N(\mathbf{L}\#\boldsymbol{\theta}^n)) = 2\omega(K) = \hat{\omega}(N(\mathbf{L}))$ and that $\hat{\omega}(T(\mathbf{L})) = 2\omega(K) + 1$.

$$S(N(\mathbf{L}\#\boldsymbol{\theta}^n)) = (-A)^{\hat{\omega}(N(\mathbf{L}\#\boldsymbol{\theta}_n))} R(N(\mathbf{L}\#\boldsymbol{\theta}^n))$$
$$= \frac{[(A^{-1}+1)S(N(\mathbf{L})) + \sigma S(T(\mathbf{L})) - A^{2\omega(K)+2}(A+1)]}{[\sigma^2(\sigma - 1)]} S(\mathcal{O}_n)$$
$$+ \frac{[-AS(N(\mathbf{L})) + \sigma S(T(\mathbf{L})) + A^{2\omega(K)}(A+1)\sigma]}{[\sigma^2(\sigma - 1)]} S(\theta^n)$$
$$+ \frac{[VS(N(\mathbf{L})) + A^{-1}(A+1)^2\sigma S(T(\mathbf{L})) + A^{2\omega(K)}W\sigma]}{[\sigma^3(\sigma - 1)]} S(\tau_n).$$

There is a slightly different evaluation that makes more apparent the symmetry of the interrelationships of the structure. The goal is to place in evidence more clearly the role of those elements of these equations which change under a mirror reflection. Those, such as invariants of planar graphs, which are unchanged under the involution taking A to A^{-1} are collected together in various ways to point out the symmetry of their contribution.

Proposition (3.12). *The following expression gives the invariant for the type of configuration shown in Figure 3.10:*

$$S(N(\mathbf{L}\#\theta^n)) = \frac{S(\mathcal{O}_n)[S(N(\mathbf{L})) - S(K) - (-A)^{2\omega(K)}]}{[(\sigma - 1)(\sigma^2 - \sigma - 1)]}$$
$$+ \frac{S(\theta^n)[S(N(\mathbf{L})) + \sigma(1 - \sigma)S(K) + \sigma^2(\sigma - 2)(-A)^{2\omega(K)}]}{[\sigma(\sigma(\sigma^2 - \sigma - 1))]}$$
$$+ \frac{S(\tau_n)[\sigma S(N(\mathbf{L})) + (1 - \sigma^2)S(K) - \sigma(-A)^{2\omega(K)}]}{[\sigma(\sigma - 1)(\sigma^2 - \sigma - 1)]}$$

Proof. One again uses the expansion

$$R(N(\mathbf{L}\#\theta^n)) = X \ R(N(\zeta\#\theta^n)) + Y \ R(N(\xi\#\theta^n)) + Z \ R(N(\chi\#\theta^n))$$

in which the terms X, Y, and Z are finite integral Laurent polynomials determined by the Yamada state expansion.

By observing that $R(N(\zeta\#\theta^n)) = R(\mathcal{O}_n)$, $R(N(\xi\#\theta^n)) = R(\theta^n)$, and that $R(N(\chi\#\theta^n)) = R(\tau_n)$ we see that the calculation is reduced to the determination of polynomials X, Y, and Z. This is accomplished by a new family of equations:

$$R(N(\mathbf{L})) = X \ R(N(\zeta\#\zeta)) + Y \ R(N(\xi\#\zeta)) + Z \ R(N(\chi\#\zeta))$$
$$R(D(\mathbf{L})) = X \ R(N(\zeta\#\xi)) + Y \ R(N(\xi\#\xi)) + Z \ R(N(\chi\#\xi))$$
$$R(\chi(\mathbf{L})) = X \ R(N(\zeta\#\chi)) + Y \ R(N(\xi\#\chi)) + Z \ R(N(\chi\#\chi)).$$

In our case, $N(\mathbf{L})$ is the double of the diagram of a knot K thereby giving a certain twisted double of the knot, the amount of twisting determined by the writhe of the diagram of K. $D(\mathbf{L})$ is a diagram of the trivial knot and $\chi(\mathbf{L})$ is the connected sum of the diagram of K with θ. The values

$$R(N(\zeta\#\zeta)) = R(N(\xi\#\xi)) = \sigma^2,$$
$$R(N(\zeta\#\xi)) = R(N(\xi\#\zeta)) = \sigma,$$
$$R(N(\chi\#\zeta)) = R(N(\zeta\#\chi)) = R(N(\chi\#\xi)) = R(N(\xi\#\chi)) = -\sigma^2 \quad \text{and}$$
$$R(N(\chi\#\chi)) = \sigma - \sigma^2 + \sigma^3$$

give the system of equations:

$$R(N(\mathbf{L})) = \sigma^2 X + \sigma Y - \sigma^2 Z$$
$$\sigma = \sigma X + \sigma^2 Y - \sigma^2 Z$$
$$R(\chi(\mathbf{L})) = R(K)(1 - \sigma) - \sigma = -\sigma^2 X - \sigma^2 Y + (\sigma - \sigma^2 + \sigma^3)Z,$$

whose solution, in this case, is:

$$X = [R(N(\mathbf{L})) - R(K) - 1]/[(\sigma - 1)(\sigma^2 - \sigma - 1)]$$
$$Y = [R(N(\mathbf{L})) + \sigma(1 - \sigma)R(K) + \sigma^3 - 2\sigma^2]/[\sigma(\sigma - 1)(\sigma^2 - \sigma - 1)]$$
$$Z = [\sigma R(N(\mathbf{L})) + (1 - \sigma^2)R(K) - \sigma)]/[\sigma(\sigma - 1)(\sigma^2 - \sigma - 1)]$$

Substitution in the first equation completes the calculation of $R(N(\mathbf{L}\#\boldsymbol{\theta}^n))$:

$$R(N(\mathbf{L}\#\boldsymbol{\theta}^n)) = \frac{R(\mathcal{O}_n)[R(N(\mathbf{L})) - R(K) - 1]}{[(\sigma - 1)(\sigma^2 - \sigma - 1)]}$$
$$+ \frac{R(\boldsymbol{\theta}^n)[R(N(\mathbf{L})) + \sigma(1 - \sigma)R(K) + \sigma^3 - 2\sigma^2]}{[\sigma(\sigma - 1)(\sigma^2 - \sigma - 1)]}$$
$$+ \frac{R(\tau_n)[\sigma R(N(\mathbf{L})) + (1 - \sigma^2)R(K) - \sigma)]}{[\sigma(\sigma - 1)(\sigma^2 - \sigma - 1)]}$$

To complete the proof of the proposition we determine the contribution of the reduced writhe, $\hat{\omega}(N(\mathbf{L}\#\boldsymbol{\theta}^n))$, in terms of the writhe of the knot K. Specifically, we find that $\hat{\omega}(N(\mathbf{L}\#\boldsymbol{\theta}^n)) = 2\omega(K) = \hat{\omega}(N(\mathbf{L}))$ and that $\hat{\omega}(\chi(\mathbf{L})) = \omega(K)$.

$$S(N(\mathbf{L}\#\boldsymbol{\theta}^n)) = (-A)^{\hat{\omega}(N(\mathbf{L}\#\boldsymbol{\theta}_n))} R(N(\mathbf{L}\#\boldsymbol{\theta}^n))$$
$$= \frac{S(\mathcal{O}_n)[S(N(\mathbf{L})) - S(K) - (-A)^{2\omega(K)}]}{[(\sigma - 1)(\sigma - 1)]}$$
$$+ \frac{S(\boldsymbol{\theta}^n)[S(N(\mathbf{L})) + \sigma(1 - \sigma)S(K) + \sigma^2(\sigma - 2)(-A)^{2\omega(K)}]}{[\sigma(\sigma(\sigma^2 - \sigma - 1)]}$$
$$+ \frac{S(\tau_n)[\sigma S(N(\mathbf{L})) + (1 - \sigma^2)S(K) - \sigma(-A)^{2\omega(K)})]}{[\sigma(\sigma - 1)(\sigma^2 - \sigma - 1)]}$$

Corollary 3.13. *The formula given in the proposition can be presented in an alternative fashion:*

$$S(N(\mathbf{L}\#\theta^n)) = \frac{S(N(\mathbf{L}))[\sigma S(\mathcal{O}_n) + S(\theta^n) + \sigma S(\tau_n)]}{[\sigma(\sigma - 1)(\sigma^2 - \sigma - 1)]}$$
$$+ \frac{S(K)[-\sigma S(\mathcal{O}_n) + \sigma(1 - \sigma)S(\theta^n) + (1 - \sigma^2)S(\tau_n)]}{[\sigma(\sigma - 1)(\sigma^2 - \sigma - 1)]}$$
$$+ \frac{A^{2\omega(K)}[-\sigma S(\mathcal{O}_n) + \sigma^2(\sigma - 2)S(\theta^n) - \sigma S(\tau_n)]}{[\sigma(\sigma - 1)(\sigma^2 - \sigma - 1)]}$$

$$= \frac{S(N(\mathbf{L}))[(\sigma^2 - \sigma - 1) + \sigma(2 - \sigma)^{n+1} + (2 - \sigma)(1 - \sigma)^n]}{(\sigma - 1)(\sigma^2 - \sigma - 1)}$$
$$+ \frac{S(K)[-(\sigma^2 - \sigma - 1) + (\sigma^3 - \sigma^2 - 2\sigma + 1)(2 - \sigma)^n + \sigma(\sigma - 2)(1 - \sigma)^n]}{[(\sigma - 1)(\sigma^2 - \sigma - 1)]}$$
$$+ \frac{A^{2\omega(K)}[-(\sigma^2 - \sigma - 1) - \sigma(2 - \sigma)^{n+1} + (\sigma^3 - 2\sigma^2 + \sigma - 1)(1 - \sigma)^n]}{[(\sigma - 1)(\sigma^2 - \sigma - 1)]}$$

Proof. The first formula is achieved by rearranging the terms of the formula in the proposition. The expressions

$$S(\tau_n) = \sigma(1 - \sigma)[(2 - \sigma)^n - (1 - \sigma)^{n-1}] \quad \text{and}$$
$$S(\mathcal{O}_n) = -\sigma(2 - \sigma)(1 - (2 - \sigma)^{n-1}) - (1 - \sigma)(1 - (1 - \sigma)^{n-1})$$
$$S(\theta^n) = \sigma(1 - \sigma)^n$$

are invoked to further modify the formula so as to make the universal role of $\sigma = A^{-1} + 1 + A$ transparent as in the second formula.

The terms $S(N(\mathbf{L}))$, $S(K)$, and $A^{2\omega(K)}$ describe the contribution of the knotting of the molecular graph to the associated invariant via these associated classical knots and links. They are shown, for this specific case, in Figure 3.14. In the case of these DNA graph models, the results are known in classical knot theory as the (2,0) and (2,1) cables of the knot, K, forming the central axis of the model.

The presence of supercoiling or twisting in the model, as is illustrated in Figures 3.5 and 3.6, can be detected means of the localization theory as well. The basic method is to employ the well-known interconvertability of supercoils and twists, shown in Figure 3.15. Thus, as we have done earlier, change all supercoils to twists and isolate all the twists in the same region as the 'rungs' of the ladder structure. To illustrate this method, consider the effect of one full twist on the relaxed unknotted graph. The invariant of the relaxed unknotted graph, shown in Figure 2.5, is $S(\mathcal{O}_n)$, while the invariant of the once twisted graph shown in Figure 3.16, is

$$A^2[(A^{-1} - A + A^2)S(\mathcal{O}_{n+1})],$$

for $n \geq 3$.

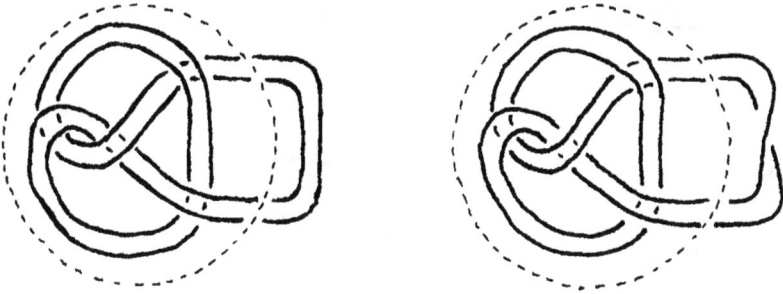

Figure 3.14 The associated classical knots and links: $S(N(\mathbf{L}))$ and $S(T(\mathbf{L}))$

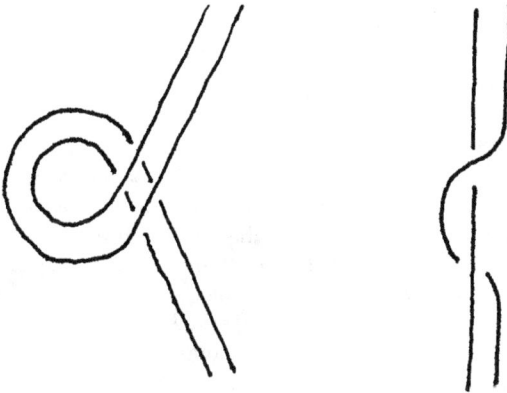

Figure 3.15 Supercoiling and twisting interconvertability

More generally, for the case of k half twists on the relaxed unknotted graph, ${}^{k}\mathcal{O}_n$, one has the following recursion formula for the polynomial invariant:

$$S({}^{k}\mathcal{O}_n) = A^2 S({}^{k-2}\mathcal{O}_n) + (A^2 - 1)\{(-A)^{k+1}S(\theta^n) - S({}^{k-1}\mathcal{O}_n)\}$$

which reduces the calculation to that of ${}^{0}\mathcal{O}_n = \mathcal{O}_n$ and ${}^{1}\mathcal{O}_n = \mu_n$, the Mobius ladder graph shown in Figures 2.5 and 3.16, whose invariants were expressed in terms of $S(\theta^n)$ and $S(\mathcal{O}_{n+1})$ in the results of Proposition (3.2).

4. SUMMARY CONCLUSIONS AND OPEN PROBLEMS

In this paper we have presented an extension of the Yamada polynomial invari-

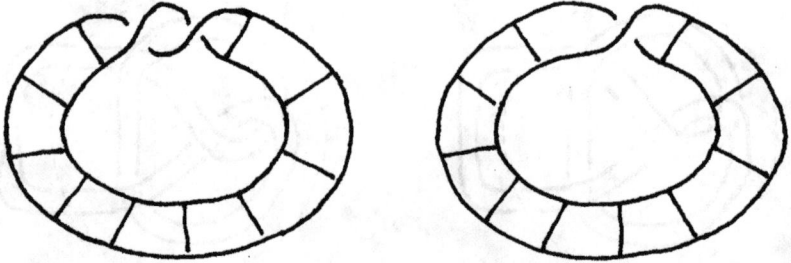

Figure 3.16 The once twisted and the Mobius ladder graphs, respectively

ant which defines a topological invariant of spatial embeddings of 3–valent graphs embellished with the choice of a certain family of circuits in the graph. This theory is sufficiently robust so as to be able to detect topological chirality in interesting situations. In addition, we have shown that the natural graph models for DNA are well adapted to this theory and that calculations can be easily made by exploiting the topological properties of the models and the multilinearity of the skein theory of these invariants. These calculations allow one to detect the knotting, linking and, supercoiling of the DNA models by means of simple calculations.

The discussion in this paper has been restricted to topological invariants associated to spatial configurations corresponding to embeddings of regular 3–valent graphs in 3–space. Yamada, [Y1], has given a definition which gives a 'rigid vertex' or 'chimerical' invariant for the θ_n graphs and Jonish–Millett, [JM2], and also, Kauffman–Vogel, [K4], have given an invariant for chimerical regular 4–valent graphs. Nevertheless, there does not exist an analogous theory for the topological spatial invariance of even the regular graphs of valence larger than 3.

Problem (4.1): What is the appropriate normalization procedure to extend the Yamada invariant to a spatial invariant of all regular k–valent graphs, $k \geq 4$?

If one restricts attention to classical knots and links, then

$$(-1)^{c(L)-1}A^{4\lambda(L)}(A^{-1}+1+A)F_G(L)(iA^2, i(A^{-1}-A)) = A^{-2\hat{\omega}(L)}R(L)(A),$$

where $\lambda(L)$ denotes the total linking number of L, i.e. the sum of the linking numbers of all pairs of distinct components of L, and $\hat{\omega}(L)$ denotes the reduced writhe of the diagram L. Yamada has also identified another important connection with earlier polynomials. If $\widetilde{K}$ is a diagram for a knot, K such that $\omega(\widetilde{K}) = 0$, i.e. an untwisted diagram, then

$$(A^{-1}+1+A)F_G^*(\widetilde{K})(A^{-2}, (A^{-1}-A)) + 1 = R(\widetilde{K})(A^{-1}) + 1$$
$$= -(A^{-1/2}+A^{1/2})V(\widetilde{K}^{(2)})(A)$$

where $\widetilde{K}^{(2)}$ is the untwisted (2,0)–cable of $\widetilde{K}$. In the previous section we have observed that it is useful to calculate the R polynomial of the (2,0) and (2,1) cables of a knot. As a result of the calculations provided in the previous section, it is clear that an analogous theory, applicable to F^* or the R polynomial invariants, would be of significant value in understanding the spatial or geometric content of these equations.

Problem (4.2): Are there simple formulae for the F^* or the R polynomials of the (2,0) and (2,1) cables of a knot analogous to the formula given above for the Jones polynomial?

The complexity of the formulae describing the presence of twist/supercoiling in the molecular graph models for the simple closed DNA as well as the presence of spatial knotting suggest the continuing need for simpler invariants in which these influences are more easily recognized.

Problem (4.3): Develop simpler polynomial invariants reflecting the supercoiling and knotting of the DNA graph models.

The fundamental issues in the creations of a state model for a topological invariant of a spatial graph appear to focus on two aspects: the evaluation of the base states associated to the model and the normalization procedure necessary to insure invariance under the type I generalized Reidemeister movement. The Yamada approach works for 3–valent graphs.

Problem (4.4): Develop alternative state model base state evaluations and normalizations to create topological invariants for spatial graphs, e.g. regular 4–valent graphs.

REFERENCES

[A] J. W. Alexander, *Topological invariants of knots and links*, Trans. Amer. Math. Soc. **20** (1928), 275–306.

[B] R. D. Brandt, W. B. R. Lickorish, and K. C. Millett, *A polynomial invariant for nonoriented knots and links*, Invent. Math. **84** (1986), 563–573.

[C] J. H. Conway, *An enumeration of knots and links and some of their algebraic properties*, Computational problems in abstract algebra (John Leech, ed.), Pergamon Press, Oxford and New York, 1969, pp. 329–358.

[F] P. Freyd, D. Yetter; Y. Hoste; W. B. R. Lickorish, K. Millett; A. Ocneanu, *A new polynomial invariant for knots and links*, Bulletin (New Series) of the American Mathematical Society **12, No. 2** (April 1985), 239–246.

[H1] C. F. Ho, *A new polynomial for knots and links—preliminary report*, Abstracts AMS **6, 4** (1985), abstract 821–57–16.

[H2] J. Hoste, *A polynomial invariant of knots and links*, Pacific J. Math **124** (1986), 295–320.

[J1] V. F. R. Jones, *A polynomial invariant for knots via von Neumann algebras*, Bull. Am. Math. Soc. **12** (1985), 103–111.

[J2] _____, *Hecke algebra representations of braid groups and link polynomials*, Ann. of Math. **126** (1987), 335–388.

[J3] _____, *On knot invariants related to some statistical mechanical models*, preprint, 1988.

[JM1]   D. Jonish, K. C. Millett, *Extrinsic topological chirality indices of molecular graphs*, Graph Theory and Topology in Chemistry **51** (1987), 82–90.

[JM2]   ———, *Isotopy invariants of graphs*, Trans. Amer. Math. Soc. (to appear).

[K0]   L. Kauffman, *The Conway polynomial*, Topology **20** (1980), 101–108.

[K1]   ———, *An invariant of regular isotopy*, Trans. Amer. Math. Soc. **318** (1990), 417–471.

[K2]   ———, *Invariants of graphs in three-space*, Trans. Amer. Math. Soc. **311** (1989), 679–710.

[K3]   ———, *State models and the Jones polynomial*, Topology **26** (1987), 395–407.

[K4]   L. Kauffman and P. Vogel, *Link polynomials and graphical calculus*, preprint, 1987.

[L1]   W. B. R. Lickorish, *Proc. Artin's Braid Group Conference*, Santa Cruz, 1986.

[L2]   ———, *Polynomials for links*, Bull. London Math. Soc. **20** (1988), 558–588.

[LM1]   W. B. R. Lickorish and K. C. Millett, *A polynomial invariant for oriented links*, Topology **26** (1987), 107–141.

[LM2]   ———, *The new polynomials for knots and links*, Mathematics Magazine **61** (1988), 3–23.

[M1]   K. C. Millett, *Stereotopological indices for a family of chemical graphs*, J. Comp. Chem. **8** (1987), 536–550.

[M2]   ———, *Configuration census, topological chirality and the new combinatorial invariants*, The Proceedings of the International Symposium on Applications of Mathematical Concepts to Chemistry, Croatica Chemica Acta **59(3)** (1986), 669–684.

[O]   A. Ocneanu, *A polynomial invariant for knots: a combinatorial and an algebraic approach*, preprint.

[P]   J. Przytycki and P. Traczyk, *Invariants of the Conway type*, Kobe J. Math. **4** (1987), 115–139.

[R]   D. Rolfsen, *Knots and Links*, Publish or Perish Press, 1976.

[Sch]   M. Scharlemann, A. Thompson, *Detecting unknotted graphs in 3-space*, preprint, 1989.

[S1]   J. Simon, *Topological chirality of certain molecules*, Topology **25(2)** (1986), 229–235.

[S2]   ———, *Molecular graphs as topological objects in space*, J. Comp. Chem. **8** (1987), 718–726.

[S3]   J. Simon, K. Wolcott, *Minimally knotted graphs in S^3*, preprint, 1989.

[W1]   E. Witten, *Quantum field theory and the Jones polynomial*, Comm. Math. Phys. **121** (1989), 351–399.

[W2]   ———, *Gauge theories and integrable lattice models*, preprint, February 1989.

[W3]   ———, *Gauge theories, vertex models, and quantum groups*, Nucl. Physics B **330** (1990), 285–346.

[Y1]   S. Yamada, *An invariant of spatial graphs*, J. Graph Theory **13** (1989), 537–551.

[Y2]   S. Yamamoto, *Knots in spatial embeddings of the complete graph on four vertices*, preprint, 1988.

DEPARTMENT OF MATHEMATICS, UNIVERSITY OF CALIFORNIA, SANTA BARBARA, CA 93106, U.S.A.

E-mail address: millett@henri.ucsb.edu

COINCIDENCE THEORY – TOPOLOGICAL & HOLOMORPHIC

KALYAN MUKHERJEA

Indian Statistical Institute, Calcutta

ABSTRACT. We draw attention to the inadequacy of standard topological methods for determining whether or not two maps $f, g : M \longrightarrow N$ between compact manifolds have a coincidence. By some simple examples we illustrate how complex-analytic methods a la Atiyah-Bott can be used to answer similar questions in the category of complex manifolds and holomorphic maps.

1. INTRODUCTION

In this note I wish to discuss a class of coincidence problems involving holomorphic maps. These are very easily treated by the natural extension of Lefschetz's method to the category of complex manifolds and holomorphic maps; yet they are difficult to deal with by extending Lefschetz's method in the topological category. Many other examples of this genre can be easily constructed. By highlighting these examples I hope to draw the attention of topological fixed-point theorists to the very interesting problems in this area. This is why section 3 contains a reasonably detailed account of the holomorphic set-up.

2. LEFSCHETZ'S METHOD – A REVIEW

Let $f, g : X \longrightarrow Y$ be maps between two spaces. The coincidence problem is to develop a criterion for deciding when these two maps have a coincidence, i.e. a point $x \in X$ such that $f(x) = g(x)$.

When X and Y are closed, oriented manifolds of the same dimension a sufficient condition is provided by Lefschetz's celebrated coincidence theorem. The following general procedure for studying coincidence theory is what I mean by the *Lefschetz method.*

Suppose we have a group-valued contravariant functor, $\mathcal{H}$, defined on the homotopy category of topological spaces and an element $\alpha \in \mathcal{H}(Y \times Y)$ which is supported by the diagonal Δ_Y. (This means that whenever $\varphi : Z \longrightarrow Y \times Y$ is a map such that $\varphi(Z) \cap \Delta_Y = \emptyset$, $\mathcal{H}(\varphi)(\alpha) = 0$.) Then, f and g (or more generally any pair of maps, φ and ψ, which are homotopic to f and g respectively) will have a coincidence whenever $\mathcal{H}(f \times g)(\alpha) \neq 0$, where $f \times g$ is the map $x \rightsquigarrow (f(x), g(x))$.

When Y is a compact differentiable manifold there is a systematic method we can follow to produce classes that are supported by the diagonal.

Let $\mathcal{H}$ be a generalized cohomology theory and let Y be any closed, differentiable manifold whose tangent bundle is orientable with respect to this theory. Let τ_Y denote the tangent bundle of Y, let $B(\tau_Y)$ and $S(\tau_Y)$ be the unit disc and sphere bundles of τ_Y with respect to a Riemannian metric and let $\tau_Y^+ = B(\tau_Y)/S(\tau_Y)$ be the 1-point compactification or *Thom space* of τ_Y. If we choose an orientation $\xi \in \mathcal{H}(B(\tau_Y), S(\tau_Y)) \cong \widetilde{\mathcal{H}}(\tau_Y^+)$ we can define the *umkehr* homomorphism

$$\Delta_! : \mathcal{H}(Y) \longrightarrow \widetilde{\mathcal{H}}(Y \times Y)$$

by the following Thom-Pontryagin construction.

The normal bundle of the diagonal submanifold $\Delta_Y = \{(x, y) \in Y \times Y \mid x = y\}$ is isomorphic to τ_Y via the homomorphism which takes a tangent vector v at $y \in Y$ to the normal vector $(v, -v)$ at $(y, y) \in Y \times Y$. Let U be any neighbourhood of Δ_Y. Then using a suitable Riemannian metric on τ_Y we can construct a diffeomorphism of pairs:

$$\theta : (B(\tau_Y), S(\tau_Y)) \longrightarrow (N, \dot{N}), \text{ such that } N \subseteq U.$$

Let $\Gamma : Y \times Y \longrightarrow \tau_Y^+$ be the map which identifies $N \setminus \dot{N}$ with the set of vectors of length less than 1 and which takes all points in the complement of $N \setminus \dot{N}$ to the point at infinity in τ_Y^+. Then $\Delta_! : \mathcal{H}(Y) \longrightarrow \widetilde{\mathcal{H}}(Y \times Y)$ is defined as the composition:

$$\mathcal{H}(Y) \xrightarrow{\Phi_\xi} \widetilde{\mathcal{H}}(\tau_Y^+) \xrightarrow{\Gamma^*} \widetilde{\mathcal{H}}(Y \times Y)$$

where Φ_ξ is the Thom isomorphism arising from the choice of the orientation ξ. If $a \in \mathcal{H}(Y)$ is any nonzero element then $\Delta_!(a) \in \widetilde{\mathcal{H}}(Y \times Y)$ is supported by the diagonal. To see this observe that since the neighbourhood U of Δ_Y was chosen arbitrarily if $\varphi : Z \longrightarrow Y \times Y$ is disjoint from the diagonal we may suppose that $\varphi(Z)$ lies in the complement of N. Then $Z \xrightarrow{\varphi} Y \times Y \xrightarrow{\Gamma} \tau_Y^+$ is the constant map and hence kills any element of $\widetilde{\mathcal{H}}(\tau_Y^+)$. We now give some examples of this construction.

Examples.

1. Let Y be a closed oriented manifold of dimension n, $\mathcal{H}$ the singular cohomology functor with a field as coefficients. (Using $\mathbb{Z}_2$ coefficients these remarks hold even if Y is nonorientable.) Then it is well-known that the pair $(B(\tau_Y), S(\tau_Y))$ can be replaced by $(Y \times Y, Y \times Y \setminus \Delta_Y)$, the homomorphism $\Gamma^* : H^*(\tau_Y) \longrightarrow H^*(Y \times Y)$ is equivalent to the homomorphism $H^*(Y \times Y, Y \times Y \setminus \Delta_Y) \longrightarrow H^*(Y \times Y)$ induced by inclusion and that $\Delta_!(1) = U_Y \in H^n(Y \times Y)$ is the *Poincaré dual* of the fundamental class of Δ_Y. Then the nonvanishing of $(f \times g)^*(U_Y)$ is a sufficient condition for f and g to have a coincidence. If X is also a closed and oriented manifold of dimension n then evaluating $(f \times g)^*(U_Y)$ on the fundamental class of X yields the well-known alternating sum of traces known as the Lefschetz Number of f and g.

This is the reason I refer to the use of the classes $\Delta_!(a)$ for detecting coincidences in our more general context as the *Lefschetz method*.

I will say that the coincidences of $f, g : X \rightarrow Y$ are *homologically invisible* if $(f \times g)^*(U_Y)$ is zero.

2. Let G be one of the stable groups $\mathbf{O}$, $\mathbf{SO}$ or $\mathbf{U}$. Corresponding to these we get the bordism theories Ω_*^G of unoriented, oriented and weakly almost complex manifolds and the corresponding cobordism theories. If Y is a closed manifold whose stable tangent bundle admits a G-reduction then $\Delta_!(1) \in \Omega_G^*(Y \times Y)$ is the cobordism class of the embedding $\Delta : Y \rightarrow Y \times Y$. (See [Q].) If $f, g : X \rightarrow Y$ are maps from a smooth compact manifold then in the generic case they will be transverse regular and the coincidence set $C_{f,g} = \{x \in X \mid f(x) = g(x)\}$ will be a closed submanifold. In this situation $(f \times g)^*(\Delta_!(1)) \in \Omega_G^*(X)$ is the cobordism class of the embedding $C_{f,g} \rightarrow X$. (Observe that the normal bundle of $C_{f,g}$ being a pull-back of the normal bundle of Δ_Y has a G-reduction.) If the dimensions of X and Y are equal then $C_{f,g}$ is a finite set of points (with orientations when $G = \mathbf{SO}$, $= \mathbf{U}$) and the classical Lefschetz number may be thought of as determining this cobordism class.

3. If $Y = \mathbb{C}P^n$ and $\mathcal{H}$ the complex K-theory functor, then $\Delta_! : K(Y) \rightarrow K(Y \times Y)$ is the direct image map and one can show that:

$$\Delta_!(1) = \sum_{k=0}^{n} t_k \sum_{i+j=n+k} \xi^i \otimes \xi^j$$

where t_k is the coefficient of x^k in the binomial expansion of $(1 + x)^{-n}$ and ξ is the class in $\tilde{K}(\mathbb{C}P^n)$ represented by $\mathbf{L}_n - 1$, $\mathbf{L}_n$ being the Hopf line bundle and 1 the trivial line bundle. In the above we have used the Künneth formula to identify $K(\mathbb{C}P^n \times \mathbb{C}P^n)$ and $K(\mathbb{C}P^n) \otimes K(\mathbb{C}P^n)$. (See [Mu] for an application of this class to the detection of homologically invisible coincidences.)

It may appear from the generality of the Lefschetz method that in any situation one would be able to detect coincidences using a suitable functor and a class supported on the diagonal. The limitation of the method is that in order to be able to make any kind of computations, the following conditions must obtain:

a) We must have an explicit understanding of Poincaré duality (with respect to the theory $\mathcal{H}$) in $Y \times Y$ or have a sufficiently faithful natural transformation from $\mathcal{H}$ to an easier theory and a Riemann-Roch type theorem to compute $\Delta_!(1)$;

b) in order to perform the last step, there should be a Künneth formula available in the cohomology theory $\mathcal{H}$.

These conditions are indeed quite stringent. For example, if $Y = \mathbb{C}P^n$ it is difficult to perform these steps except when $\mathcal{H}$ is complex K-thory or, what is in effect the same thing, complex cobordism.

As we will show, even this will fail to detect coincidences of very simple and geometrically interesting pairs of maps. For instance, the complex K-theory of $\mathbb{C}P^n$ is generated by a line bundle and so the above computation of $\Delta_!(1)$ is useless for detecting coincidences of maps $f, g : X \rightarrow \mathbb{C}P^n$ if $H^2(X)$ is trivial.

Example.

For each $n \geq 1$, let $h : \mathbf{S}^{2n+1} \to \mathbb{C}P^n$ be the Hopf fibration. Let $X = \mathbf{S}^{2n+1} \times \mathbf{S}^{2n+1}$ and let $H : X \to \mathbb{C}P^n \times \mathbb{C}P^n$ be the map, $H(x_1, x_2) = (h(x_1), h(x_2))$. Let π_1, π_2 be the projections of $\mathbb{C}P^n \times \mathbb{C}P^n$ on the first and second factors respectively and for $i = 1, 2$, consider the maps $\alpha_i = \pi_i \circ H$. As explained above $(f \times g)^*(\Xi) = 0$ for any $\Xi \in K(\mathbb{C}P^n \times \mathbb{C}P^n)$ and any maps $f, g : X \to \mathbb{C}P^n \times \mathbb{C}P^n$. And yet, any two maps, f and g, homotopic respectively to α_1 and α_2 will have coincidences. To see this, observe that $\alpha_1 \times \alpha_2 : \mathbf{S}^{2n+1} \times \mathbf{S}^{2n+1} \to \mathbb{C}P^n \times \mathbb{C}P^n$ is a Serre fibration and that the identity map of X covers $\alpha_1 \times \alpha_2$. Hence there is a map $\varphi : X \to X$ homotopic to the identity which covers $f \times g$. If f and g do not have a coincidence, $f \times g$ will not be surjective and hence φ will not be surjective either. But this is absurd since φ, being homotopic to the identity map, has degree equal to 1.

In the next section we will indicate how the holomorphic version of the Lefschetz method can be used to detect the coincidences of holomorphic maps in this situation.

3. HOLOMORPHIC COINCIDENCE THEORY

The subject, at least the fixed-point theory, started with Atiyah and Bott's Lefschetz fixed-point formula for elliptic complexes [A-B]. Their basic philosophy is easily explained. If $f : X \to X$ is a smooth map of a smooth manifold and f is transverse to the diagonal then the set F of fixed-points of f is finite and to each fixed-point $p \in F$ one can attach the index $i(p; f) = \text{sign}(\det(I - df(p)))$. Then it is classically known that the Lefschetz Number $L(f) = \sum_{p \in F} i(p; f)$. Their viewpoint is that if the tangent bundle of X has more refined structural properties (for instance if X is a complex manifold or a spin manifold) and f in an appropriate way respects this structure (say f is holomorphic when X is complex) then one should define a more general Lefschetz trace formula and relate the generalized Lefschetz Number to the local behaviour of f in a neighbourhood of F.

For topological fixed-point theorists it is easier to follow Toledo's elegant reformulation [To] or the account in the book of Griffiths and Harris, [G-H, pp.419–426]. Toledo's paper is quite lucid and rather than give an incomprehensibly condensed version thereof, we will give below some of the basic definitions and indicate his approach to fixed-point and coincidence theory in the holomorphic context.

Let M be a compact complex manifold of *complex dimension* n and let $(z_1, \ldots, z_n)$ be holomorphic local coordinates in an open set, $U \subseteq M$. Then the complexified tangent space, $TM \otimes \mathbb{C}$, is a $2n$-dimensional complex vector bundle. At each point, $p \in U$, one has a basis consisting of the tangent vectors:

$$\partial_j = \frac{\partial}{\partial z_j} = \frac{1}{2}\left(\frac{\partial}{\partial x_j} - \sqrt{-1}\frac{\partial}{\partial y_j}\right),$$
$$\overline{\partial}_j = \frac{\partial}{\partial \overline{z}_j} = \frac{1}{2}\left(\frac{\partial}{\partial x_j} + \sqrt{-1}\frac{\partial}{\partial y_j}\right),$$

where, $z_j = x_j + \sqrt{-1}y_j$. Let dz_j and $d\overline{z}_j$ are the differentials dual to ∂_j and $\overline{\partial}_j$ respectively. A smooth $\mathbb{C}$-valued form which can be written as a linear combination of forms of the type, $dz_{i_1} \wedge \ldots \wedge dz_{i_p} \wedge d\bar{z}_{j_1} \wedge \ldots \wedge d\bar{z}_{j_q}$, is called a form of type (p, q) and the module of (p, q)-forms is denoted $\Lambda^{p,q}(M)$. The exterior derivative operator, d, splits into $d = \partial + \bar{\partial}$, where ∂ (resp. $\bar{\partial}$) takes a (p, q)-form to a $(p + 1, q)$-form (resp. $(p, q + 1)$-form) and one has that $\bar{\partial}^2 = 0$. Since $\bar{\partial}$ is the Cauchy-Riemann operator, a $(p, 0)$–form, η, is called a *holomorphic p-form* if $\bar{\partial}\eta = 0$.

The cohomology groups,

$$H^{p,q}(M) = \frac{\text{kernel}\{\bar{\partial} : \Lambda^{p,q}(M) \to \Lambda^{p,q+1}(M)\}}{\text{image}\{\bar{\partial} : \Lambda^{p,q-1}(M) \to \Lambda^{p,q}(M)\}}$$

are known as the Dolbeaut cohomology groups of M and are isomorphic to the qth cohomology of M with the sheaf, $\Omega^p(M)$, of germs of holomorphic p-forms as coefficients.

When M is connected, quite trivially, one has the isomorphisms, $H^{0,0}(M) \cong H^0(M; \mathbb{C}) \cong \mathbb{C}$ and $H^{n,n}(M) \cong H^{2n}(M; \mathbb{C}) \cong \mathbb{C}$. But in general there is no known relationship between the Dolbeaut cohomoloogy groups and the singular or the de Rham cohomology groups of a complex manifold. In fact, one point to which we wish to draw attention is that the Dolbeaut cohomology is an especially useful tool for detecting coincidences of holomorphic maps defined on non-Kähler manifolds.

There is a duality pairing analogous to Poincaré duality for the Dolbeaut cohomology groups. The wedge product followed by integration on pairs of forms,

$$(\alpha, \beta) \rightsquigarrow \int_M \alpha \wedge \beta$$

gives rise to a nonsingular bilinear map:

$$H^{p,q}(M) \otimes H^{n-p,n-q}(M) \to \mathbb{C}$$

which is known as Serre duality.

When M is a Kähler manifold, i.e. when there is a $(1,1)$-form, ω, such that $d\omega = 0$, Hodge theory tells us among other things that there is an isomorphism, $H^k(M, \mathbb{C}) \cong \bigoplus_{p+q=k} H^{p,q}(M)$. So for Kähler manifolds, the Dolbeaut groups may be computed with only a little more effort than that required to compute ordinary cohomology.

Another consequence of Hodge theory is that if M_1, M_2 are compact, conneted, complex manifolds and $\pi_i : M_1 \times M_2 \to M_i$ the projections, then

$$\bigoplus_{p,q,r,s} \pi_1{}^* \otimes \pi_2{}^* : \bigoplus_{p,q,r,s} H^{p,q}(M_1) \otimes H^{r,s}(M_2) \to H^{*,*}(M_1 \times M_2)$$

is an isomorphism; this is the Künneth formula for Dolbeaut cohomology.

Suppose that E, B, F are compact, complex manifolds and that $\pi : E \to B$ is a *holomorphic* locally trivial fibre bundle. Then there is an analogue of the Serre spectral sequence which is useful sometimes in computing the Dolbeaut cohomology groups of a non-Kähler E if B, F are both Kähler. (See the appendix by A. Borel in [Hi]).

Finally recall that a map $f : M \to N$ between complex manifolds is said to be *holomorphic* if for each point $x \in M$ and for complex-analytic charts $\varphi : U \to \mathbb{C}^m$ and $\psi : V \to \mathbb{C}^n$ centered at $x \in M$ and $f(x) \in N$ respectively, the map $\psi \circ f \circ \varphi^{-1} : \mathbb{C}^m \to \mathbb{C}^n$ is holomorphic. If $f : M \to N$ is holomorphic then the usual contravariant transformation of differential forms induced by f preserves the type of the forms and so there is an induced map $f^* : \Lambda^{p,q}(N) \to \Lambda^{p,q}(M)$; this in turn induces homomorphisms $f^* : H^{p,q}(N) \to H^{p,q}(M)$. The Atiyah-Bott fixed-point theorem for holomorphic maps relates the alternating traces of some of these endomorphisms of $H^{*,*}(M)$ with the local behaviour of f near its fixed points. (See [G-H].)

We now describe Toledo's approach to fixed- and coincidence-point theory a la Atiyah and Bott. Toledo constructs classes in $H^{*,*}(N \times N)$ (we will refer to them as *analytical Thom classes*) which have the following property, analogous to being supported by the diagonal:

Given any open set V in $N \times N$ which is disjoint from the diagonal, an analytical Thom class can be represented by a $\bar\partial$-closed form in $N \times N$ which is $\bar\partial$-exact in V.

This implies that if ξ is such a class and F, $G : M \to N$ holomorphic maps such that $(F \times G)^*(\xi) \in H^{*,*}(M) \neq 0$ then F and G have a coincidence. In the fixed-point context Toledo shows that by choosing a suitable analytical Thom class ξ and integrating $(F \times 1_M)^*(\xi)$ over M, one obtains a certain alternating sum of traces of maps induced by F on the Dolbeaut groups of M. On the other hand, the explicit formula for these classes in a neighbourhood of the diagonal enables him to identify these alternating sums with the behaviour of the map near an isolated fixed point. These yield the various holomorphic versions of the Atiyah-Bott-Lefschetz fixed-point formulae.

The account above shows that the holomorphic theory is an appropriate extension of the Lefschetz method to an analytical context.

We now describe two kinds of analytical Thom classes which we will use in our applications of these ideas.

Let N be a compact, connected, complex manifold and let $\{\omega_{p,q;\alpha}\}_{\alpha\in\Lambda_{p,q}}$ be a basis for $H^{p,q}(N)$ and let $\{\eta_{n-p,n-q;\alpha}\}_{\alpha\in\Lambda_{p,q}}$ be the basis of $H^{n-p,n-q}(N)$ dual to the ω's obtained from Serre duality.

(i) The class, ξ_Δ, given by

$$\xi_\Delta = \sum_{p,q;\alpha\in\Lambda^{p,q}} (-1)^{p+q}\omega_{p,q;\alpha} \otimes \eta_{n-p,n-q;\alpha}$$

is an analytical Thom class. In fact, it represents the *cohomological* class U_Y (with $\mathbb{C}$ as coefficient field) discussed in Example 1 in Section 2. Topologists will, of course, immediately see the analogy of this with Lefschetz's diagonal formula. The analysis of this class in the context of fixed-point theory yields only the classical

Lefschetz theorem. Although the example we give is rather trivial, we will see that when the domain is a non-Kähler manifold even this class can detect homologically invisible coincidences.

(ii) The class ξ_Δ^0 defined by:

$$\xi_\Delta^0 = \sum_{q;\alpha \in \Lambda^{0,q}} (-1)^q \omega_{0,q;\alpha} \otimes \eta_{n,n-q;\alpha}.$$

is an analytical Thom class whose study leads to the holomorphic versions of the fixed-point formula.

We will use these classes to detect coincidences which, if not invisible, are quite difficult to detect by topological methods.

4. THE EXAMPLES

The manifolds, $X = S^{2n+1} \times S^{2n+1}$, admit a class of complex structures discovered by Calabi and Eckmann [C-E]. With respect to these structures, the map:

$$H : X \to \mathbb{C}P^n \times \mathbb{C}P^n$$

is a holomorphic fibre bundle with a torus, $S^1 \times S^1$, as fibre. Any choice of a complex structure for this torus gives a complex structure on X. The Dolbeault cohomology of X is computed in [Hi] using the Borel spectral sequence and may be described as follows. Recalling that $\pi_i \circ H = \alpha_i$;

$$H^{*,*}(X) \cong \mathbb{C}[x_{1,1}]/(x_{1,1}^{n+1}) \otimes E(z_{1,0}, z_{n,n+1})$$

where $x_{1,1} = \alpha_i^*(\Omega)$ $(i = 1,2)$ and $\Omega \in H^{1,1}(\mathbb{C}P^n)$ is the standard generator of $H^{1,1} \cong H^2$, viz.:

$$\Omega = \frac{\sqrt{-1}}{2} \partial\bar{\partial} \log(|Z_0|^2 + \cdots + |Z_n|^2).$$

The subscripts indicate the type of the classes, and E denotes an exterior algebra on the generators enclosed in parentheses.

Now $\mathbb{C}P^n$ is a Kähler manifold with Kahler class, Ω, and the analytical Thom classes have the form:

$$\xi_\Delta = \sum_{i+j=n} \Omega^i \otimes \Omega^j \text{ and } \xi_\Delta^0 = 1 \otimes \Omega^n.$$

We begin with the trivial:

Proposition 4.1. *The maps α_1, $\alpha_2 : X \to \mathbb{C}P^n$ have a coincidence.*

Proof. $(\alpha_1 \times \alpha_2)^*(\xi_\Delta) = (n+1)x_{1,1}^n \neq 0$. $\square$

Remark. Of course every point (x, x) is a coincidence of α_1 and α_2. The point we wish to make is that these coincidences are homologically invisible and yet the class ξ_Δ (which in fixed-point theory only yields homological information) can detect these coincidences.

Proposition 4.2. *If $F : X \to \mathbb{C}P^n$ is any holomorphic map, then F has a coincidence with α_2.*

Proof. $(F \times \alpha_2)^*(\xi_\Delta^0) = x_{1,1}^n \neq 0$. □

If we use a little more information about the complex structure of X, then we can prove more.

Theorem 4.3. *If F, $G : X \to \mathbb{C}P^n$ are holomorphic maps then they will have a coincidence provided at least one of them is nonconstant.*

Proof. Theorem V of [C-E] asserts that if ψ is any meromorphic function on X then it is constant along the fibres of the fibration: $H : X \to \mathbb{C}P^n \times \mathbb{C}P^n$. Now suppose that $F : X \to \mathbb{C}P^n$ is a nonconstant holomorphic map and let p, p' be two points lying in the same fibre. If $F(p) \neq F(p')$, then by composing F with Z_i/Z_j for suitable homogeneous coordinates Z_i and Z_j we will obtain a meromorphic function which violates the above result of [C-E]. Hence F factors through $\mathbb{C}P^n \times \mathbb{C}P^n$ via H. If $F = F' \circ H$ and F is nonconstant then $F'^*(\Omega) \neq 0$. Let $F'^*(\Omega) = a \cdot 1 \otimes \Omega + b \cdot \Omega \otimes 1$. Since $\Omega^{n+1} = 0$ and the $1 \otimes \Omega$ and $\Omega \otimes 1$ commute either a or b must be zero. It follows that $(G \times F)^* \xi_\Delta^0 \neq 0$. □

5. CONCLUDING REMARKS

1. As mentioned before, in [Mu] there is an example in which $\Delta_!(1) \in K(\mathbb{C}P^n \times \mathbb{C}P^n)$ is used to detect coincidences (on maps defined on a product of Lens spaces and projective spaces) which are not detectable using ordinary cohomology. These coincidences are also detected by the class ξ_Δ^0; the proof reads almost the same as for 4.2. The computation of $\Delta_!(1)$ is a highly nontrivial exercise, even if it is viewed as a routine application of Riemann-Roch. The proof using holomorphic methods is far simpler.

2. Since there is no "homotopy axiom" in Dolbeaut cohomology, we cannot as in the topological situation obtain information concerning coincidences of homotopy classes.

3. The kind of situation which we have illustrated is of course special; but many examples of this genre may be easily constructed. For example, suppose that G is a compact, connected, simply-connected Lie group and $T \subseteq G$ a maximal torus. Then $G \times G$ can be given a homogeneous complex structure which is not Kähler [Wa]. On the other hand, G/T is in a natural way an algebraic variety and hence is Kähler. There is an abundance of holomorphic maps between these manifolds and though in some cases I can use holomorphic methods to detect coincidences between such maps I do not know how to apply the topological version of the Lefschetz method to detect coincidences of maps from $G \times G$ to G/T. Recently H. Pittie has developed methods (as yet unpublished) of computing the Dolbeaut cohomology of compact Lie groups which admit complex structures and many results should be obtainable by using his techniques in conjunction with the methods I have illustrated in this note.

Finally, I would like to observe that whenever the domain has dimension greater than that of the range manifold, coincidences that are homologically invisible are very difficult to detect. This is why the methods discussed above are of particular interest. It would also be very interesting to have some new topological methods which could be used in such situations.

REFERENCES

[A-B] M. Atiyah and R. Bott, *A Lefschetz Fixed-Point Theorem for Elliptic Complexes I*, Annals of Math. **86** (1967), 374–407.

[C-E] E. Calabi and B. Eckmann, *A Class of Compact, Complex Manifolds which are not Algebraic*, Annals of Math. **58** (1953), 494–500.

[G-H] P. Griffiths and J. Harris, *Principles of Algebraic Geometry*, John Wiley, New York, 1978.

[Hi] F. Hirzebruch, *Topological Methods in Algebraic Geometry*, (Third Ed.), Springer-Verlag, Heidelberg, Berlin, New York, 1966.

[Mu] K. Mukherjea, *New Methods in Coincidence Theory*, Proc. Amer. Math. Soc. **34** (1972), 615–620.

[Q] D. Quillen, *Elementary Proofs of some Results of Cobordism Theory using Steenrod Operations*, Advances in Mathematics **7** (1971), 29–56.

[To] D. Toledo, *On the Atiyah-Bott Formula for Isolated Fixed Points*, J. Diff. Geom. **8** (1973), 401–436.

[Wa] H. C. Wang, *Closed Manifolds with Homogeneous Complex Structure*, Amer. J. Math. **76** (1954), 1–32.

STAT.-MATH. DIVISION, INDIAN STATISTICAL INSTITUTE, 203 BARRACKPORE TRUNK ROAD, CALCUTTA 700035, INDIA

ENERGY FUNCTIONALS OF KNOTS

JUN O'HARA

ABSTRACT. We define *energy functionals* on the space of embeddings from S^1 into $\mathbb{R}^3$ and show the finiteness of knot types under bounded value of those functionals.

0. INTRODUCTION

In this paper we define the *energy functionals of knots*, characterized as functionals on the space of embeddings from S^1 into $\mathbb{R}^3$ that are well-behaved for embeddings and that blow up for curves with self-intersections. The main motivation is to define a suitable functional on the space of embeddings from S^1 into $\mathbb{R}^3$ or S^3. The *canonical positions of knots* are characterized as the embeddings attaining the minimum (or minimal) of that functional in their ambient isotopy classes. The canonical positions of knots are 'symmetric' or 'beautiful' in some sense. We have not yet reached a complete solution.

This paper also concerns the *finite* number of ambient isotopy classes of knots in $\mathbb{R}^3$ that can have any given number as upper bound for the energy functionals. An energy functional for PL (piecewise linear) knots is given. We use this in a computer simulation to get the approximated canonical positions of knots by moving given PL knots along the gradient of the functional. This can be illustrated by the computer graphic video by Ahara ([A]) that the author demonstrated at the conference at Hawaii in August 1990. A table of some figures from that video is annexed at the end of this paper.

Sections 1 and 2 are brief reviews of [O3]. We refer the reader to [O1], [O2], and [O3] for proofs.

The author thanks deeply Prof. K. Ahara for making programs and video tapes of computer simulations and the figures of the table.

Throughout this paper we work with the fixed domain of functionals, $\mathcal{E}$, given by

$$\mathcal{E} = \{f : S^1 = \mathbb{R}/\mathbb{Z} \to \mathbb{R}^3; \ f \text{ is an embedding of class } C^2 \text{ such that}$$
$$|f'(t)| = 1 \text{ for all } t \in S^1\}.$$

1. Energy functionals

In this section we define the energy functional, give some examples and state some basic properties.

Definition 1.1. *A functional* $e : \mathcal{E} \to \mathbb{R}$ *is an energy functional if the following two conditions are satisfied:*

(1) *For any* $x, y \in S^1$ *with* $x \neq y$, *and for any real number* M, *there exists a positive number* ϵ *such that* $|f(x) - f(y)| \leq \epsilon$ $(f \in \mathcal{E})$ *implies* $e(f) \geq M$.

(2) *If* $f(S^1)$ *and* $g(S^1)$ *are congruent* $(f, g \in \mathcal{E})$, *then* $e(f) = e(g)$ *holds.*

Put the arc length between $x, y \in S^1 = \mathbb{R}/\mathbb{Z}$ by $\delta(x, y) = \min\{|x - y|, 1 - |x - y|\}$.

Definition 1.2. *For* $f \in \mathcal{E}$, $x, y \in S^1$ *and for* $0 \leq j \leq 2$, *define* $\phi(f; j; x, y)$ *as follows:*

$$\phi(f; j; x, y) = \int_{\delta(x,y)^{-1}}^{|f(x)-f(y)|^{-1}} \xi^{j-1} \, d\xi \qquad \text{if } x \neq y,$$

$$= \begin{cases} \dfrac{1}{j}\left(\dfrac{1}{|f(x)-f(y)|^j} - \dfrac{1}{\delta(x,y)^j}\right) & \text{if } j \neq 0 \text{ and } x \neq y, \\[2ex] \log \dfrac{\delta(x,y)}{|f(x)-f(y)|} & \text{if } j = 0 \text{ and } x \neq y, \end{cases}$$

$$\phi(f; j; x, x) = \begin{cases} 0 & \text{if } j \neq 2, \\[1ex] \dfrac{|f''(x)|^2}{24} & \text{if } j = 2. \end{cases}$$

Proposition 1.3. *Let* $\mathcal{E}$ *be endowed with* C^2-*topology.*

(1) *The map* $\phi(\ ; \ ; \ , \ ) : \mathcal{E} \times [0, 2] \times S^1 \times S^1 \to \mathbb{R}$ *is non-negative, and is continuous outside* $\mathcal{E} \times \{j = 2\} \times \{(x, x) \in S^1 \times S^1; x \in S^1\}$.

(2) *The map* $\phi(f; 2; \ , \ ) : S^1 \times S^1 \to \mathbb{R}$ *is bounded for each* $f \in \mathcal{E}$, *and is continuous if* f *is of class* C^4.

Definition 1.4. *Let* $f \in \mathcal{E}$. *For* $0 \leq j \leq 2$ *and* $1 \leq p \leq \infty$, *define the functional* $e_j{}^p$ *to be the* L^p *norm of the map* $\phi(f; j; ,) : S^1 \times S^1 \to \mathbb{R}$:

$$e_j{}^p(f) = \|\phi(f; j; ,)\|_{L^p(S^1 \times S^1)}$$

$$= \begin{cases} \left\{\int_0^1 \int_0^1 \phi(f; j; x, y)^p \, dx dy\right\}^{\frac{1}{p}} & \text{if } p < \infty \\[2ex] \text{ess } \sup_{x, y \in S^1} \phi(f; j; x, y) & \text{if } p = \infty \end{cases}$$

Proposition 1.5. *The functional* $e_j{}^p(f)$, *defined in 1.4, is positive and continuous in* j, p, *and* f.

Theorem 1.6. *The functional $e_j{}^p (0 \le j \le 2; 1 \le p \le \infty)$ is an energy functional if and only if $p \ge \frac{2}{j}$ $(0 < j \le 2)$ or $p = \infty$.*

Example 1.7. *The energy $E : \mathcal{E} \to \mathbb{R}$ in [O2] is given by the following:*

$$E(f) = \frac{1}{2} \int_0^1 \lim_{\epsilon \downarrow 0} \left(\int_{x+\epsilon}^{1+x-\epsilon} \frac{dy}{|f(x) - f(y)|^2} - \frac{2}{\epsilon} \right) dx$$

$$= \frac{1}{2} \int_0^1 \int_0^1 |f(x) - f(y)|^{-2} - \pi^2 \{\sin \pi(y - x)\}^{-2} dx dy$$

$$= e_2{}^1(f) - 2.$$

Example 1.8. *Gromov's distortion ([G] page 113) of a knot is given by the following:*

$$\text{Distor}(f) = \sup_{x,y \in S^1, x \ne y} \frac{\delta(x, y)}{|f(x) - f(y)|}$$

$$= \exp(e_0{}^\infty(f)).$$

Definition 1.9. *If we have $p \ge \frac{2}{j}$ $(0 < j \le 2)$ or $p = \infty$, then we call $e_j{}^p(f)$ the (j,p)-energy of f, where $f \in \mathcal{E}$.*

Definition 1.10. *An embedding $f_0 : S^1 \to \mathbb{R}^3$ is in canonical position with respect to the (j, p)-energy if $e_j f^p$ attains a relative minimum at f_0 as f varies within the ambient isotopy class of f_0.*

We state some basic properties of $e_j{}^p$.

Theorem 1.11. *If $p \ge 2j^{-1}$ $(0 < j \le 2)$ or $p = \infty$, then for a given positive number b there exists a positive number $K = K(j, p, b)$ such that if we have $e_j{}^p(f) \le b$ $(f \in \mathcal{E})$, then*

$$|f(x) - f(y)| \ge K\delta(x, y)$$

for all $x, y \in S^1$.

Proof. If $j = 0$, set $K(0, \infty, b) = \exp(-b)$.
   If $j > 0$, set

$$K(j, p, b) = \frac{1}{8} \exp \left\{ -4 \left(\frac{jb}{1 - (\frac{2}{3})^j} \right)^{\frac{2}{j}} \right\}.$$

The desired properties for $K(j, p, b)$ follow easily.

Theorem 1.12. *If $p > 2j^{-1}$ $(0 < j \le 2)$, then for a given positive number b there exist positive numbers $A = A(j, p, b)$ and $q = q(j, p)$ such that if $e_j{}^p(f) \le b$ $(f \in \mathcal{E})$, then*

$$|f(x) - f(y)| \ge (1 - A\delta(x, y)^q)\delta(x, y)$$

for all $x, y \in S^1$.

Proof. Set

$$A(j, p, b) = 2^{\frac{7}{3}}(j+1)^{\frac{1}{3}}b^{\frac{1}{3}},$$

$$q(j, p) = \frac{pj - 2}{3p}.$$

This A and q have the desired property.

Let $\tau(S^1)$ be a planar simple closed curve including a straight line segment. Put a small nontrivial "knotted part" in its center. Keeping the "knotted part" similar, make it shrink to a point as indicated in Fig. 1.13.0. (From each knot in this process, we can obtain the corresponding knot that belongs to $\mathcal{E}$ by suitable similarity and reparametrization.) We investigate the behavior of the value of $e_j{}^p$ under this process.

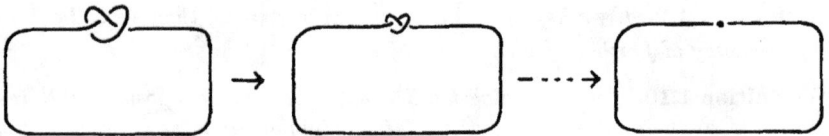

FIG. 1.13.0

Theorem 1.13. *Under this process:*
  (1) *If* $p > 2j^{-1}$ $(0 < j \le 2)$, *then the value of* $e_j{}^p$ *explodes to* $+\infty$.
  (2) *If* $p = 2j^{-1}$ $(0 < j \le 2)$ *or* $p = +\infty$ $(j = 0)$, *then the value of* $e_j{}^p$ *converges to some constant, which is greater than* $e_j{}^p(\tau)$.
  (3) *If* $p < 2j^{-1}$ $(0 < j \le 2)$ *or* $p < +\infty$ $(j = 0)$, *then the value of* $e_j{}^p$ *converges to* $e_j{}^p(\tau)$.

2. FINITENESS OF KNOT TYPES.

Definition 2.1 ([K]). *Let* $f : S^1 \to \mathbb{R}^3$ *be an embedding of class* C^2 *such that* $|f'(t)| = 1$ *for all* $t \in S^1$. *For* $x \in S^1$ *define* $\chi_f(x)$ *to be the smallest critical value of the map* $S^1 \ni y \mapsto |f(x) - f(y)| \in \mathbb{R}$ *that is different from zero. Then Kuiper's self-distance of* f *is given by*

$$sd(f) = \inf_{x \in S^1} \chi_f(x).$$

Proposition 2.2 ([O2]). *Given a positive number r. Then there exist only finitely many ambient isotopy classes of knots which can have r as lower bound for Kuiper's self-distance.*

The combination of Theorems 1.11 and 1.12 yields:

Theorem 2.3. *If $p > 2j^{-1}(0 < j \leq 2)$, then for a given positive number b there is a positive number $r = r(j, p, b)$ such that if $e_j{}^p(f) \leq b$ $(f \in \mathcal{E})$, then $sd(f) \geq r$.*

Corollary 2.4. *If $p > 2j^{-1}(0 < j \leq 2)$, then for a given positive number b there exist only finitely many ambient isotopy classes of knots which can have b as upper bound for the (j, p)-energy.*

We can obtain an algorithm of calculating an upper bound of the number of knot types with $e_j{}^p \leq b$, where $p > 2j^{-1}(0 < j \leq 2)$.

We make some remarks on Theorems 2.3 and Corollary 2.4 about the cases when $p \leq 2j^{-1}$ $(0 < j \leq 2)$ or $j = 0$. Recall that if $p < 2j^{-1}$ $(0 < j \leq 2)$ or $p < \infty$ $(j = 0)$ then $e_j{}^p$ is not an energy functional.

Theorem 1.13 implies that Theorem 2.3 does not hold if $p \leq 2j^{-1}$ $(0 < j \leq 2)$ or $j = 0$.

On page 114 of [G] Gromov raised the following question:

> "Does every ambient isotopy class of knots in $\mathbb{R}^3$ have a representative with distortion ≤ 100?"

We remark that there exist infinitely many knot types with bounded value of Gromov's distortion, or equivalently, $e_0{}^\infty$.

Theorem 2.5. *If $p < 2j^{-1}$ $(0 < j \leq 2)$ or $j = 0$, then there is a positive number $B = B(j, p)$ such that there exist infinitely many ambient isotopy classes of knots (including infinitely many ambient isotopy classes of prime knots) which can have B as upper bound for $e_j{}^p$. (We can take $B(0, \infty) \leq 100$.)*

3. PL-VERSION OF THE ENERGY FUNCTIONAL E.

In this section we define a PL version of the energy functional $E = e_2{}^1 - 2$, $\hat{E}$, which we use in a computer simulation of moving given PL knots along the gradient of $\hat{E}$ to get approximated canonical positions of knots with respect to E.

Proposition 3.1. ([O1]) *For any $x \in S^1$*

$$E(f) = \lim_{n \to +\infty} \left(\frac{1}{n^2} \sum_{0 \leq i < j \leq n-1} |f(x + \frac{i}{n}) - f(x + \frac{j}{n})|^{-2} - \frac{\pi^2}{6}n \right).$$

Fix a natural number $n (\geq 3)$. Define $\mathbb{D}_n$ and L_n by

$$\mathbb{D}_n = \{ \mathbb{X} = (X_1, X_2, \cdots, X_n) \in (\mathbb{R}^3)^n \, ; \, X_i = (x_i, y_i, z_i) \in \mathbb{R}^3, \, X_i \neq X_j \text{ if } i \neq j \},$$

$$L_n = \{ \mathbb{X} \in \mathbb{D}_n; \, |X_i - X_{i-1}| = \frac{1}{n} \text{ for all } i \},$$

where the suffixes are taken mod n. Then $\mathbb{L}_n$ is $2n$-dimensional open submanifold in $\mathbb{R}^{3n}$. Define the function $\hat{E} : \mathbb{D}_n \to \mathbb{R}$ by

$$\hat{E}(\mathbb{X}) = \frac{1}{n^2} \sum_{i<j} \frac{1}{|X_i - X_j|^2} - \frac{\pi^2}{6} n = \frac{1}{2} \frac{1}{n^2} \sum_{i \neq j} \frac{1}{|X_i - X_j|^2} - \frac{\pi^2}{6} n,$$

and denote the restriction of $\hat{E}$ to the space $\mathbb{L}_n$ by the same letter $\hat{E}$. Then $\hat{E} + \frac{\pi^2}{6} n$ is 'electric potential energy' of a polygonal knot $\mathbb{X}$ each vertex of which is equally charged with 'electrons' with total amount 1 whose 'Coulomb's force' is proportional to the cubic inverse of the distance. $\hat{E} = \hat{E}|_{\mathbb{L}_n}$ is the PL version of the energy functional $E = e_2{}^1 - 2$ in the sense of Proposition 3.1. (Restriction of the domain of $\hat{E}$ to the space $\mathbb{L}_n$ corresponds to the condition that $|f'(t)| = 1$ for all $t \in S^1$ of the domain of E, $\mathcal{E}$.)

Define

$$-\mathrm{grad}_{\mathbb{X}}^{\mathbb{D}_n}(\hat{E}) = -\left(\frac{\partial \hat{E}}{\partial x_1}, \cdots, \frac{\partial \hat{E}}{\partial z_n} \right)$$
$$= (U_1, U_2, \cdots, U_n)$$

then we have

$$U_i = \frac{2}{n^2} \sum_{i \neq j} \frac{X_i - X_j}{|X_i - X_j|^4}.$$

Put

$$V_i = X_i - X_{i-1} \ (\bmod n),$$
$$\mathbb{V}_1 = (V_1, O, \cdots, O, -V_1),$$
$$\mathbb{V}_2 = (-V_2, V_2, O, \cdots, O),$$
$$\vdots$$
$$\mathbb{V}_n = (O, \cdots, O, -V_n, V_n).$$

Then the tangent space of $\mathbb{L}_n$ at $\mathbb{X}$, $T_{\mathbb{X}}(\mathbb{L}_n)$, is the orthogonal complement in $\mathbb{R}^{3n}$ of the n-dimensional subspace spanned by n vectors $\mathbb{V}_1, \mathbb{V}_2, \cdots, \mathbb{V}_n$.

Proposition 3.2. ([F], [O1])

(1) *The orthogonal projection $\pi : \mathbb{R}^{3n} \to T_{\mathbb{X}}(\mathbb{L}_n)$ is given by the following formula;*

$$\pi(\mathbb{W}) = \mathbb{W} - \mathbb{W}({}^t\mathbb{V}_1 \cdots {}^t\mathbb{V}_n) \left[\begin{pmatrix} \mathbb{V}_1 \\ \vdots \\ \mathbb{V}_n \end{pmatrix} ({}^t\mathbb{V}_1 \ldots {}^t\mathbb{V}_n) \right]^{-1} \begin{pmatrix} \mathbb{V}_1 \\ \vdots \\ \mathbb{V}_n \end{pmatrix},$$

for $\mathbb{W} \in \mathbb{R}^{3n}$.

(2) *The gradient of $\hat{E} = \hat{E}|_{\mathbb{L}_n} : \mathbb{L}_n \to \mathbb{R}$ at $\mathbb{X}$ is given by*

$$\mathrm{grad}_{\mathbb{X}}^{\mathbb{L}_n}(\hat{E}) = \pi(\mathrm{grad}_{\mathbb{X}}^{\mathbb{D}_n}(\hat{E})).$$

4. OPEN PROBLEMS.

We finish with a few questions which we were unable to settle. Suppose $p \geq 2j^{-1}$ $(0 < j \leq 2)$ or $p = \infty$.

(a) Show the existence of the canonical positions for each ambient isotopy class of knots. Are the critical points of the functional $e_j{}^p$ nondegenerate?

(b) Are there only finitely many ambient isotopy classes of knots which can have bounded (j, p)-energy when $p = 2j^{-1}$ $(j \neq 0)$?

(c) Show that the standard planar circle gives the minimum with respect to the (j, p)-energy $e_j{}^p$.

(d) How many canonical positions of the trivial knot are there? One or infinitely many? (Computer simulation implies that there seems to be at least two canonical positions of trefoil.)

Find a suitable energy functional such that there are no canonical positions of the trivial knot except for the standard planar circle. (Hatcher showed in page 606 of [H] that the space of smoothly embedded unknotted circles in S^3 deformation retracts onto the space of great circles in S^3.)

(e) Show that there is a positive number $\alpha = \alpha(j, p)$ such that $\alpha > e_j{}^p(\iota)$, where $\iota(S^1)$ is the standard planar circle, and that if $f(S^1)$ is a non-trivial knot then we have $e_j{}^p(f) \geq \alpha$.

(f) Is the inverse of Kuiper's self-distance an energy functional?

REFERENCES

[A] Ahara, K., *video : Energy of a knot*, (not a paper).

[B-O] Buck, G. and Orloff, J., *Computing canonical conformations for knots*, preprint.

[B-S] Buck, G. and Simon, J., *Knots as dynamical systems*, preprint.

[F] Fukuhara, S., *Energy of a knot*, A Fête of Topology, Y. Matsumoto, T. Mizutani, and S. Morita, ed., Academic Press, 1987, pp. 443–452.

[G] Gromov, M., *Filling Riemannian manifolds*, J. Differential Geom. **18** (1983), 1–147.

[Gu] Gunn, C., *program : Linkmover*, at Geometry Supercomputer Project, (not a paper).

[H] Hatcher, A., *A proof of the Smale conjecture, Diff(S^3)* $\simeq O(4)$, Ann. of Math. **117** (1983), 553–607.

[K] Kuiper, N.H., private note.

[O1] O'Hara, J., *Energy of a knot*, Master Thesis, Univ. of Tokyo, 1986, in Japanese.

[O2] _____ , *Energy of a knot*, Topology **30** (1991), 241–247.

[O3] _____ , *Family of energy functionals of knots*, to appear, Topology and its Applications.

[S] Sakuma, M., *Energy of geodesic links in S^3*, preprint.

COMPUTER SIMULATIONS

In this table we present five examples of computer simulations: two of the trivial knot, two of the trefoil and one of the figure eight knot, which illustrate how piecewise linear knots with 50 vertices transform themselves in the space $\mathbb{L}_{50}$ to decrease $\hat{E} = \hat{E}|_{\mathbb{L}_{50}}$ (see § 3). Strictly speaking, the flow we used in this computer simulation along which to move PL knots is obtained from $-\mathrm{grad}_{\mathbf{X}}^{\mathbb{D}_{50}}(\hat{E})$, but is not $-\mathrm{grad}_{\mathbf{X}}^{\mathbb{L}_{50}}(\hat{E})$ itself. The final figure of each example is the (probably stable) critical point of that flow.

The numbers beside each figure are its values of $\hat{E}$ and Gromov's distortion D.

Remark. Suppose $\iota(S^1)$ is the standard planar circle and $\mathbb{X}_{50}$ denotes the regular 50-gon in a plane. Then

$$E(\iota) = 0. \qquad\qquad \mathrm{Distor}(\iota) = \frac{\pi}{2} = 1.571.$$
$$\hat{E}(\mathbb{X}_{50}) = -0.1410. \qquad \mathrm{Distor}(\mathbb{X}_{50}) = 1.570.$$

1. Example of the trivial knot

$$\hat{E} = 45.89 \quad D = 14.97$$

$\downarrow$

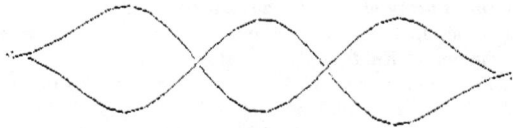

$$\hat{E} = 29.57 \quad D = 8.758$$

$\downarrow$

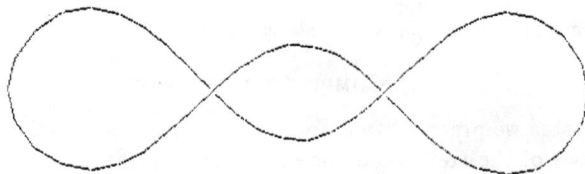

$$\hat{E} = 18.33 \quad D = 8.574$$

$\downarrow$

$\hat{E} = 11.78 \ D = 6.405$

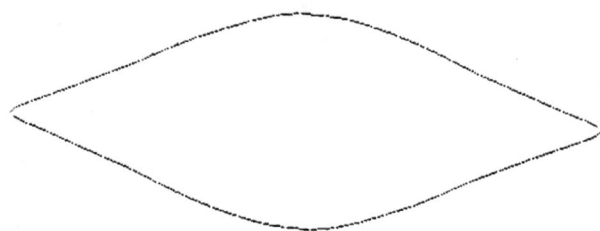

$\hat{E} = 4.80 \ D = 3.748$

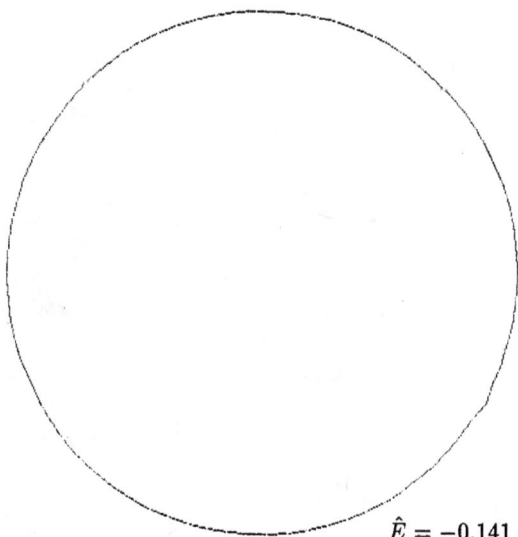

$\hat{E} = -0.141 \ D = 1.586$

JUN O'HARA

2. Example of the trivial knot

$$\hat{E} = 12.32 \quad D = 6.787$$

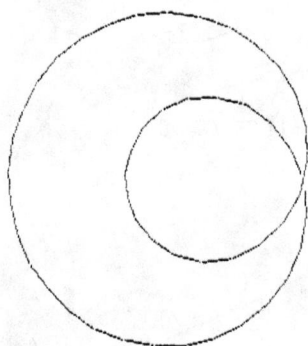

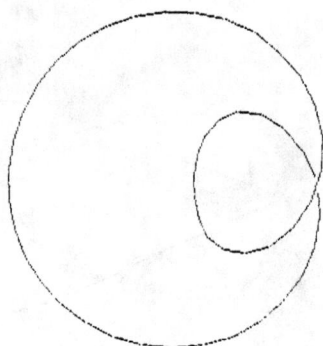

$$\hat{E} = 7.96 \quad D = 4.370$$

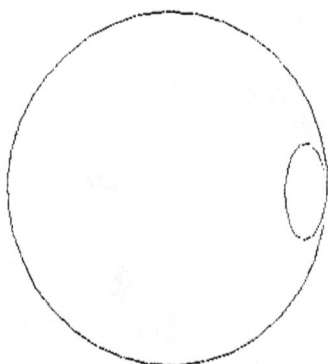

$$\hat{E} = 3.51 \quad D = 2.841$$

$$\hat{E} = 0.871 \quad D = 2.034$$

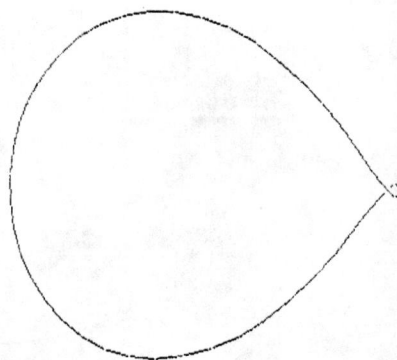

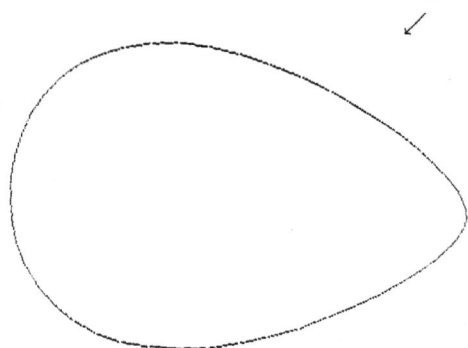

$\hat{E} = 0.042 \ D = 1.738$

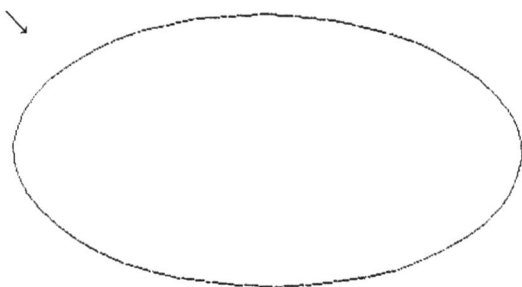

$\hat{E} = -0.141 \ D = 1.579$

3. Example of the trefoil

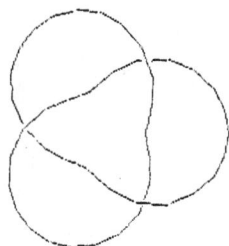

$\hat{E} = 48.14 \ D = 9.789$

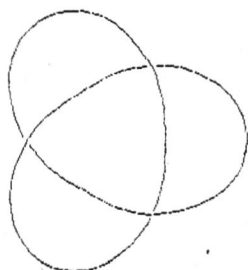

$\hat{E} = 38.82 \ D = 9.810$

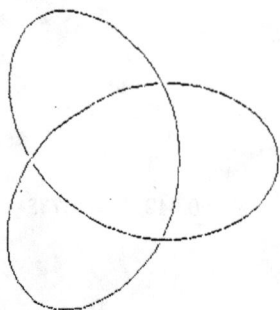

$\hat{E} = 34.99 \ D = 8.063$

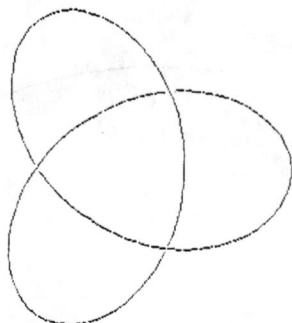

$\hat{E} = 34.78 \ D = 8.009$

4. Example of the trefoil

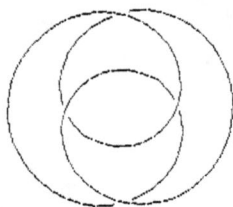

$\hat{E} = 56.41 \ D = 10.76$

↓

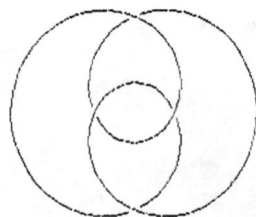

$\hat{E} = 45.54 \ D = 13.98$

↓

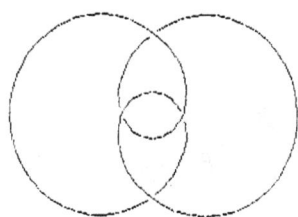

$\hat{E} = 39.15 \ D = 17.50$

$\downarrow$

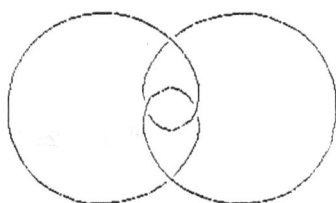

$\hat{E} = 36.38 \ D = 20.57$

$\downarrow$

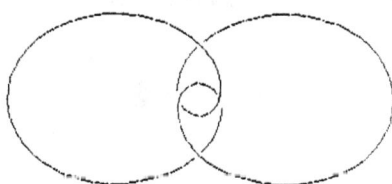

$\hat{E} = 34.73 \ D = 25.60$

5. Example of the figure eight knot

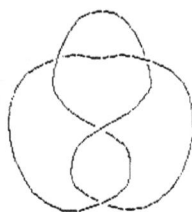

$\hat{E} = 81.85 \ D = 16.05$

$\downarrow$

$\downarrow$

$\hat{E} = 57.63 \quad D = 22.09$

$\downarrow$

$\hat{E} = 51.83 \quad D = 25.72$

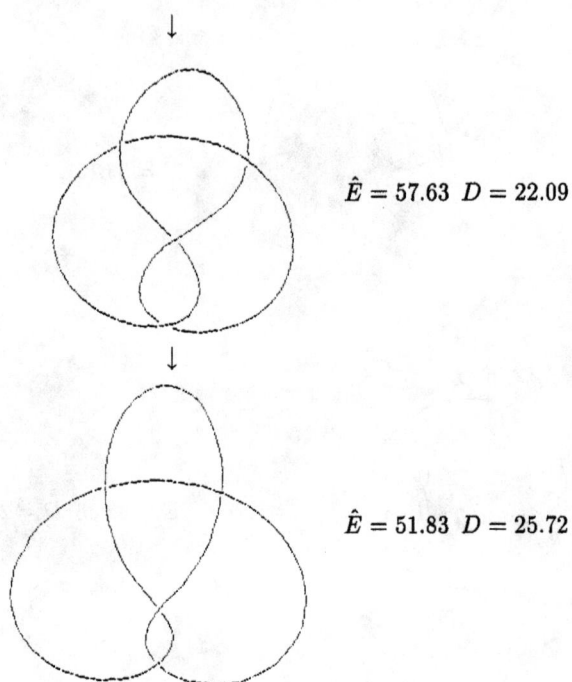

DEPARTMENT OF MATHEMATICS, TOKYO METROPOLITAN UNIVERSITY, 1-1 MINAMIOHSAWA, HACHIOHJI-SHI, TOKYO 192-03, JAPAN

BASIC RELATIVE NIELSEN NUMBERS

XUEZHI ZHAO

Liaoning University

ABSTRACT. Let $f : (X, A) \longrightarrow (X, A)$ be a selfmap of a pair of compact polyhedra. We shall introduce some basic relative Nielsen numbers of f, which will construct all the existing relative Nielsen numbers. Using the method in the relative Nielsen theory, we obtain all of the distribution of the minimal fixed point set for maps in the homotopy class of a given map.

1. INTRODUCTION

Let $f : (X, A) \longrightarrow (X, A)$ be a selfmap of a pair of compact polyhedra. To estimate the number and location of the fixed points of such a map, several variants of relative Nielsen number have been defined in [3], [4] and [5]. It was shown that $N(f; X, A)$, $\widetilde{N}(f; X, A)$ and $N(f; X - A)$ are respectively lower bounds for the number of fixed points on X, on $\mathrm{Cl}(X - A)$ and on $X - A$ for all maps in the homotopy class of f.

In this paper, we shall introduce some basic relative Nielsen numbers, which play the role of "particle" in the relative Nielsen theory. And every relative Nielsen number we have known can be written as a partial sum of these basic relative Nielsen numbers. It is known that the location of the fixed points of such a map f on X is not arbitrary. Using the methods in [3] and [5], we shall obtain all of the distribution of the minimal fixed point set for maps in the homotopy class of a given map.

This paper is organized as follows. We describe the relative Nielsen theory and give the definition of the basic relative Nielsen numbers in Section 2. We discuss some properties of these basic relative Nielsen numbers in Section 3. Finally, Section 4 is concerned with the minimal fixed point sets of relative maps.

This paper requires knowledge of some results and proofs from [3], [4] and [5]. Background material about Nielsen fixed point theory can be found in [1] and [2].

I want to thank the referee for his help.

The author gratefully acknowledges the support of K. C. Wong Education Foundation, Hong Kong

2. Basic relative Nielsen numbers

Let $f : (X, A) \longrightarrow (X, A)$ be a selfmap of a pair of compact polyhedra. We write $\overline{f} : A \longrightarrow A$ for the restriction of f to A and $\widehat{A} = \bigcup A_k$ for the disjoint union of all components of A which are mapped by f into themselves.

As we know, X and all the components of $\widehat{A}$ have universal coverings

$$p : \widetilde{X} \longrightarrow X$$
$$p_k : \widetilde{A}_k \longrightarrow A_k \qquad k = 1, 2, \ldots, n \ .$$

Thus, fixed point classes of f and f_k are respectively in one-to-one correspondence with lifting classes of f and f_k [2; Ch. II, Sec. 1]. For example, a lifting class $[\widetilde{f}]$ of f corresponds to fixed point class $p\,\mathrm{Fix}\,\widetilde{f}$, which is the projection of the fixed point set of $\widetilde{f}$. Hence the fixed point classes we are concerned with in this paper maybe empty, and fixed point classes labelled by different lifting classes will be regarded as different ones even if all of them are empty.

For each $f_k : A_k \longrightarrow A_k$, which is the restriction of f to A_k, there is a natural correspondence from lifting classes of f_k to those of f (see [5, p190]). It is denoted by $i_{k,FPC}$, and we have

Theorem 2.1. ([5, Proposition 2.1]) *Let $p_k\,\mathrm{Fix}\,\widetilde{f}_k$ be a non-empty fixed point class of f_k. Then $p_k\,\mathrm{Fix}\,\widetilde{f}_k$ is contained in $p\,\mathrm{Fix}\,\widetilde{f}$ if and only if $i_{k,FPC}([\widetilde{f}_k]) = [\widetilde{f}]$.* $\square$

Hence, we shall say $p_k\,\mathrm{Fix}\,\widetilde{f}_k$ is contained in $p\,\mathrm{Fix}\,\widetilde{f}$ if and only if $i_{k,FPC}([\widetilde{f}_k]) = [\widetilde{f}]$ even if $p_k\,\mathrm{Fix}\,\widetilde{f}_k$ is empty. Thus we have

Theorem 2.2. *Let $p_k\,\mathrm{Fix}\,\widetilde{f}_k$ be a fixed point class of f_k. Then it is contained in a unique fixed point class $p\,\mathrm{Fix}\,\widetilde{f}$ of f, where $i_{k,FPC}([\widetilde{f}_k]) = [\widetilde{f}]$.* $\square$

As in [3] and [4], fixed point classes of each f_k are also said to be fixed point classes of $\overline{f} : A \longrightarrow A$. For a non-empty fixed point class $\mathbb{F}$ of f, $\mathbb{F} \cap A$ is empty or the union of fixed point classes of $\overline{f}$([3]). We say $\mathbb{F}$ assumes its index in A if

$$\mathrm{ind}(X, f, \mathbb{F}) = \mathrm{ind}(A, \overline{f}, \mathbb{F} \cap A),$$

where $\mathrm{ind}(A, \overline{f}, \mathbb{F} \cap A)$ is well defined as the sum of the indices of the non-empty fixed point classes of $\overline{f}$ which are contained in $\mathbb{F} \cap A$ ([4]). Since the index of an empty fixed point class is zero, we can generalize the definition above to all fixed point classes of f, i.e. a fixed point class $\mathbb{F}$ of f is said to be assume its index in A if its index is equal to the sum of the indices of the (empty and non-empty) fixed point classes of $\overline{f}$ which are contained in $\mathbb{F}$. It is easy to see that an empty fixed point class of f assumes its index in A.

Thus, for an arbitrary fixed point class $\mathbb{F}$ of f, there are four basic questions as follow.

　　i) Is $\mathbb{F}$ essential?

ii) Does $\mathbb{F}$ contain an essential fixed point class of $\overline{f}$?

iii) Does $\mathbb{F}$ contain an inessential fixed point class of $\overline{f}$?

iv) Does $\mathbb{F}$ assume its index in A?

Hence, the set of fixed point classes of f has been divided into 16 "groups". We write $N_{ijkl}(f; X, A)$, $i,j,k,l, = 0,1$, for the numbers of fixed point classes in these "groups", where, for each group, the negative or positive answer to a basic question are denoted by the subscript 0 or 1. For example, $N_{1010}(f; X, A)$ is defined as the number of essential fixed point classes of f which do not contain any essential fixed point classes of $\overline{f}$, contain inessential fixed point classes of $\overline{f}$ and do not assume their indices in A. $N_{ijkl}(f; X, A)$, $i,j,k,l = 0,1$, are said to be *basic relative Nielsen numbers*.

3. PROPERTIES OF THE BASIC RELATIVE NIELSEN NUMBERS

In this section, we shall discuss some properties of the basic relative Nielsen numbers.

Theorem 3.1. (homotopy invariance) *If two maps $f, g : (X, A) \longrightarrow (X, A)$ are homotopic, i.e. they are connected by a homotopy of the form $H : (X \times I, A \times I) \longrightarrow (X, A)$, then $N_{ijkl}(f; X, A) = N_{ijkl}(g; X, A)$, $i,j,k,l = 0,1$.*

Proof. Let $\mathbb{F}$ and $\mathbb{G}$ be respectively fixed point classes of f and g, where $\mathbb{F}$ and $\mathbb{G}$ are H-related. Recall from [3] and [5] that a fixed point class of f is said to be common if it contains an essential fixed point class of $\overline{f}$ and to be weakly common if it contains a fixed point class of $\overline{f}$. Hence, it suffices to show

i) $\mathbb{F}$ is essential if and only if $\mathbb{G}$ is essential,

ii) $\mathbb{F}$ is common if and only if $\mathbb{G}$ is common,

iii) $\mathbb{F}$ is weakly common if and only if $\mathbb{G}$ is weakly common,

iv) $\mathbb{F}$ assumes its index in A if and only if $\mathbb{G}$ assumes its index in A.

As we know, i) is equivalent to the homotopy invariance of Nielsen number. ii), iii), and iv) are respectively contained in the proof of [3, Theorem 3.3], [5, Theorem 2.6] and [4, Theorem 4.1]. Thus, this theorem holds. $\square$

Similar to [3, Theorem 3.4 and Theorem 3.5], we can show that these basic relative Nielsen numbers have the commutativity and homotopy type invariance. We leave the details to the reader.

By the definition of the basic relative Nielsen numbers, we have

Theorem 3.2.

$$R(f) = \sum_{i,j,k,l=0}^{1} N_{ijkl}(f; X, A),$$

$$N(f) = \sum_{i,j,k=0}^{1} N_{1ijk}(f; X, A),$$

where $R(f)$ is the Reidemeister number of f. $\square$

Theorem 3.3. $N_{1ijk}(f; X, A)$ and $N_{01ij}(f; X, A)$ are all non-negative integers, and

$$N_{1ijk}(f; X, A) \leqslant N(f),$$
$$N_{i1jk}(f; X, A) \leqslant N\left(\overline{f}\right), \qquad i, j, k = 0, 1.$$

Proof. Since, by the definition, basic relative Nielsen numbers are non-negative integers or infinite, the first inequality follows directly from Theorem 3.2. Since each fixed point class of $\overline{f}$ is contained in a unique fixed point class of f, the number $N\left(\overline{f}\right)$ of essential fixed point classes of $\overline{f}$ is not less than the number of fixed point classes of f which contain essential fixed point classes of $\overline{f}$, i.e.

$$\sum_{i,j,k=0}^{1} N_{i1jk}(f; X, A) \leqslant N\left(\overline{f}\right).$$

Thus, the second inequality is true. $\square$

Theorem 3.4. *Let* $\mathbb{F}$ *be a fixed point class of* f *which does not contain any essential fixed point classes of* $\overline{f}$. *Then* $\mathbb{F}$ *is essential if and only if it does not assume its index in* A.

Proof. Since $\mathbb{F}$ does not contain any essential fixed point classes of $\overline{f}$, the sum of the indices of the fixed point classes of $\overline{f}$ which are contained in $\mathbb{F}$ is zero. Hence, $\mathbb{F}$ does not assume its index in A if and only if its index is not equal to zero, i.e. $\mathbb{F}$ is essential. $\square$

Corollary 3.5.

$$N_{1011}(f; X, A) = N_{1001}(f; X, A) = N_{0010}(f; X, A) = N_{0000}(f; X, A) = 0. \quad \square$$

From this corollary, we know that there exist actually 12 basic relative Nielsen numbers. These, together with the Reidemeister number $R\left(\overline{f}\right)$ and the Nielsen number $N\left(\overline{f}\right)$ of f, will construct all the relative Nielsen numbers we know. Here are some formulae.

Theorem 3.6.

$$N\left(f, \overline{f}\right) = \sum_{i,j=0}^{1} N_{11ij}(f; X, A),$$
$$N(f; X, A) = N\left(\overline{f}\right) + N_{1010}(f; X, A) + N_{1000}(f; X, A).$$

Proof. By the definition, $\sum_{i,j=0}^{1} N_{11ij}(f; X, A)$ is the number of essential fixed point classes of f which contain essential fixed point classes of $\overline{f}$. Since essential fixed point classes are non-empty, the essential fixed point classes of f and $\overline{f}$ we are

concerned with here can be regarded as those defined via path homotopies in [3].
Thus, the first equality follows from the definition of $N\left(f, \overline{f}\right)$.

From the definition of $N(f; X, A)$ and Theorem 3.2,

$$N(f; X, A) = N\left(\overline{f}\right) + N(f) - N\left(f, \overline{f}\right)$$

$$= N\left(\overline{f}\right) + \sum_{i,j,k=0}^{1} N_{1ijk}(f; X, A) - \sum_{i,j=0}^{1} N_{11ij}(f; X, A)$$

$$= N\left(\overline{f}\right) + \sum_{i,j=0}^{1} N_{10ij}(f; X, A)$$

By Corollary 3.5, we get the second equality. $\square$

Theorem 3.7.

$$\widetilde{N}(f; X, A) = \sum_{i,j,k=0}^{1} N_{ijk0}(f; X, A),$$

$$\widetilde{n}(f; X, A) = \sum_{i,j=0}^{1} N_{i1j0}(f; X, A).$$

Proof. Recall from [4] that the Nielsen number of the complement $\widetilde{N}(f; X, A)$ is the number of fixed point classes of f which do not assume their indices in A and the Nielsen number of the boundary $\widetilde{n}(f; X, A)$ is the number of fixed point classes of f which do not assume their indices in A and are common. Since a fixed point class of which does not assume its index in A is non-empty, these two equalities follow from the definition of basic relative Nielsen numbers. $\square$

Recall from [5] that the Nielsen number for the complement $N(f; X - A)$ is the number of fixed point classes of f which are essential and are not weakly common. We have

Theorem 3.8. $N(f; X - A) = N_{1000}(f; X, A)$. $\square$

From the formulae, we have

Theorem 3.9.

$$N(f) - N\left(f, \overline{f}\right) = \widetilde{N}(f; X, A) - \widetilde{n}(f; X, A)$$

$$= N(f; X, A) - N\left(\overline{f}\right)$$

$$= N(f; X - A) + N_{1010}(f; X, A)$$

$$= N_{1000}(f; X, A) + N_{1010}(f; X, A).   \square$$

From Theorem 3.3 and Corollary 3.5, basic relative Nielsen numbers, except $N_{0001}(f; X, A)$ and $N_{0011}(f; X, A)$, are non-negative integers. But it is known that

$R(f)$, which is the sum of all the basic relative Nielsen numbers, is a non-negative integer or infinite. Two examples showing that $N_{0001}(f; X, A)$ and $N_{0011}(f; X, A)$ can be infinite follow.

Example 3.10. Let $X = S^1 \times I$ and $A = S^1 \times \{0\}$, and let $f : (X, A) \longrightarrow (X, A)$ be identity map.

It is easy to see the $\overline{f}$ as well as f has infinitely many fixed point classes, all of them are inessential. Since $\pi_1(A) \xrightarrow{i} \pi_1(X)$ is surjective, it follows from [5, Theorem 4.1] that every fixed point class of f is weakly common and is not common. Thus, $N_{0011}(f; X, A)$ is infinite, and the other basic relative Nielsen numbers are zero.

Example 3.11. Let $X = S^1$ and A be a single point in X, and let $f : (X, A) \longrightarrow (X, A)$ be identity map.

Since $\overline{f}$ has only one fixed point class, which is essential, and f is the same as the map $\overline{f}$ in Example 3.10, we have $N_{0100}(f; X, A) = 1$, $N_{0001}(f; X, A)$ is infinite and the others are zero.

4. Minimal fixed point sets

In this section, we shall give the upper and lower bounds for the number of fixed points on $\operatorname{Int} A$, $\operatorname{Bd} A$, A, $X - A$ and $\operatorname{Cl}(X - A)$ for all maps in the homotopy class of f which have minimal fixed point sets on X. Under some assumptions, we shall give all the possible distributions. Homotopies of f are maps of the form $H : (X \times I, A \times I) \longrightarrow (X, A)$.

Theorem 4.1. If $f : (X, A) \longrightarrow (X, A)$ has $N(f; X - A)$ fixed points on $X - A$, then it has at least $N(\overline{f}) + N_{1010}(f; X, A)$ fixed points on A, and at least $\widetilde{n}(f; X, A) + N_{1010}(f; X, A)$ fixed points on $\operatorname{Bd} A$.

Proof. It is known that $N(f; X, A)$ and $\widetilde{N}(f; X, A)$ are respectively the lower bounds for the number of fixed points on X and $\operatorname{Cl}(X - A)$, and that

$$N(f; X, A) = N(\overline{f}) + N_{1010}(f; X, A) + N(f; X - A),$$
$$\widetilde{N}(f; X, A) = \widetilde{n}(f; X, A) + N_{1010}(f; X, A) + N(f; X - A).$$

Since f has $N(f; X - A)$ fixed points on $X - A$, it has at least $N(\overline{f}) + N_{1010}(f; X, A)$ fixed points on A, and at least $\widetilde{n}(f; X, A) + N_{1010}(f; X, A)$ fixed points on $\operatorname{Cl}(X - A) - (X - A) = \operatorname{Bd} A$. $\square$

Using Theorem 3.9, a relative Nielsen number on a subspace also gives an upper bound for the number of fixed points on its complementary space for all maps in the given homotopy class which have minimal fixed point sets on X. Hence, we have

Theorem 4.2. If $f : (X, A) \longrightarrow (X, A)$ has $N(f; X, A)$ fixed points on X, then f has

    i) at most $N(\overline{f}) - \widetilde{n}(f; X, A)$ fixed points on $\operatorname{Int} A$,

ii) at least $\widetilde{n}(f; X, A)$ and at most $N\left(\overline{f}\right) + N_{1010}(f; X, A)$ fixed points on Bd A,

iii) at least $N\left(\overline{f}\right)$ and at most $N\left(\overline{f}\right) + N_{1010}(f; X, A)$ fixed points on A,

iv) at least $N(f; X - A)$ and at most $N_{1010}(f; X, A) + N(f; X - A)$ fixed points on $X - A$,

v) at least $\widetilde{N}(f; X, A)$ and at most $N(f; X, A)$ fixed points on $Cl(X - A)$. □

As for the realization of these upper and lower bounds, we have

Theorem 4.3. *Let (X, A) be a pair of compact polyhedra such that*

i) *X is connected,*

ii) *X-A has no local cut point and is not a 2-manifold,*

iii) *every component of A is a Nielsen space with a non-empty interior,*

iv) *A can be by-passed in X.*

Let $f : (X, A) \longrightarrow (X, A)$ be a selfmap. If $\widetilde{n}(f; X, A) \leqslant k_1 \leqslant N\left(\overline{f}\right)$ and $0 \leqslant k_2 \leqslant N_{1010}(f; X, A)$, then we can homotope f to a map $g : (X, A) \longrightarrow (X, A)$ with $N(f; X, A)$ fixed points on X, of which $N\left(\overline{f}\right) - k_1$ lie on Int A, $N(f; X - A) + k_2$ lie on $X - A$, and therefore $k_1 + N_{1010}(f; X, A) - k_2$ lie on Bd A.

Proof. As in the proof of [4, Theorem 5.1], f is homotopic to a map $f' : (X, A) \longrightarrow (X, A)$ with $N(f; X, A)$ fixed points on X, $N\left(\overline{f}\right)$ fixed points on A, $\widetilde{N}(f; X, A)$ fixed points on $Cl(X - A)$ and k_1 fixed points on Bd A, in which the number k_1 of fixed points of f' on Bd A can be attained by moving extra $k_1 - \widetilde{n}(f; X, A)$ fixed points on Int A to Bd A in Step 1.

Using the method in the proof of [5, Theorem 3.8], we can move $N_{1010}(f; X, A) - k_2$ fixed points on $X - A$ to Bd A, which are in the essential and weakly common fixed point classes of f'. Thus, we get a map $g : (X, A) \longrightarrow (X, A)$ homotopic to f with $N(f; X, A)$ fixed points on X, of which $N\left(\overline{f}\right) - k_1$ lie on Int A, $k_1 + N_{1010}(f; X, A) - k_2$ lie on Bd A and $N(f; X - A) + k_2$ lie on $X - A$. So the theorem holds. □

Remark 4.4. It is easy to see that Theorem 4.3 has summarized the minimum theorems of the existing relative Nielsen numbers, for, it will be [3, Theorem 6.2], if $k_1 = N\left(\overline{f}\right)$ and $k_2 = N_{1010}(f; X, A)$; be [4, Theorem 5.1], if $k_1 = \widetilde{n}(f; X, A)$ and $k_2 = N_{1010}(f : X, A)$; be [5, Theorem 3.9], if $k_1 = \widetilde{n}(f; X, A)$ and $k_2 = 0$.

Remark 4.5. All the results in this paper, except Theorem 4.3, can be generalized to the admissible maps of pairs of ANR's as in [4].

REFERENCES

1. R. F. Brown, *The Lefschetz Fixed Point Theorem*, Scott, Foresman and Co., Glenview, Ill., 1971.

2. Boju Jiang, *Lectures on Nielsen Fixed Point Theory*, Contemporary Mathematics **14**, Amer. Math. Soc., Providence, Rhode Island, 1983.

3. H. Schirmer, *A relative Nielsen number*, Pacific J. Math **122** (1986), 459-473.

4. H. Schirmer, *On the location of fixed points on pairs of spaces*, Topology and its Applications **30** (1988), 253-266.

5. Xuezhi Zhao, *A relative Nielsen number for the complement*, in "Topological Fixed Point Theory and Applications", Lecture Notes in Math., Vol. 1411, Springer, Berlin, Heidelberg, 1989, pp. 189-199.

AMS Subject Classification (1980): 55M20, 54H25

DEPARTMENT OF MATHEMATICS, LIAONING UNIVERSITY, SHENYANG 110036, P. R. CHINA.